Elephant Seals

Elephant Seals:
Population Ecology,
Behavior, and
Physiology

EDITED BY

Burney J. Le Boeuf and Richard M. Laws

UNIVERSITY OF CALIFORNIA PRESS

Berkeley   Los Angeles   London

University of California Press
Berkeley and Los Angeles, California

University of California Press
London, England

Library of Congress Cataloging-in-Publication Data
Elephant seals : population ecology, behavior, and physiology / edited
by Burney J. Le Boeuf and Richard M. Laws.
    p. cm.
  Includes bibliographical references and index.
  ISBN 0-520-08364-4 (cloth)
  1. Elephant seals. I. Le Boeuf, Burney J. II. Laws, Richard M.
QL737.P64E44 1994
599.74'8—dc20   93-38142
                                                           CIP

Printed in the United States of America

1 2 3 4 5 6 7 8 9

The paper used in this publication meets the minimum requirements of American National
Standard for Information Sciences–Permanence of Paper for Printed Library Materials, ANSI
Z39.48-1984 ∞

To those who provided funds for the conference:
Gordon Reetz and William Lang of Minerals Management Service,
John Twiss and Robert Hofman of the Marine Mammal Commission, and
George A. Malloch via the Gerena Macgowan Trust

CONTENTS

PREFACE

"If you like superlatives, you will love this animal," a pitchman for elephant seals might say. The list is long and far-ranging. It is the largest seal and one of the most sexually dimorphic marine mammals. It is extremely polygynous by comparison with any other large vertebrate. Females fast while lactating, and the largest breeding males fast for more than one hundred days during the breeding season. Elephant seals dive deeper and longer than any other pinniped, and they spend more time submerged during their long aquatic wanderings than most whales. No other large vertebrate has come so close to extinction as the northern elephant seal did one hundred years ago and made such a rapid recovery.

Although these superlatives make the two elephant seal species in the genus *Mirounga* ideal subjects for a variety of scientific studies, some of their more mundane attributes are responsible for much of the attention by scientists. Elephant seals breed on open beaches where they are plainly visible. They are unafraid of humans and do not flee when approached or disturbed. Consequently, when sleeping, they can be easily marked or tagged individually. Tagging at weaning, one month after birth, yields a cohort of known-age animals by sex. This is critical for long-term behavior studies and studies of growth, aging, and survival and defines the age and maturation variables in physiological studies in the laboratory. To put this in perspective, a biologist can identify, sex, and age more elephant seals in one afternoon than a student of killer whales might accomplish in a month, if at all. Arrivals and departures on rookeries for breeding and molting are predictable, which has facilitated instrumentation of individuals for the study of diving and at-sea behavior.

Owing in large part to their ease of study, the rudiments of elephant seal natural history, the basis for a good monograph, were provided by the first systematic observations and studies conducted in the 1940s and 1950s. It was in elephant seal teeth that annual growth layers were first demonstrated as a reliable method of age determination of mammals. George Bartholomew pioneered studies of the northern elephant seal, *M. angustirostris*, in California and Mexico and one of us (RML) concurrently conducted an intensive study of southern elephant seals, *M. leonina*, at South Georgia and other rookeries in Antarctic waters. During the next three decades, this information base expanded rapidly as a result of fundamental behavioral, physiological, and population studies conducted by scientists from a number of countries at most places where elephant seals breed. One measure of the ease and depth of study of this animal is that lifetime reproductive success of several cohorts has been estimated by measuring reproductive success annually in identifiable males and females throughout their lifetimes, something that has been accomplished in only a few large vertebrates. Studies of at-sea behavior began in 1983, made possible by the development of small self-contained diving instruments. By 1990, elephant seals were one of the most thoroughly studied and well-known marine mammals.

When there is sufficient knowledge and intellectual activity in a field, it becomes interesting and worthwhile to survey this knowledge, and holding a conference has numerous advantages. In 1990, Gordon Reetz, representing the Minerals Management Service, asked one of us (BJL) to organize and coordinate such a conference. The impetus for the international conference on elephant seals held in Santa Cruz, California, on May 20-21, 1991, was to review and update the status of both the northern and southern elephant seal populations and present current research findings in life history, diving and foraging behavior, and physiological ecology. Because it was apparent that the population of the northern species was expanding rapidly while major colonies of the southern species were in long-term decline, it was hoped that causes of colony decline might be suggested from an examination of behavioral, reproductive, and survival data. Second, there was a great deal of exciting research being done on diving and foraging behavior and energy requirements during feeding and breeding on diverse rookeries, and we thought it would be stimulating to meet, present our findings, and discuss progress in this fast-developing area of research. Third, we aimed to examine factors affecting survival and reproductive success. For both species, we wanted the principal researchers from around the world to present their most recent findings on these topics.

Like the conference, this volume is intended to provide a detailed summary of current knowledge of certain aspects of elephant seal life. It is not an attempt to present all that is known about elephant seals. Everything that was presented orally at the conference is reported here except for two talks. Additional information, acquired after the talks were given, has been added to many of the chapters, and an introductory chapter was written to provide background information on natural history.

This volume has 40 authors, 39 more than the typical monograph. The advantage of a multiauthor monograph over the more typical single-author volume is breadth of coverage, depth of treatment in each chapter, and multiple perspectives on the same issues. Little effort was made to hold authors to a uniform format or writing style. We hope that the information presented in this volume will be of interest to students of animal behavior, ecology, physiology, and marine science, as well as wildlife managers and administrators linked with government, fisheries, and petroleum assessment and development, that is, people who are concerned with the impacts of activities in our oceans on the animals that live in them.

In this volume, the first section on population ecology contains chapters addressing the history and current status of both species, the impact of southern elephant seals on the Antarctic ecosystem, and possible causes of the decline of some colonies. In the second section, the results of long-term studies of juvenile survivorship, diet, and breeding strategies of the northern elephant seal are presented, basic behavioral and life history data that are intended to elucidate the causes of population growth and decline. The third section deals with at-sea behavior of both species. In chapter 12, Roger D. Hill presents the theory that made it possible to determine the migratory paths and foraging areas of seals treated by Marthan N. Bester and Ian S. Wilkinson (chap. 5), Burney J. Le Boeuf (chap. 13) and Brent S. Stewart and Robert L. DeLong (chap. 16). Other chapters review recent findings in this fast-developing field and explore new directions for elucidating foraging behavior and physiological constraints on diving, such as early development, measurement of swimming speed, and analysis of dive types. The volume closes with a consideration of key elements of foraging economics in southern elephant seals, endocrine changes during development, breath-hold performance during sleep on land and underwater, and the role of hormones in fuel regulation during fasting, a key aspect of the life history strategy of both species.

We are grateful to Gordon Reetz, Carol Fairfield, James Lang, Mark Pierson, and others from the Minerals Management Service for support in acquiring the finances that enabled us to bring this international group together for a stimulating exchange of information in a pleasant setting and for the funding to initiate this volume. We thank John Twiss and Bob Hofman of the Marine Mammal Commission for providing a small grant to help defray travel costs for participants from out of the country. BJL acknowledges the financial assistance of George A. Malloch and the Gerena MacGowan Estate and grants from the National Science Foundation. RML acknowledges a grant from the Royal Society. We thank Daniel Costa and

Christophe Guinet for their oral presentations at the conference and James Estes, Gerald Kooyman, and Joanne Reiter for chairing sessions and acting as discussants. Communication with authors and transmittal of manuscripts to the publisher were facilitated by Gigi Nabors and Marie McCullough. We are grateful to the science editor at the University of California Press, Elizabeth Knoll, for taking on this project and speeding it along.

CONTRIBUTORS

Sarah G. Allen, Point Reyes Bird Observatory, 4990 Shoreline Highway, Stinson Beach, CA 94970

George A. Antonelis, National Marine Mammal Laboratory, NOAA, National Marine Fisheries Service, 7600 Sand Point Way, Seattle, WA 92115

Tom A. Arnbom, British Antarctic Survey, Natural Environment Research Council, High Cross, Madingley Road, Cambridge CB3 OET, United Kingdom, and Department of Zoology, Stockholm University, Stockholm, Sweden

Tomohiro Asaga, Tokyo University of Fisheries, 4-5-7 Kohnan Minato-ku, Tokyo 108, Japan

Marthan N. Bester, Mammal Research Institute, University of Pretoria, Pretoria 0002, South Africa

Ian L. Boyd, British Antarctic Survey, Natural Environment Research Council, High Cross, Madingley Road, Cambridge CB3 OET, United Kingdom

Michael M. Bryden, Department of Veterinary Anatomy, University of Sydney, NSW 2006, Australia

Harry R. Burton, Australian Antarctic Division, Channel Highway, Kingston, Tasmania 7050, Australia

Michael A. Castellini, Institute of Marine Sciences, University of Alaska, Fairbanks, AK 99775

C. Chambers, Sea Mammal Research Unit, Natural Environment Research Council, High Cross, Madingley Road, Cambridge CB3 OET, United Kingdom

Walter L. Clinton, 5844 NE 75th St., No. A312, Seattle, WA 98115

Daniel P. Costa, Department of Biology and Institute of Marine Sciences, University of California, Santa Cruz, CA 95064

Daniel E. Crocker, Department of Biology, University of California, Santa Cruz, CA 95064

Robert L. DeLong, National Marine Mammal Laboratory, NOAA, National Marine Fisheries Service, 7600 Sand Point Way, Seattle, WA 92115

Charles J. Deutsch, Department of Biology, University of California, Santa Cruz, CA 95064

Michael A. Fedak, Sea Mammal Research Unit, Natural Environment Research Council, High Cross, Madingley Road, Cambridge CB3 OET, United Kingdom

Clifford H. Fiscus, National Marine Mammal Laboratory, Alaska Fisheries Science Center, National Marine Fisheries Service, 7600 Sand Point Way, Seattle, WA 92115

J. Harwood, Sea Mammal Research Unit, Natural Environment Research Council, High Cross, Madingley Road, Cambridge CB3 OET, United Kingdom

Roger D. Hill, Wildlife Computers, Woodinville, WA 98072

Mark A. Hindell, Department of Zoology, University of Tasmania, P.O. Box 252c, Hobart, Tasmania 7001, Australia

Harriet R. Huber, National Marine Mammal Laboratory, NOAA, National Marine Fisheries Service, 7600 Sand Point Way, Seattle, WA 92115

Ronald J. Jameson, U.S. Fish and Wildlife Service, National Ecology Research Center, Piedras Blancas Research Station, P.O. Box 70, San Simeon, CA 93452

Vicky Lee Kirby, Department of Biology, University of California, Santa Cruz, CA 96064

Richard M. Laws, St. Edmund's College, University of Cambridge, Cambridge CB3 0BN, United Kingdom

Burney J. Le Boeuf, Department of Biology and Institute of Marine Sciences, University of California, Santa Cruz, CA 96064

Mark S. Lowry, Southwest Fisheries Science Center, National Marine Fisheries Service, La Jolla, CA 92038

T. S. McCann, British Antarctic Survey, Natural Environment Research Council, High Cross, Madingley Road, Cambridge CB3 OET, United Kingdom

B. J. McConnell, Sea Mammal Research Unit, Natural Environment Research Council, High Cross, Madingley Road, Cambridge CB3 OET, United Kingdom

Patricia Morris, Department of Biology, University of California, Santa Cruz, CA 95064

Yasuhiko Naito, National Institute of Polar Research, 9-10, Kaga 1-Chome, Itabashi-ku, Tokyo 173, Japan

Nadav Nur, Point Reyes Bird Observatory, 4990 Shoreline Highway, Stinson Beach, CA 94970

C. Leo Ortiz, Department of Biology, University of California, Santa Cruz, CA 96064

Joanne Reiter, Institute of Marine Sciences, University of California, Santa Cruz, CA 96064

Haruo Sakurai, National Institute of Polar Research, 9-10, Kaga 1-Chome, Itabashi-ku, Tokyo 173, Japan

David J. Slip, Australian Antarctic Division, Channel Highway, Kingston, Tasmania 7050, Australia

Brent S. Stewart, Hubbs-Sea World Research Institute, 1700 South Shores Road, San Diego, CA 92109

William J. Sydeman, Point Reyes Bird Observatory, 4990 Shoreline Highway, Stinson Beach, CA 94970

Philip H. Thorson, Department of Biology, University of California, Santa Cruz, CA 95064, and Hubbs-Sea World Research Institute, 1700 South Shores Road, San Diego, CA 92109

Ian S. Wilkinson, Sea Fisheries Research Institute, Private Bag X2, Rogge Bay 8012, South Africa

Pamela K. Yochem, Hubbs-Sea World Research Institute, 1700 South Shores Road, San Diego, CA 92109

ONE

Elephant Seals:
An Introduction to the Genus

Burney J. Le Boeuf and Richard M. Laws

ABSTRACT. The two species of elephant seals, *M. angustirostris* and *M. leonina*, inhabit different parts of the world, in opposite hemispheres and contrasting climatic zones, they breed at different times of the year, and they differ in size and elaboration of male secondary sexual characteristics. The northern species is lacking in genetic variation, apparently the result of an extreme population bottleneck caused by sealing in the last century; the population has increased rapidly in this century. The southern species has decreased in overall number over the last half century. There are three (or four) stocks, with little genetic interchange; the major stock appears to be stable, others have decreased, and one population is increasing. Despite these differences and thousands of years of separation without intermingling, the two species are remarkably similar in morphology, behavior, and life history traits serving reproduction and the diving pattern.

The aim of this chapter is to provide information on the origins, nomenclature, and key features in the ecology and life history of elephant seals that will serve as background for the specialized chapters that follow. Research on both species has been both intensive and extensive. The more accessible northern species has been particularly closely studied at Año Nuevo Island, where large cohorts of known-age marked animals provide longitudinal data, and intensive year-round studies of behavior have been made. Research on the southern species has been concentrated on the breeding season because the remote and rigorous environment in which it lives has made year-round studies more difficult. Nevertheless, a recent review remarked that it "is one of the most exhaustively and widely studied pinnipeds" (Ling and Bryden 1992).

While the northern species' breeding range is comparatively limited, with extensive interchange between colonies, the southern elephant seal has a number of separate populations, with limited interbreeding, over a circumpolar range and in a broad range of latitudes (Laws, this volume). Its population ecology is more varied.

ORIGINS

Elephant seals are the largest in size of the 34 extant species of pinnipeds (King 1983). There are two species in the genus *Mirounga*, the northern elephant seal, *M. angustirostris*, and the southern elephant seal, *M. leonina*. The genus is in the family Phocidae, the true seals, as distinct from the other two families in the suborder Pinnipedia: Otariidae, the fur seals and sea lions, and Odobenidae, the walrus.

Northern elephant seals are distributed along the west coast of North America from mid-Baja California, Mexico, to the eastern Aleutian Islands in Alaska (Stewart et al., this volume). The distribution of southern elephant seals is circumpolar on island and mainland sites, from the Antarctic continent to Patagonia, but concentrated on subantarctic islands (Laws, this volume).

The origins of elephant seals, like all pinnipeds, are obscure, but there is general agreement that phocids originated in the Tethyan-Mediterranean part of what is now Asia during the middle Miocene, about 15 to 20 million years ago (King 1964; Hendey 1972; Hendey and Repenning 1972; Repenning 1980). J. L. Davies (1958), drawing heavily on J. E. King (1956), argues that Antarctic seals derived from ancestral monachinoid seals that spread southward to the Caribbean at the time of seawater cooling in the Miocene and thus took advantage of the absence of the Central American isthmian barrier and entered the East Pacific cold-water route to the south. C. A. Repenning (1980) also thinks that phocid seals first invaded the south during the Miocene, approximately 10 million years ago. The earliest fossil records of seals in Argentina and South Africa date from this time.

J. L. Davies (1958) and Q. B. Hendey (1972) argue that the elephant seal genus developed in the Antarctic and the present species, *M. leonina*, colonized most of the anti-Boreal zone. They reason that elephant seals reinvaded the Northern Hemisphere by retracing the ancestral route along the west coast of South America during a Pleistocene glacial age, a time when passage to the Caribbean was closed. When the rewarming of the seas occurred, the group in the lower California region, which we now know as *M. angustirostris*, was cut off from the main elephant seal population to the south. According to Hendey (1972:108), "*Mirounga angustirostris* can thus be regarded as a relict species, surviving in isolation far from the origins of the genus."

K. T. Briggs and G. V. Morejohn (1976) present a different interpretation. They argue that the presence of fossil elephant seals in California and the relative primitiveness of the northern species (more complicated teeth and reduced sexual dimorphism) are inconsistent with a putative southern origin. They propose that the genus originated in the subtropics from unspecialized ancestors of modern monk seals. Ancestors of *Mirounga* entered the Pacific by way of the Central American Seaway. Uplift of the Tertiary Central American Seaway and poleward retreat of elephant seal populations, due to climatic conditions, led to geographic isolation, speciation, and the present distribution of the two species. According to this interpretation, the northern species is the older of the two. The southern species either evolved directly from the northern congener or they shared a common ancestor.

The poor fossil record does not permit an adequate test of these competing hypotheses. It is also difficult to say how long the two species have been separated. Perhaps the separation was as recent as the last major glaciation, which ended about 5 to 10 thousand years ago, or possibly it began as long ago as the early Pleistocene. Elephant seals were in California 100 to 130 thousand years ago as revealed by fossils found in the San Diego formation (Miller 1971), about the time of the beginning of the last glaciation. Other fossil material from California may be from 1 to 4 million years old, but whether the material is definitely *Mirounga* is in question (C. A. Repenning, pers. comm.). That the two species are considered distinct and differ in many ways argues for a separation long enough to have permitted the evolution of different structures and behaviors. The distance separating the breeding ranges of the two species today is more than 8,000 km, and there is no evidence of intermingling.

NOMENCLATURE

Whether the ancestors of elephant seals headed north after entering the Pacific or headed south to the Antarctic, or took both directions, remains moot, but it is clear that the name for the genus has a southern origin. *Mirounga* is from *miouroung*, an old Australian Aboriginal name for elephant seals. A more fitting name, *Macrorhinus*, was proposed by Cuvier in 1824, but the name had already been given to an insect, so *Mirounga*, suggested by J. E. Gray (1827), prevailed.

The species names have even more frivolous origins (King 1964; Le Boeuf 1989). T. Gill (1866) proposed *angustirostris*, or narrow nose, for northern elephant seals on the coasts of western North America, to differentiate them from *leonina* in the Antarctic. He was impressed by the "peculiarly narrowed and pronounced snout" of the only northern specimen he had for examination, a female skull. This is ironic because the most outstanding feature of the animal is the elaborate nasal appendage of the male, not the subtle anatomical character of the female skull. Linnaeus's name for the southern species, *leonina*, was based on Lord Anson's erroneous description (Linnaeus 1758); the southern species bears no resemblance to a lion or a sea lion.

Even among the common names, there are errors and inconsistencies that have come into common usage. Calling adult elephant seals "bulls" and "cows" and the newborns "pups" is the legacy of sealers. Seals breed on "rookeries," a term first used to describe the breeding grounds of gregarious birds, and polygynous male seals defend "harems" of females.

APPEARANCES

Elephant seals are extremely sexually dimorphic, fully adult males being up to ten times larger than adult females (figs. 1.1 and 1.2). Males also have distinctive secondary sexual characteristics besides size which females lack—an enlarged proboscis and thick skin on the sides and underside of the neck, ruddy and more rugose in the northern species. The short pelage of adults varies from gray to brown, except for immediately following a drastic annual molt when the pelage takes on a gray or silver hue. The elephant seals share with the monk seals, *Monachus* sp., a drastic molt process unlike that of any other mammal, during which the hairs are shed attached by their roots to large sheets of molted epidermis (Ling and Bryden 1992; Worthy et al. 1992). Pups have a woolly black natal pelage that is molted at 4 to 6 weeks of age and replaced by a silver-gray coat (Laws 1956*a*). The new hair pushes the old coat upward and out. The new hair is short and flat but grows to a length of about 12 to 14 mm during the two weeks after the old coat has been shed. (The duration and location of molting are discussed below.) The fusiform body has been modified by selection to reduce drag in the water; the male penis and testicles and female mammary glands are internalized when not in use.

SIMILARITIES AND DIFFERENCES BETWEEN THE TWO SPECIES

The close relationship between the two species is at once obvious to the casual observer from the similarity in size, appearance, and movements. Behavior of the two species is, on the whole, remarkably similar on land and at sea. For example, the social organization of males competing for access to females during the breeding season is similar. We describe these similarities in more detail in later sections.

Laws (1956*a*) described the behavior of a small population of southern elephant seals in the South Orkney Islands, breeding on sea ice in spring and present under fast ice, in which they kept open breathing holes, in early winter. Similar behavior occurs in the South Shetland Islands, and elephant seals have been seen hauled out on pack ice (R. M. Laws, unpubl. observ.). This is in extreme contrast to some subtropical breeding locations of the northern species, for example, desert islands on the west

Fig. 1.1. A northern elephant seal bull bellows a threat vocalization to a competitor. An adult female is in the foreground, and others are in the background. Photograph by Franz Lanting.

Fig. 1.2. Two southern elephant seal bulls fighting at Península Valdez, Argentina. Photograph by Burney J. Le Boeuf.

coast of Baja California, Mexico. The dive depths and dive durations during transit and foraging are remarkably similar despite occurring in disparate parts of the ocean (Le Boeuf et al. 1988; Le Boeuf, this volume; Stewart and DeLong 1991; Stewart and DeLong, this volume; Hindell 1990; Fedak et al., this volume).

Differences between the species have not been subjected to special study; we describe the most obvious. The southern male is longer and heavier than its northern counterpart. Southern bulls in harems weigh 1,500 to 3,000 kg, with maximum weights reaching 3,700 kg (Ling and Bryden 1981). The largest northern males weigh 2,300 kg (Deutsch, Haley, and Le Boeuf 1990). Females of tne two species do not appear to differ significantly in mass. Southern females range widely in mass from 350 to 800 kg shortly after giving birth, with most of them in the range 400 to 600 kg (Fedak et al., this volume). Northern females have a postpartum mass ranging from 360 to 710 kg (Deutsch et al., this volume). The mean mass of northern pups, however, may exceed that of southern pups at weaning, 131 kg versus 121 kg (Le Boeuf, Condit, and Reiter 1989; Deutsch et al., this volume; Fedak et al., this volume), perhaps owing to a difference in suckling duration (Le Boeuf, Whiting, and Gantt 1972; Laws 1953b; McCann 1980). However, at South Georgia, mean male weaning mass in four years ranged from 118.7 to 137.2 kg (SCAR 1991), and more southerly populations of *M. leonina* have higher weaning weights; for example, R. M. Laws (1953b) found weaning weights of about 200 kg at Signy Island in 1948–1949, and H. Burton (pers. comm.) reports high weaning weights at King George Island.

Laws (1953b) and R. Carrick, S. E. Csordas, and S. E. Ingham (1962) report southern bulls measuring 6.2 m, and others refer to males of 7.62 to 9.14 m; the early records, however, probably included hind flippers and are not comparable (see Scheffer 1958). C. M. Scammon (1874) reports a northern male that measured 6.71 m, but the longest northern bull measured in recent times was 5.03 m long (B. Le Boeuf, unpubl. data). The average lengths of males seem to be less disparate across species; Laws (1960) estimated 4.72 m for southern males, and mean estimates for northern males range from 4.33 m (Deutsch et al., this volume) to 4.48 m (Clinton, this volume).

Some morphological differences between the species beg explanation. For example, the southern elephant seal can bend its body backward into a U-shape over a much greater angle than the northern elephant seal. Perhaps because of this difference, southern males also seem to be able to rear up higher during fighting than northern males. Paradoxically, despite greater sexual dimorphism in skull characteristics in southern elephant seals, the fleshy exterior of the northern species is more sexually dimorphic. The proboscis of the northern male is larger and the integumentary neck and chest shield are more highly developed than these features in the southern male (Murphy 1914; Laws 1953*b*).

Developmental differences in early life deserve further study. Southern pups are weaned at 22 to 23 days of age (Laws 1953*b*; McCann 1980; Campagna, Lewis, and Baldi 1993); northern pups are weaned at 24 to 28 days, nursing duration increasing with the age of the mother (Reiter, Panken, and Le Boeuf 1981). Some southern elephant seal pups begin molting while still suckling; a small percentage molt in utero (Laws 1953*b*, 1956*a*; Carrick et al. 1962; Le Boeuf and Petrinovich 1974). For pups at Signy Island, Laws (1953*b*) reports that the molt lasts about 24 days and is completed by 30 to 38 days of age. The molt in northern elephant seal pups is equally long but does not begin until after weaning, at about 28 days of age; the process begins slightly later in males than in females (Reiter, Stinson, and Le Boeuf 1978).

In the southern elephant seal, weaned pups fast for an average of 37 days, and the duration of the postweaning fast increases linearly with increasing weight at weaning. Pups continue to fast until they reach a lower weight threshold of about 70% of weaning weight (Wilkinson and Bester 1990). The duration of the fast in the Northern Hemisphere varies with the date of weaning. Pups weaned early fast for a mean of 73.5 ± 7.6 days; pups weaned late in the season fast for a mean of 55.6 ± 13.2 days (Reiter, Stinson, and Le Boeuf 1978). As in the south, the duration of the fast is positively correlated with weaning mass, and at departure from the rookery, pups have lost an average of 25 to 30% of their weaning weight. Some pups of both species remain near the rookery for up to 2 to 2 1/2 months before going to sea for the first time (Reiter, Stinson, and Le Boeuf 1978; Condy 1979).

The threat vocalizations of males are quite distinct, evidently a divergence that has come about with geographic separation (Le Boeuf and Petrinovich 1974). The threat call of northern males is composed of 3 to 20 discrete expulsive, low-frequency bursts that are emitted at a relatively constant rate, with some individuals adding a longer embellishment to the beginning or end of the call. The mean duration of the call is 6.8 ± 2.5 seconds. In contrast, the threat vocalizations of southern males are more than twice as long (mean = 19.1 ± 8.3 seconds) and are composed of long roars that vary in pitch and loudness, the result of a sound being produced during inspiration as well as during expiration.

The near-annihilation of the northern elephant seal population by sealers in the last century, compared to the less destructive sealing efforts in the larger and more widely distributed elephant seal populations in the Southern Hemisphere, has apparently resulted in differences in genetic variation. The extreme population bottleneck experienced by the northern species—a reduction to less than 100 individuals breeding on only one island in the late 1880s—is interpreted as being responsible for lack of allozyme diversity and reduced DNA sequence diversity in two mtDNA regions, relative to southern elephant seals (Bonnell and Selander 1974; Hoelzel et al. 1993).

TERRESTRIAL HABITAT

The traditional rookeries and haul-out sites of elephant seals are islands or remote continental shores. Preferred breeding and resting sites are usually located on gradually sloping, sandy beaches or sand spits. Given similar accessibility, elephant seals settle first on beaches with a fine sandy substrate, next on pebbles, and as a last resort on boulders or rocky shores. A sandy substrate is ideal for the caterpillarlike movements of these large animals. Moreover, in both species, flipping damp sand, loose dirt, or small damp pebbles on their backs aids in temperature regulation on warm days (Laws 1956a; White and Odell 1971; Heath and Schusterman 1975).

THE ANNUAL CYCLE ON LAND

The annual cycle of the two species is similar except that the time scale is shifted (fig. 1.3). *M. leonina* breeds in the austral spring (early September to mid-November), while *M. angustirostris* breeds in the northern winter (December to February). Temporal intervals between reproductive events are similar (Carrick et al. 1962; Le Boeuf and Petrinovich 1974). There are slight shifts in peak events, such as the maximum number of females during the breeding season, among rookeries distributed widely across latitudes. This is especially well documented in the southern elephant seal (Laws 1956a; Condy 1979; McCann 1985; Hindell and Burton 1988). As one proceeds from northern rookeries such as Peninsula Valdez, Argentina, to southern ones like the islands of King George and South Georgia (across 22 degrees of latitude or about 1,736 km), the peak number of females is delayed by 22 days, from October 3 to October 25 (McCann 1985; Campagna, Lewis, and Baldi 1993). Variation in the onset of the season in the northern species, from one rookery to the next, has not been reported, but differences, if they exist, are thought to be slight.

The annual cycle of northern elephant seals at Año Nuevo, California, seems to be typical of that of other colonies of seals in the Northern Hemisphere (with the possible exception of numbers during the female and juvenile molt; see fig. 1.3a) but differs in some respects from southern colonies, not only in the shift in seasons but in the timing of haul-out in different age and sex groups. The annual cycle is described below for Año Nuevo and for Macquarie Island.

Northern Species

It is useful to divide the annual cycle into four terrestrial phases: breeding season, female and juvenile molt, male molt, and juvenile haul-out. The relative numbers present on the rookery during each phase is shown in figure 1.3a.

Breeding Season: December to Mid-March. The breeding season at Año Nuevo begins in early December with the arrival of the adult males. Usually, the older bulls arrive first, and all serious competitors are on the rookery by the end of December. Concurrent with the arrival of adult males, there is a rapid decline in the number of juveniles, 1 to 4 years old, that previously predominated on the rookery. Pregnant females begin arriving in mid-December, reach a peak during the period from January 26 to February 2, and then their numbers decline until all of them have returned to sea by the end of the first week in March (fig. 1.4). Younger males begin leaving the rookery in late February, but the larger bulls remain on the rookery until the end of March, long after the last female has departed.

The pupping period is from about the third week in December to the end of the first week in February. Copulations occur from the first week in January through the first week in March, with February 14 being the peak day of copulation frequency.

Female and Juvenile Molt: Mid-March through May. As the last females wean their pups and return to sea, the first females that gave birth early in December begin returning from sea in mid-March to molt, a process that takes about one month to complete. This influx of adult females continues for about two months. The adult females are joined by juveniles, 1 to 4 years old, of both sexes. The highest number of animals present on the rookery are seen in late April. Present in the spring but declining in number are adult males, all of whom return to sea by the end of March, and the newly weaned pups, 80% of whom leave the rookery by the end of April. Rather suddenly, in early to late May, there is a rapid decline in total numbers on the rookery.

Male Molt: June through August. The lowest number of animals are observed on the rookery in June, July, and August (fig. 1.3a), when breeding-age males molt. There is a tendency for the younger pubertal males, 5 to 6 years old, to arrive in early summer, and they are followed by the older males in late summer (fig. 9.1, this volume). A few young of the year and 1½-year-olds are observed early in this period, but they make up less than 5% of the total number of seals in residence. Juvenile numbers begin to increase in August, and in some years, they outnumber molting males by the end of the month.

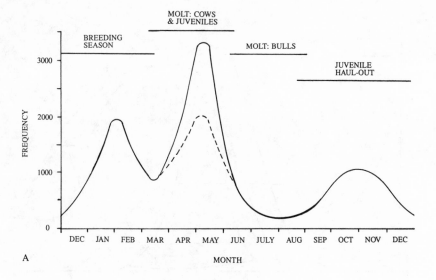

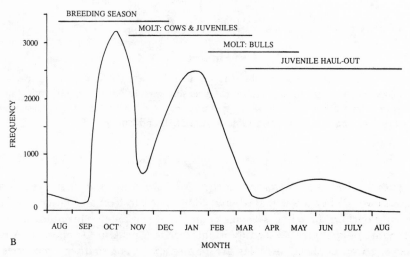

Fig. 1.3. A schematic representation of the annual cycle of northern and southern elephant seals showing the relative number of animals present throughout the year. The times of breeding and other events during the annual cycle are indicated by horizontal bars. *(A)* The relative number of northern elephant seals on a rookery such as Año Nuevo Island during the late 1970s (solid line). The dashed line indicates that for some rookeries (e.g., San Nicolas and San Miguel), the number of animals during the molt of cows and juveniles is nearly the same as the number observed during the breeding season (Le Boeuf and Bonnell 1980). *(B)* The relative number of southern elephant seals present on a typical rookery during the annual cycle (based on data from Laws 1956a and Hindell and Burton 1988).

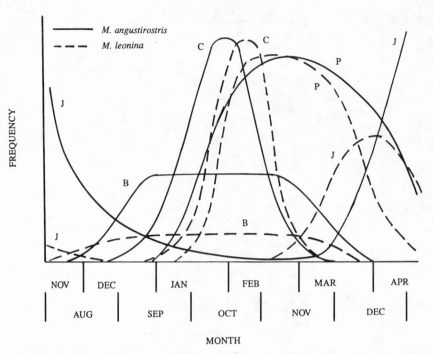

Fig. 1.4. A schematic diagram showing the increase and decrease in categories of elephant seals present during the breeding season on Año Nuevo Island (solid line; adapted from Le Boeuf 1972) and a typical rookery of the southern elephant seal (dashed line; adapted from Laws 1956a and Hindell and Burton 1988).

Juvenile Haul-Out: September through November. By the first week in September, all but a few males have gone to sea to feed and fatten up for the approaching breeding season. Young of the year and juveniles up to four years of age increase steadily in number to a peak in October. At Año Nuevo, many of these juveniles are immigrants from southern rookeries. Some yearlings exhibit a pathological skin and pelage called "scabby molt," but as a rule, normal molting does not occur during this period. Pubertal, subadult males begin to arrive on the rookery during the middle of November, a time when juveniles are decreasing in number.

The schematic representation of the total number of elephant seals present on Año Nuevo throughout the year (fig. 1.3a) shows the relative number of animals present in each phase of the annual cycle. The temporal patterning of peaks and troughs in total animals, as well as the relative size of the peaks to each other, has varied little at this rookery during the last two decades. Elephant seals are in residence throughout the year, but their numbers fluctuate greatly because animals of both sexes and different age groups are moving in and out predictably.

The pattern in figure 1.3a resembles that observed at other northern elephant seal colonies, such as San Nicolas, San Miguel, and Santa Barbara islands (Odell 1974; Bonnell et al. 1979), during the same period that Año Nuevo was monitored (Le Boeuf 1981). The difference is that the female-juvenile molt numbers are equal to or only slightly higher than breeding season numbers. This is because animals from southern California rookeries may haul out to molt in a different place from where they breed, showing up to be counted on northern California rookeries like Año Nuevo. (Similarly, some South Georgia elephant seals haul out to molt at Signy Island, Elephant Island, and South Shetland Islands; Macquarie Island animals haul out on the Antarctic continent. Also, bulls haul out to molt in limited numbers at Año Nuevo compared with colonies of the southern species.)

Southern Species

Information for the southern species is drawn from R. M. Laws (1956*a*), R. Carrick et al. (1962), and M. A. Hindell and H. R. Burton (1988) (fig. 1.3b).

Breeding Season: August to Mid-December.   At Macquarie Island, the adult males arrive in early August, and all serious competitors are present by mid-September (fig. 1.4). Small numbers of juveniles (females of 1 to 4 years; males of 1 to 6 years) haul out in winter but decline to near zero by the end of August. Pregnant females begin to arrive in mid-September, reaching a peak about October 16, subsequently declining so that all have returned to sea by the end of November. Adult bulls remain until mid-December. The pupping period is from late September to the end of October. Copulations occur from mid-October to mid-November, with peak frequency in early November.

Female and Juvenile Molt: January and February.   Mature females begin returning from sea about the beginning of January, to molt for about one month, reaching peak numbers about the end of January; all have returned to sea by the end of February. The juvenile haul-out begins much earlier, in mid-November, reaches a peak about mid-December, and ends in late January.

Male Molt: November to April.   Subadult males (6 to 8 years old) molt from early November to early March, with a peak at the end of December. Adult males haul out for molt from early February until May, with a peak in mid-March. The lowest numbers are observed on the rookeries in winter, July to August, representing a haul-out of adult males and juveniles.

Juvenile Haul-Out: Mid-March to the End of August.   By the end of August, males have completed their molt and gone to sea to feed and lay down blubber for the breeding season. At Macquarie Island, there is a winter haul-out of juveniles from April to August, with peak numbers in early June less than half of the peak molting haul-out. At Macquarie, juveniles are present ashore for a greater proportion of the year than any other population component. At Signy Island and South Georgia, there is no significant winter haul-out.

At Signy Island, because the breeding population is small, the highest numbers are ashore molting in the last week of January, representing immigrants from South Georgia. But at Macquarie Island and South Georgia, the largest numbers are present in the breeding season.

Thus, the general haul-out cycles for northern and southern elephant seals are broadly similar in the breeding season, but there are significant differences in detail throughout the rest of the year. The overall breeding season is longer in the southern species, mainly due to earlier haul-out of the bulls, but the pupping season is more synchronous with a more pronounced peak in numbers of cows; weaned pups depart to sea earlier (fig. 1.4). The female and juvenile molt is slightly shorter and later at Macquarie Island than at Año Nuevo. The period when males are ashore for the molt is twice as long at Macquarie as at Año Nuevo, starting earlier and ending earlier at Macquarie; also, few adult males molt at Año Nuevo (i.e., only the males that breed there). The juvenile haul-out is similar in the two locations, although slightly longer and earlier at Macquarie. At South Georgia and Signy Island, south of the Antarctic Convergence, it would be expected to be earlier than at Año Nuevo or Macquarie owing to lower air temperatures (and pack ice around Signy Island) from April onward. The limited data confirm this (Laws 1956a).

Some colony and species differences in the shape of the annual cycle of total animals are to be expected given that there are differences in molting locations, in the time that food is abundant, and in the distance of prey from the rookery. By and large, the presence of elephant seals on traditional rookeries and hauling grounds is predictable by age and sex.

TIME AT SEA

It is useful to consider the time at sea from the individual point of view. This varies with age and sex (fig. 13.1, this volume). In northern elephant seals, pups make their first contact with the water within two weeks of weaning at about six weeks of age, learning to swim and dive in shallow water near the rookery, until they depart on their first trip to sea at 3½ months of age (Reiter, Stinson, and Le Boeuf 1978; Thorson and Le Boeuf, this volume). After 4 to 5 months at sea, they return to the rookery for about one month in the fall, then return to sea for another 4 to 5 months. Both sexes take two trips to sea per year until they undergo puberty. When they come of reproductive age, the fall haul-out is skipped, and the seal remains at sea for 11 months of the year. Thereafter, the adult pattern is adopted. Females make a postlactation trip to sea lasting a mean of 72.6 ± 5.0 days, followed by a one-month molt on land, followed by another 8-month trip to sea over the course of pregnancy. Thus, adult females are at sea for 10 months of the year (Le Boeuf et al. 1988; Le Boeuf, this volume). Adult males also make two trips to sea per year but spend less total time at sea, 8 months, than females (Le Boeuf, this volume; Stewart and DeLong, this volume). Southern elephant seals appear to hold to a similar schedule at sea (Laws 1956a; Hindell and Burton 1988).

Research on the migratory paths and location of foraging areas of both species is in progress (Hindell, Slip, and Burton 1991; DeLong, Stewart, and Hill 1992; Bester and Pansegrouw 1992; McConnell, Chambers, and Fedak 1992; Le Boeuf et al. 1993) and is treated in several chapters in this volume (chaps. 5, 13, 16, and 20).

REPRODUCTION

Social Organization and Mating Behavior
Bulls are belligerent when they arrive on the rookery at the start of the breeding season, threatening and fighting with each other in areas where females will settle to give birth. When pregnant females begin arriving, they seek each other out for safety from the sexually aggressive males and gather in groups, or harems. The result of the fierce, bloody encounters between males is a dominance hierarchy at each breeding area (Le Boeuf and Peterson 1969; Le Boeuf 1974; McCann 1981). This social structure effectively reduces access to the grouped females to a few of the highest-ranking males in the area. One male, the alpha or beachmaster, dominates all other males and keeps them away from females with the species-specific threat vocalization delivered with the head and neck elevated, braced up with the foreflippers. The second-ranked male keeps all other males away from females, too, but defers to the top-ranked male, and so it goes with decreasing rank. Social rank is directly associated with access to females; the alpha bull situates himself in the middle of the group of females. Although small harems of 100 females or less may be ruled by one bull, as the female group increases in number, it becomes too time- and energy-consuming for the alpha male to keep all other males out. Other males, next in rank, take up peripheral positions in the harem. In small colonies, the harems are smaller and discrete, each controlled by a single dominant bull, with subordinate bulls positioned around but outside the harem. In very large colonies, like

some beaches at South Georgia, there may be a nearly continuous mass of breeding females stretching for a kilometer or more, without individual harems. Alpha males and subordinate bulls are then positioned at intervals among the females (Laws 1956a).

Mating Success of Males

The ultimate result of this power structure is that few males mate, and male mating success is directly related to social rank in the dominance hierarchy. At Año Nuevo, as few as 5 out of 180 males were responsible for up to 92% of the copulations observed with up to 470 females during a breeding season; one male held the alpha position in a large harem and dominated mating for four consecutive years, inseminating an estimated 200 females (Le Boeuf 1974). In the southern elephant seal, M. N. Bester and I. S. Wilkinson (this volume) recorded that "dominant bulls controlling the harems achieved over 98% of all matings."

A study of lifetime reproductive success at Año Nuevo confirms that there is great variance in male reproductive success (Le Boeuf and Reiter 1988). In a sample of 91 male pups, only 19 reached breeding age. Three males were extremely successful, inseminating an estimated 121, 97, and 63 females, respectively; 5 other males apparently inseminated 69 females, and the remaining 11 surviving males (as well as the 72 nonsurvivors) failed to mate. The males most successful at mating achieved high ranks in the dominance hierarchies associated with harems.

Although males undergo puberty at about 5 years of age in the northern species and 4 to 5 years of age in the southern species, they do not achieve high rank until they are older, at least eight years of age, and larger (Laws 1956a; McCann 1985; Clinton, this volume; Deutsch et al., this volume). Prime breeding years for northern males are at age 9 to 12, males are in decline at age 13, and 14 years is the maximum life span (Le Boeuf and Reiter 1988). The majority of breeding bulls in the southern species are also 9 to 12 years old, but the maximum life span is 20 years (Laws 1953a; McCann 1985). Despite the predominance of adult males in mating and male-male competition, four other categories of subadult males (SAMs) are present during the breeding season: SAM1 = 4 years old, SAM2 = 5 years old, SAM3 = 6 years old, and SAM4 = 7 years old (Le Boeuf 1974; Cox and Le Boeuf 1977; Clinton, this volume).

The Reproductive Cycle of Females

There is strong circumstantial evidence that most females of both species mate for the first time at sea (Laws 1956a, 1956b, this volume; B. J. Le Boeuf, unpubl. observ.). At Año Nuevo, only a few identifiable (marked) virgin females mate on land each year, late in the breeding season; the vast majority must mate at sea because they are not observed elsewhere on land, either at this time or later in the season. Studies on the southern species demonstrated that in parous females, after copulations and fertilization, development is suspended and the blastocyst remains free in the uterus for about 4½ months; implantation occurs at the end of the summer molt, and active embryonic growth occupies about 7½ months (Laws 1956b).

Within 6 days after a pregnant northern elephant seal arrives on the rookery during the breeding season, she gives birth to a single pup (this period in the southern elephant seal is 5 days). She keeps it near her and nurses it daily for 27 days (23 days in the southern elephant seal), all the while remaining in the harem and fasting from food and water. During the last 3 to 5 days of nursing, she copulates (about 4 days in the southern elephant seal). Four weeks after giving birth, 34 days after arriving on the rookery (28 days in the southern elephant seal), and on her last day of estrus, the female weans her pup by returning to sea (Le Boeuf, Whiting, and Gantt 1972; Laws 1953b, 1956a, 1956b).

Age at Primiparity and Natality.   In expanding colonies in northern California, females give birth for the first time at age 3 to 6, with 4 years of age being the mean age at primiparity (Reiter, Panken, and Le Boeuf 1981; Huber 1987; Le Boeuf and Reiter 1988). High local density of breeding females is correlated with deferred maturity (Huber et al. 1991). Most females produce a pup annually until they die at a maximum age of about 20 years; however, skipping a year has been observed in some rookeries, especially following first breeding (Huber 1987; Huber et al. 1991).

At South Georgia, *M. leonina* females also give birth for the first time at age 3 to 6, the majority at age 4. At Macquarie Island, there was deferment of the range and mean by one year (McCann 1980; Carrick et al. 1962). The reason for the difference is not clear, although body growth rates are lower at Macquarie Island. M. A. Hindell and G. J. Little (1988) report two known-age females observed suckling pups on Macquarie Island at age 23.

Female Reproductive Success.   Female weaning success increases with age up to at least age 8 (Reiter, Panken, and Le Boeuf 1981; Le Boeuf and Reiter 1988; Huber 1987; Huber et al. 1991; Sydeman et al. 1991; Deutsch et al., this volume). Reproductive experience may have a positive (Reiter, Panken, and Le Boeuf 1981) or a negative effect on subsequent weaning success (Sydeman et al. 1991; Sydeman and Nur, this volume). An important factor in weaning success is female mass, which increases with age (Deutsch, this volume). At Año Nuevo, older, larger females dominate younger, smaller females, displacing them or their pups and preempting areas in the harem that are less prone to nursing disruptions caused by peripheral males or high water (Christenson and Le Boeuf 1978).

Preweaning Pup Mortality. The major cause of pup mortality on the rookery and lack of weaning success is the trauma-starvation syndrome that begins with mother-pup separation and ends with starvation or lethal injury inflicted by adults (Le Boeuf and Briggs 1977). Trauma results from crushing by breeding bulls, or from orphaned pups being bitten while attempting to suckle alien females. Three classes of variables affect mother-pup separation and increase pup mortality either directly or indirectly: female aggression; the number, distribution, and density of females; and winter storms that flood the harems at peak season.

An important variable affecting the maximum pup mortality rate is the size of the breeding unit in relation to the topography of the breeding site. At Año Nuevo, if mothers and their pups can move inland to higher ground when surf and high water threatens, such as occurs on large island beaches or mainland breeding sites, the annual pup mortality rate is usually less than 10% of pups born. If, however, there is no fallback position, due to wave cut platforms or high cliffs, the pup mortality rate may rise to 75 to 100% (Le Boeuf and Condit 1983; Stewart and Yochem 1991). On northern elephant seal island colonies, where the pup mortality rate has been monitored closely, the annual rate is usually in the range of 10 to 40% (Le Boeuf and Reiter 1991; Huber, Beckham, and Nisbet 1991). In the southern species, it has been recorded at 2 to 6% on average (Laws 1953*b*; Condy 1978). It is higher in unfavorable conditions. Thus, at South Georgia, in some situations on land at the beginning of the season, pups melted deep holes in the snow, which prevented them from sucking and resulted in mortality as high as 30%. Mortality (80%) was associated with ice-breeding colonies at Signy Island due to breakup of the fast ice in storms (Laws 1953*b*).

Nursing and Adoption. Most females nurse their own pups exclusively until weaning; however, some females that lose their own pups adopt orphans and raise them as they would their own. In a study by M. L. Riedman and B. J. Le Boeuf (1982) at Año Nuevo, 5% of the orphans reunited with their mothers, 27% were adopted or frequently cared for by foster mothers, and 68% remained orphaned and died. Since young, inexperienced females most often lost their pups, they were the ones most likely to foster an orphan. Usually, they adopted pups that were the same age as those they had lost. Adoption of a single pup was most common, but pupless females also attempted to steal suckling pups, adopted a weaned pup, adopted two pups, or indiscriminately nursed any orphan that approached. Nursing between unrelated cows and pups has been observed in southern elephant seals, but it is unusual. Orphaned pups that attempt to steal milk are usually unsuccessful (McCann 1982).

At Año Nuevo, a few "superweaners" may be produced when pups are suckled by two "mothers" or are adopted by a pupless female after being weaned by their own mother (Reiter, Stinson, and Le Boeuf 1978; Le Boeuf 1981). These weanlings may attain twice the mass of normal weanlings. In addition, some weanlings attempt to steal milk from nursing females by stealth or perseverance; the majority of these milk thieves are males.

Philopatry and Site Fidelity. Like most other pinnipeds and many terrestrial mammals, most female elephant seals give birth on the rookery where they were born in roughly the same site from year to year. Seventy-one percent of the females born at Año Nuevo during the 1970s returned to give birth there for the first time (Reiter, Panken, and Le Boeuf 1981). The rest moved to adjacent rookeries as their birth site became crowded (Le Boeuf, Ainley, and Lewis 1974). Similarly, 70% of females monitored returned to give birth at the original site the following year. Movement to a new site was associated with failing to wean a pup or high density of breeding females. Similarly, D. G. Nicholls (1970) reported that at Macquarie Island, 77% of branded cows up to 11 years old were found breeding within 4 km of their birth site, and Hindell and Little (1988) reported two 23-year-old cows breeding within 1 km of their birth site.

FASTING: A KEY LIFE HISTORY TRAIT

Fasting on land is an integral part of the life history strategy of elephant seals (Bartholomew 1970). Consider the breeding bull. Clumping of females into groups enhances the potential for polygyny (Emlen and Oring 1977). An alpha bull that can keep others away from a group of females will mate with many females and sire many offspring, provided that he remains near the females night and day keeping competitors away. If he has to make periodic trips to sea to feed, other males will mate in his absence. Clearly, the mating success of a male that dominates his competitors will depend on the length of his uninterrupted tenure on the rookery during the mating season. Breeding males fast for over 100 days while fighting for social rank and competing for mates (Le Boeuf 1974; Wilkinson and Bester 1990; Deutsch et al., this volume; Clinton, this volume).

Females fast while nursing, which provides numerous advantages. By not having to feed while nursing, as sea lions and fur seals do, an elephant seal transfers a great deal of milk energy to her pup in a short time (Costa et al. 1986; Fedak et al., this volume). This frees the mother to prepare for future offspring, to forage on her own without encumbrance.

Fasting is imposed on weaned pups by the departure of the mother. They fast while learning to swim and dive before embarking on a trip to sea to find food (Reiter, Stinson, and Le Boeuf 1978; Thorson and Le Boeuf, this volume).

To reap the benefits of fasting, elephant seals must be able to accumulate sufficient energy during their long trips to sea, they must have an enormous capacity to store fat, and they must economize this energy store while on land. Evidently, they have no problem satisfying the first two requirements (Laws 1953*b*; Le Boeuf et al. 1988; Le Boeuf, this volume; Fedak et al., this volume). We emphasize some behaviors that minimize energy expenditure on land.

When elephant seals are not fighting, mating, nursing, or learning water skills, as in the case of newly weaned pups, they sleep. Sleep does not follow a diel pattern. It occurs at all hours but especially during the heat of the day. Among breeding males and nursing females, it is clearly opportunistic. Moreover, elephant seals undergo long apneas during sleep lasting up to 25 minutes (Blackwell and Le Boeuf 1993; Castellini, this volume). This aperiodic pattern of breathing is one of several ways to economize water and save energy (see also Huntley, Costa, and Rubin 1984). Since a major avenue of water loss is through respiration, reducing the number of exhalations by breath holding during sleep reduces water loss. This means a reduction in lipid metabolism, the sole source of water for a fasting seal (Ortiz, Costa, and Le Boeuf 1978).

Besides serving as an energy reserve and a source of water during long fasts, the fat stores, or blubber, provide insulation against the cold both on land and at sea. Conversely, overheating as a result of active movement is minimized in seals by a countercurrent system in the uninsulated flippers.

FORAGING

Elephant seals exhibit long, deep, and nearly continuous diving at sea (Le Boeuf et al. 1986, 1988, 1989; Hindell, Slip, and Burton 1991; Boyd and Arnbom 1991; Stewart and DeLong 1991). Foraging occurs during these journeys that take them long distances from their rookeries. Prey, foraging locations, and the unusual diving pattern of both species are subjects of much current interest and intensive study (e.g., Hindell, Burton, and Slip 1991; DeLong, Stewart, and Hill 1992; McConnell et al. 1992; Bester and Pansegrouw 1992); these topics are treated in detail in several chapters in this volume (chaps. 5, 11, 13, 14, 15, 16, and 20).

PREDATION

The major predators on elephant seals are white sharks, *Carcharodon carcharias*, and killer whales, *Orca orcinus* (Laws 1953*b*; Ainley et al. 1981; Le Boeuf, Riedman, and Keyes 1982). Shark attacks on northern elephant seals have been observed throughout the breeding range. White sharks kill elephant seals near the surface, victimizing both sexes and all age groups, including the largest bulls. At the Farallon Islands, near the mouth of San Francisco Bay, white shark predation is restricted to juveniles during the fall haul-out and is localized in the zone of shallow water ($<$ 11 m) within 0.4 km of the island (Klimley et al. 1992). At Año Nuevo, peak attacks on elephant seals occur in December and January, and breeding age males are most frequently victimized (Le Boeuf, Riedman, and Keyes 1982). The contribution of white shark predation to total mortality at sea is unclear, but the diving behavior of elephant seals in the high-risk zone, moving to and from the Año Nuevo rookery over the continental shelf, suggests selection for predator avoidance (Le Boeuf and Crocker 1993). Predation by killer whales has been documented less frequently than predation by sharks (Laws 1953b; M. Pierson, pers. observ.). Leopard seals, *Hydrurga leptonyx*, will occasionally take a southern elephant seal but cannot be classed as an important predator (Laws 1953b).

The cookiecutter shark of the genus *Isistius* bites out circular chunks of skin and blubber the size of a tennis ball from northern elephant seals in the southern part of their range, but since the craterlike injuries are not lethal, this is more a matter of parasitism than predation (Le Boeuf, McCosker, and Hewitt 1987).

Historically, man has been the major predator on both species of elephant seals, killing them for their oil. The history of sealing in the Southern Hemisphere is reviewed by Laws (this volume) and W. N. Bonner (1982); the decline in numbers and the virtual annihilation of northern elephant seals due to sealing is reviewed by G. A. Bartholomew and C. L. Hubbs (1960) and B. C. Busch (1985) and addressed in chapter 2 of this volume.

REFERENCES

Ainley, D. G., C. S. Strong, H. P. Huber, T. J. Lewis, and S. H. Morrell. 1981. Predation by sharks on pinnipeds at the Farallon Islands. *Fishery Bulletin* 78: 941–945.

Bartholomew, G. A. 1970. A model for the evolution of pinniped polygyny. *Evolution* 24: 546–559.

Bartholomew, G. A., and C. L. Hubbs. 1960. Population growth and seasonal movements of the northern elephant seal, *Mirounga angustirostris*. *Mammalia* 24: 313–324.

Bester, M. N., and H. M. Pansegrouw. 1992. Ranging behaviour of southern elephant seal cows from Marion Island. *South African Journal of Science* 88: 574–575.

Blackwell, S. B., and B. J. Le Boeuf. 1993. Developmental aspects of sleep apnoea in northern elephant seals. *Journal of Zoology, London* 231: 437–447.

Bonnell, M. L., B. J. Le Boeuf, M. O. Pierson, D. H. Dettman, and G. D. Farrens. 1979. *Summary Report 1975-1978, Marine Mammal and Seabird Surveys of the Southern*

California Bight Area. III. Pinnipeds. Bureau of Land Management, Department of the Interior, Contract AAa550-CT7-36.

Bonnell, M. L., and R. K. Selander. 1974. Elephant seals: Genetic consequences of near extinction. *Science* 184: 908–909.

Bonner, W. N. 1982. *Seals and Man: A Study of Interactions.* Seattle: University of Washington Press.

Boyd, I., and T. Arnbom. 1991. Diving behaviour in relation to water temperature in the southern elephant seal: Foraging implications. *Polar Biology* 11: 259–266.

Briggs, K. T., and G. V. Morejohn. 1976. Dentition, cranial morphology and evolution in elephant seals. *Mammalia* 40: 199–222.

Busch, B. C. 1985. *The War Against the Seals.* Montreal: McGill-Queen's University Press.

Campagna, C., M. Lewis, and R. Baldi. 1993. Breeding biology of southern elephant seals in Patagonia. *Marine Mammal Science* 9: 34–47.

Carrick, R., S. E. Csordas, and S. E. Ingham. 1962. Studies on the southern elephant seal, *Mirounga leonina* (L.). IV. Breeding and development. *CSIRO Wildlife Research* 7: 161–197.

Carrick, R., S. E. Csordas, S. E. Ingham, and K. Keith. 1962. Studies on the southern elephant seal, *Mirounga leonina* (L.). III. The annual cycle in relation to age and sex. *CSIRO Wildlife Research* 7: 119–160.

Christenson, T. E., and B. J. Le Boeuf. 1978. Aggression in the female northern elephant seal, *Mirounga angustirostris. Behaviour* 64: 158–172.

Condy, P. R. 1978. The distribution and abundance of southern elephant seals, *Mirounga leonina* (Linn.), at Marion Island. *South African Journal of Antarctic Research,* 8: 42–48.

———. 1979. Annual cycle of the southern elephant seal *Mirounga leonina* (Linn.) at Marion Island. *South African Journal of Zoology* 14: 95–102.

Costa, D. P., B. J. Le Boeuf, A. C. Huntley, and C. L. Ortiz. 1986. The energetics of lactation in the northern elephant seal. *Journal of Zoology, London* 209: 21–33.

Cox, C. R., and B. J. Le Boeuf. 1977. Female incitation of male competition: A mechanism in sexual selection. *American Naturalist* 111: 317–335.

Davies, J. L. 1958. The Pinnipedia: An essay in zoogeography. *Geographical Review* 48: 474–493.

DeLong, R. L., B. S. Stewart, and R. D. Hill. 1992. Documenting migrations of northern elephant seals using day length. *Marine Mammal Science* 8: 155–159.

Deutsch, C. J., M. P. Haley, and B. J. Le Boeuf. 1990. Reproductive effort of male northern elephant seals: Estimates from mass loss. *Canadian Journal of Zoology* 68: 2580–2593.

Emlen, S. T., and L. W. Oring. 1977. Ecology, sexual selection, and the evolution of mating systems. *Science* 197: 215–223.

Gill, T. 1866. On a new species of the genus *Macrorhinus. Proceedings of the Chicago Academy of Science* 1: 33–34.

Gray, J. E. 1827. *Mammalia.* Vol. 5 of *The Animal Kingdom Arranged in Conformity with Its Organization,* by the Baron (G.) Cuvier, with additional descriptions by Edward Griffith and others. 16 vols. London: George B. Whittaker.

Heath, M. A., and R. J. Schusterman. 1975. "Displacement" sand flipping in the northern elephant seal *(Mirounga angustirostris). Behavioral Biology* 14: 379–385.

Hendey, Q. B. 1972. The evolution and dispersal of the Monachinae (Mammalia: Pinnipedia). *Annals of the South African Museum* 59: 99–113.

Hendey, Q. B., and C. A. Repenning. 1972. A Pliocene phocid from South Africa. *Annals of the South African Museum* 59: 71–98.

Hindell, M. A. 1990. Population Dynamics and Diving Behaviour of a Declining Population of Southern Elephant Seals. Ph.D. dissertation, University of Queensland, Australia.

Hindell, M. A., and H. R. Burton. 1988. Seasonal haul-out patterns of the southern elephant seal (*Mirounga leonina* L.) at Macquarie Island. *Journal of Mammalogy* 698: 81–88.

Hindell, M. A., H. R. Burton, and D. J. Slip. 1991. The foraging areas of southern elephant seals, *Mirounga leonina*, inferred from water temperature data. *Australian Journal of Marine and Freshwater Research* 42: 115–128.

Hindell, M. A., and G. J. Little. 1988. Longevity, fertility and philopatry of two female southern elephant seals (*Mirounga leonina*) at Macquarie Island. *Marine Mammal Science* 4: 168–171.

Hindell, M. A., D. J. Slip, and H. R. Burton. 1991. The diving behaviour of adult male and female southern elephant seals, *Mirounga leonina* (Pinnipedia: Phocidae). *Australian Journal of Zoology* 39: 595–619.

Hoelzel, A. R., J. Halley, C. Campagna, T. Arnbom, B. J. Le Boeuf, S. J. O'Brien, K. Ralls, and G. A. Dover. 1993. Elephant seal genetic variation and the use of simulation models to investigate historical population bottlenecks. *Journal of Heredity* 84: 443–449.

Huber, H. R. 1987. Natality and weaning success in relation to age of first reproduction in northern elephant seals. *Canadian Journal of Zoology* 65: 1311–1316.

Huber, H. R., C. Beckham, and J. Nisbet. 1991. Effects of the 1982–83 El Niño on northern elephant seals on the South Farallon Islands, California. In *Pinnipeds and El Niño: Responses to Environmental Stress*, ed. F. Trillmich and K. Ono, 219–233. Berlin: Springer Verlag.

Huber, H. R., A. C. Rovetta, L. A. Fry, and S. Johnston. 1991. Age-specific natality of northern elephant seals at the South Farallon Islands, California. *Journal of Mammalogy* 72: 525–534.

Huntley, A. C., D. P. Costa, and R. D. Rubin. 1984. The contribution of nasal countercurrent heat exchange to water balance in the northern elephant seal, *Mirounga angustirostris*. *Journal of Experimental Biology* 113: 447–454.

King, J. E. 1956. The monk seal genus *Monachus*. *Bulletin of the British Museum of Natural History* (Zool.) 3: 203–256.

———. 1964. *Seals of the World*. London: British Museum of Natural History.

———. 1983. *Seals of the World*. 2d ed. Ithaca: Cornell University Press.

Klimley, A. P., S. D. Anderson, P. Pyle, and R. P. Henderson. 1992. Spatiotemporal patterns of white shark (*Carcharodon carcharias*) predation at the South Farallon Islands, California. *Copeia* 3: 680–690.

Laws, R. M. 1953a. A new method of age determination for mammals with special reference to the elephant seal, *Mirounga leonina* Linn. *Falkland Islands Dependencies Survey, Scientific Reports* 3: 1–11.

———. 1953b. The elephant seal (*Mirounga leonina* Linn.). I. Growth and age. *Falkland Islands Dependencies Survey, Scientific Reports* 8: 1–62.

————. 1956a. The elephant seal (*Mirounga leonina* Linn.). II. General, social and reproductive behavior. *Falkland Islands Dependencies Survey, Scientific Reports* 13: 1–88.

————. 1956b. The elephant seal (*Mirounga leonina* Linn.). III. The physiology of reproduction. *Falkland Islands Dependencies Survey, Scientific Reports* 15: 1–66.

————. 1960. The southern elephant seal (*Mirounga leonina* Linn.) at South Georgia. *Norsk Hvalfangst-Tidende* 49: 466–476, 520–542.

Le Boeuf, B. J. 1972. Sexual behavior in the northern elephant seal, *Mirounga angustirostris. Behaviour* 41: 1–26.

————. 1974. Male-male competition and reproductive success in elephant seals. *American Zoologist* 14: 163–176.

————. 1981. Elephant seals. In *The Natural History of Año Nuevo*, ed. B. J. Le Boeuf and S. Kaza, 327–374. Pacific Grove, Calif.: Boxwood Press.

————. 1989. The wars of the noses. *BBC Wildlife* 7: 144–149.

Le Boeuf, B. J., D. G. Ainley, and T. J. Lewis. 1974. Elephant seals on the Farallons: Population structure of an incipient breeding colony. *Journal of Mammalogy* 55: 370–385.

Le Boeuf, B. J., and M. L. Bonnell. 1980. Pinnipeds of the California Islands: Abundance and distribution. In *The California Islands: Proceedings of a Multidisciplinary Symposium*, ed. D. M. Power, 475–493. Santa Barbara, Calif.: Santa Barbara Museum of Natural History.

Le Boeuf, B. J., and K. T. Briggs. 1977. The cost of living in a seal harem. *Mammalia* 41: 167–195.

Le Boeuf, B. J., and R. C. Condit. 1983. The high cost of living on the beach. *Pacific Discovery* 12–14.

Le Boeuf, B. J., R. C. Condit, and J. Reiter. 1989. Parental investment and the secondary sex ratio in northern elephant seals. *Behavioral Ecology and Sociobiology* 25: 109–117.

Le Boeuf, B. J., D. P. Costa, A. C. Huntley, and S. D. Feldkamp. 1988. Continuous, deep diving in female northern elephant seals, *Mirounga angustirostris. Canadian Journal of Zoology* 66: 446–458.

Le Boeuf, B. J., D. P. Costa, A. C. Huntley, G. L. Kooyman, and R. W. Davis. 1986. Pattern and depth of dives in northern elephant seals, *Mirounga angustirostris. Journal of Zoology, London* 208: 1–7.

Le Boeuf, B. J., and D. E. Crocker. 1993. Diving behavior of elephant seals: Implications for predator avoidance. Paper presented at a symposium on the biology of the white shark, held at the Bodega Marine Laboratory, University of California, Davis, March 4–7.

Le Boeuf, B. J., D. E. Crocker, S. B. Blackwell, P. A. Morris, and P. H. Thorson. 1993. Sex differences in diving and foraging behavior of northern elephant seals. In *Marine Mammals: Advances in Behavioural and Population Biology*, ed. I. Boyd, 149–178. Symposia of the Zoological Society of London no. 66. London: Oxford University Press.

Le Boeuf, B. J., J. E. McCosker, and J. Hewitt. 1987. Crater wounds on northern elephant seals: The cookiecutter shark strikes again. *Fishery Bulletin* 85: 387–392.

Le Boeuf, B. J., Y. Naito, A. C. Huntley, and T. Asaga. 1989. Prolonged, continuous, deep diving by northern elephant seals. *Canadian Journal of Zoology* 67: 2514–2519.

Le Boeuf, B. J., and R. S. Peterson. 1969. Social status and mating activity in elephant seals. *Science* 163: 91–93.

Le Boeuf, B. J., and L. F. Petrinovich. 1974. Elephant seals: Interspecific comparisons of vocal and reproductive behavior. *Mammalia* (Paris) 38: 16–32.

Le Boeuf, B. J., and J. Reiter. 1988. Lifetime reproductive success in northern elephant seals. In *Reproductive Success*, ed. T. H. Clutton-Brock, 344-362. Chicago: University of Chicago Press.

———. 1991. Biological effects associated with El Niño Southern Oscillation, 1982–83, on northern elephant seals breeding at Año Nuevo, California. In *Pinnipeds and El Niño: Responses to Environmental Stress*, ed. F. Trillmich and K. Ono, 206–218. Berlin: Springer Verlag.

Le Boeuf, B. J., M. L. Riedman, and R. S. Keyes. 1982. White shark predation on pinnipeds in California coastal waters. *Fishery Bulletin* 80: 891–895.

Le Boeuf, B. J., R. J. Whiting, and R. F. Gantt. 1972. Perinatal behavior of northern elephant seal females and their young. *Behaviour* 43: 121–156.

Ling, J. K., and M. M. Bryden. 1981. Southern elephant seal—*Mirounga leonina*. In *Handbook of Marine Mammals*. 2. *Seals*, ed. S. H. Ridgway and R. J. Harrison, 297–327. London: Academic Press.

———. 1992. *Mirounga leonina*. *Mammalian Species* 391: 1–8.

Linnaeus, C. 1758. *Systema naturae*. *Regnum animale*. Vol. 1., pt. 1. 10th ed. rev. Stockholm: Laurentii Salvii.

McCann, T. S. 1980. Population structure and social organization of southern elephant seals, *Mirounga leonina* (L.). *Biological Journal of the Linnaean Society* 14: 133–150.

———. 1981. Aggression and sexual activity of male southern elephant seals, *Mirounga leonina*. *Journal of Zoology* 195: 295–310.

———. 1982. Aggressive and maternal activities of female southern elephant seals (*Mirounga leonina*). *Animal Behaviour* 30: 268–276.

———. 1985. Size, status and demography of southern elephant seal (*Mirounga leonina*) populations. In *Sea Mammals of South Latitudes: Proceedings of a Symposium of the 52d ANZAAS Congress in Sydney—May 1982*, ed. J. K. Ling and M. M. Bryden, 1–17. Northfield: South Australian Museum.

McConnell, B. J., C. Chambers, and M. A. Fedak. 1992. Foraging ecology of southern elephant seals in relation to bathymetry and productivity of the Southern Ocean. *Antarctic Science* 4: 393–398.

Miller, W. 1971. Pleistocene vertebrates of the Los Angeles Basin and vicinity (exclusive of Rancho La Brea). *Bulletin of the Los Angeles County Museum of Natural History* 10: 35.

Murphy, R. C. 1914. Notes on the sea elephant, *Mirounga leonina* (Linn.). *Bulletin of the American Museum of Natural History* 33: 63–79.

Nicholls, D. G. 1970. Dispersal and dispersion in relation to birth site of the southern elephant seal, *Mirounga leonina* (L.), of Macquarie Island. *Mammalia* 34: 598–616.

Odell, D. K., 1974. Seasonal occurrence of the northern elephant seal, *Mirounga angustirostris*, on San Nicolas Island, California. *Journal of Mammalogy* 55: 81–95.

Ortiz, C. L., D. P. Costa, and B. J. Le Boeuf. 1978. Water and energy flux in elephant seal pups fasting under natural conditions. *Physiological Zoology* 51: 166–178.

Reiter, J., K. J. Panken, and B. J. Le Boeuf. 1981. Female competition and reproductive success in northern elephant seals. *Animal Behaviour* 29: 670–687.

Reiter, J., N. L. Stinson, and B. J. Le Boeuf. 1978. Northern elephant seal development: The transition from weaning to nutritional independence. *Behavioral
Ecology and Sociobiology* 3: 337–367.

Repenning, C. A. 1980. Warm water life in cold ocean currents. *Oceans* 13: 18–24.

Riedman, M. L., and B. J. Le Boeuf. 1982. Mother-pup separation and adoption in
northern elephant seals. *Behavioral Ecology and Sociobiology* 11: 203–215.

Scammon, C. M. 1874. *The Marine Mammals of the North-Western Coast of North America, Described and Illustrated, Together with an Account of the American Whale-Fishery.*
San Francisco: J. H. Carmany and Co.

SCAR. 1991. Report of the workshop on southern elephant seals, Monterey, California, May 22–23, 1991.

Scheffer, V. 1958. *Seals, Sea Lions and Walruses.* Stanford: Stanford University Press.

Stewart, B. S., and R. L. DeLong. 1991. Diving patterns of northern elephant seal
bulls. *Marine Mammal Science* 7: 369–384.

Stewart, B. S., and P. K. Yochem. 1991. Northern elephant seals on the southern
California Channel Islands and El Niño. In *Pinnipeds and El Niño: Responses to Environmental Stress*, ed. F. Trillmich and K. Ono, 234-243. Berlin: Springer Verlag.

Sydeman, W. J., H. R. Huber, S. D. Emslie, C. A. Ribic, and N. Nur. 1991. Age-
specific weaning success of northern elephant seals in relation to previous breeding experience. *Ecology* 72: 2204–2217.

White, F. N., and D. K. Odell. 1971. Thermoregulatory behavior of the northern
elephant seal, *Mirounga angustirostris*. *Journal of Mammalogy* 52: 758–774.

Wilkinson, I. S., and M. N. Bester. 1990. Duration of post-weaning fast and local
dispersion in the southern elephant seal, *Mirounga leonina*, at Marion Island. *Journal of Zoology, London* 222: 591–600.

Worthy, G. A. J., P. A. Morris, D. P. Costa, and B. J. Le Boeuf. 1992. Moult
energetics of the northern elephant seal (*Mirounga angustirostris*). *Journal of Zoology,
London* 227: 257–265.

PART I

Population Ecology

TWO

History and Present Status of the Northern Elephant Seal Population

Brent S. Stewart, Pamela K. Yochem, Harriet R. Huber, Robert L. DeLong, Ronald J. Jameson, William J. Sydeman, Sarah G. Allen, and Burney J. Le Boeuf

ABSTRACT. The northern elephant seal, *Mirounga angustirostris*, was presumed extinct by 1892 owing primarily to commercial harvesting for their blubber oil that began in the early 1800s. A small, residual breeding colony survived, however, and with legal protection from further hunting, it grew rapidly through the early 1900s. Immigrants steadily colonized other island and mainland sites in Baja California and California so that by 1991 seals were breeding on fifteen islands and at three mainland beaches. Sixty-four percent of 28,164 northern elephant seal pups born in 1991 were produced on two southern California Channel Islands, San Miguel and San Nicolas. The entire elephant seal population was estimated to number around 127,000 in 1991 and was apparently still increasing by more than 6% annually. The remarkable demographic vitality and sustained population increase of northern elephant seals has evidently been unaffected by the species' low genetic variability and contrasts with recent declines of some populations of the more genetically polymorphic southern elephant seal, *M. leonina*.

> *Few, if any, living species today have been so deeply scored, so driven to the very brink of extermination—L. M. Huey (1930)*

Numerous terrestrial and marine species, like the northern elephant seal, experienced great population reductions in the nineteenth and twentieth centuries. But the single remarkable fact about the history of the northern elephant seal population is that despite only narrowly averting extinction, it rebounded with an unparalleled, century-long period of exponential increase (see, e.g., Reeves, Stewart, and Leatherwood 1992 and McCullough and Barrett 1992 for reviews of trends in pinnipeds and other vertebrates). Here we briefly review the population reduction and document its impressive recovery. We focus on number of births as an index of growth during the past three decades and estimate current population size.

PREHISTORY

Northern elephant seals lived in California waters by the late Pleistocene, evidently derived from monachine ancestors (*Callophoca* group) that entered the Pacific Ocean from the Caribbean through the Central American Seaway in the early Pliocene (Hendey 1972; Barnes and Mitchell 1975; Repenning, Ray, and Grigorescu 1979; de Muizon 1982). Little is known about their distribution during the Pleistocene when dynamic eustatic changes (Orr 1967; Vedder and Howell 1980) both greatly increased and decreased shoreline habitat available to pinnipeds, but archaeological remains show that elephant seals were in southern California waters when humans colonized the region over 15,000 years ago (e.g., Walker and Craig 1979; Snethkamp 1987; Bleitz 1993). Relatively large numbers of aboriginals lived on most of the California islands through the early nineteenth century, using the diverse marine resources on and near the islands for food, clothing, and housing; elephant seals and other pinnipeds were particularly important to aboriginal subsistence (Meighan 1959; Reimnan 1964; Glassow 1980; Stewart et al. 1993).

COMMERCIAL EXPLOITATION

Elephant seal, sea otter, whale, and fur seal hunters operated on and around the California islands from the early 1800s through the 1860s (Scammon 1870, 1874; Ogden 1933, 1941), but they left few records of northern elephant seal harvests. By 1850, northern elephant seals were scarce (Scammon 1870, 1874); it was not until 1866 that northern and southern elephant seals were scientifically recognized as taxonomically distinct (Gill 1866; but see Stewart and Huber 1993).

What we know to be incontrovertible about northern elephant seals in the early and mid-1800s is the following. Their distribution and abundance prior to 1840 is unknown. A few northern elephant seals were killed by sealers at Islas Los Coronados in 1840 and 1846, at Santa Barbara Island in May 1841, and at Cedros and Guadalupe islands in 1846 (Busch 1985). Scammon made a disappointing sealing expedition along the California coast in 1852; during a 5-month period he collected about 350 barrels of oil (Scammon 1874), probably the equivalent of around 100 to 200 adult elephant seals (see Busch 1985). Another 10-month expedition in 1857 met with even less success. Between 1865 and 1880, only a few elephant seals were reported at Isla de Guadalupe and Islas San Benito. Because all were killed as they were encountered, the species was considered extinct by the late 1870s (Townsend 1885). But in 1880, a small herd was discovered on the Baja California mainland south of Isla Cedros, at Bahia San Cristobal (fig. 2.1). Over the next four years, all 335 seals that were seen were killed by

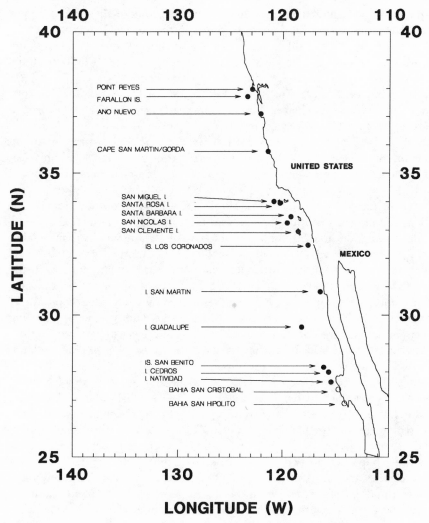

Fig. 2.1. Location of northern elephant seal colonies in 1991 (closed circles) and other alleged historical rookeries (open circles) in U.S. and Mexican waters.

the crews of six ships that visited the beach regularly, mostly in autumn. Three years later, in 1883, 80 elephant seals were found and killed at Isla de Guadalupe, and 4 were killed there in 1884. The species was again considered extinct, and no elephant seals were seen until May 1892, when C. H. Townsend and A. W. Anthony discovered 9 at Isla de Guadalupe; 7 of them were killed for the Smithsonian's museum collection (Townsend 1912; Anthony 1924). "This action was considered justifiable at the time, as the species was considered doomed to extinction by way of the sealer's trypot and few if any specimens were to be found in the museums of North America" (Anthony 1924: 146). The species was again presumed extinct, for the third time. But small numbers continued to show up at Isla de Guadalupe through 1911, and museum collectors continued to kill them: 4 in 1904 (Townsend 1912) and 14 of 40 on May 26, 1907. "This was a severe stroke dealt to a struggling species, but the appetite of science must be satisfied" (Huey 1930: 189). Townsend returned to Isla de Guadalupe on March 2, 1911, and killed 10 more seals but this time left 125 alive on the beach; on his return voyage to San Diego he searched for elephant seals at Bahia San Cristobal, Islas San Benito, and Isla Cedros but found none (Townsend 1912).

G. A. Bartholomew and C. L. Hubbs (1960), based on their interpretations of published counts of seals from the early 1900s, estimated that the total population in 1890 numbered fewer than 100 animals and speculated that it may have been as small as 20. The actual number of elephant seals that were present during the population bottleneck (or bottlenecks) in the 1800s and early 1900s is unknown because of the following flaws in sightings reports: (1) in most cases, the dates of sightings were not reported; (2) many sightings for which dates were provided were during the nonbreeding season; and (3) the age and sex composition of the seals observed was not determined. This information is vital because the number of seals on land, as well as the composition with respect to age and sex, varies greatly with time of year (Bartholomew 1951; Le Boeuf and Bonnell 1980; Stewart 1989). For example, when Townsend (1912) visited Isla de Guadalupe on March 2, 1911, and left 125 seals alive, it would have been at the end of the breeding season. At this time, some adult males should have been present, but nearly all females should already have returned to sea, leaving their weaned pups behind. Townsend noted that the herd consisted mostly of large males and immature animals of various sizes but that there were more than 15 adult females and 6 newborn young present. The photographs he took, however, show that most of the other "immature" seals were weaned pups, and it is likely that most of the seals ashore were actually molted pups-of-the-year (i.e., about 2 months old). He, like other early authors, also concluded erroneously that early March was the beginning of the breeding season, rather than the end, which emphasizes just how little was known about the natural history of elephant seals before George Bartholomew began his pioneering work on the species in the 1940s (e.g., Bartholomew 1952).

Regardless of whether the bottleneck population numbered in the tens or perhaps low hundreds, the important point is that the thousands of elephant seals alive today are all descendants of that small remnant herd.

INITIAL RECOVERY: 1900–1965

Northern elephant seals bred only at Isla de Guadalupe from the late 1890s through the 1920s. The colony grew steadily, despite sporadic poaching and scientific collecting (Bartholomew and Hubbs 1960). That early period of increase was chronicled by W. Rothschild (1908, 1910), C. M. Harris (1909), C. H. Townsend (1912), A. W. Anthony (1924) and L. M. Huey (1924, 1925, 1927, 1930) and thoroughly reviewed by Bartholomew and Hubbs (1960). On July 12, 1922, when mostly adult males were ashore molting, 264 seals were counted; a few months later, the Mexican government declared Isla de Guadalupe a biological reserve, and the seals were afforded protection from harassment and poaching (Hanna 1925). From that time on, elephant seals expanded their range; K. W. Radford, R. T. Orr, and C. L. Hubbs (1965) reviewed observations of seasonal migrants during the early 1900s along the coast from San Diego to southeastern Alaska. Other sightings were reviewed by Bartholomew and Hubbs (1960); seals were first seen on Islas San Benito in 1918, San Miguel Island in 1925, Los Coronados and Santa Barbara Island in 1948, San Nicolas Island in 1949, and Año Nuevo Island in 1955. Breeding evidently began in the 1930s at Islas San Benito, in the early 1950s at San Miguel, San Nicolas, and Santa Barbara islands (Bartholomew and Boolootian 1960; Odell 1974; Stewart 1989), and in 1961 at Año Nuevo Island (Radford, Orr, and Hubbs 1965).

From published and available unpublished counts, Bartholomew and Hubbs (1960) estimated that the total population numbered approximately 13,000 in 1957 and approximately 15,000 in 1960, with about 91% of the population residing at Isla de Guadalupe, 8% at Islas San Benito, and 1% on the Channel Islands.

RECENT TRENDS AND PRESENT STATUS: 1965–1991

Documentation of the population's recovery improved as more became known of the seasonal patterns of terrestrial abundance in the 1950s and 1960s. Table 2.1 lists births at each rookery from 1958 through 1991. The methods used varied slightly among colonies (see appendix 2.1), but all yielded estimates of births either from combined direct counts of suckling, weaned, and dead pups or derived from corrected counts of adult females made during peak breeding season (late January). Most pup counts were made on foot in February, after most births had occurred but before pups had left the rookeries. Some Mexican beaches with difficult access were surveyed from boats. The data for the three islands of Islas San Benito are combined in table 2.1 because of their closeness to each other; data for Año

TABLE 2.1 Northern elephant seal births at California and Mexico rookeries.

	Mexico					California								
	IN	IC	ISB	IG	ILC	SCLI	SBI	SNI	SMI	SRI	SMG	AN	SFAR	PR
1958									84[a,b]					
1959								48[a]						
1961												12[c]		
1962												23[c]		
1963												32[c]		
1964							11[d,b]	83[d,b]	878[d,b]			60[c]		
1965		0[l]	800[l]	3,668[l]								75[h]		
1966								136[e]				100[h]		
1967												143[h]		
1968								171[f]	1,624[g]			188[i]		
1969				7,104[s]				193[j]				247[i]		
1970		49[s]	1,578[s]	5,520[s]				210[j]				301[i]		
1971					9[3]			354[j]				310[i]		
1972					6[4]	0[k]	28[k,b]	428[k,b]	2,482[g]			367[i]	1[l]	
1973					18[3]				3,088[g]			395[i]	2[l]	
1974					28[3]							480[i]	17[m]	
1975			815[s]	6,058[s]	34[3]				3,547[g]			605[i,n]	35[m]	
1976									4,014[g]			690[i,n]	60[m]	
1977	12[5]	112[s]	1,359[s]	5,642[s]		1[p]	74[o]	693[o]	4,760[g]			814[q]	104[m]	
1978			1,727[s]	5,552[s]	45[4]				5,447[g]			994[q]	133[m]	
1979									5,531[b]	0[b]		1,173[q]	184[m]	
1980			1,752[s]	5,011[s]	58[6]			1,187[r]	6,475[b]	0[b]		1,353[q]	232[m]	
1981								1,546[r]	7,272[b]	0[b]	3[u]	1,225[q]	292[m]	1[t]
1982				4,760[s]		5[b]		2,312[r]	7,376[b]	0[b]	18[u]	1,469[q]	367[m]	1[t]

Year																
1983							0[b]	25[b]		1,906[b]	8,301[b]	0[b]	7[u]	1,566[q]	475[m]	2[t]
1984					166[7]		3[b]	80[b]		2,074[b]	9,342[b]	0[b]	25[u]	1,685[q]	447[x]	12[t]
1985				49[6]			2[b]	97[b]		2,583[b]	10,459[b]	2[v]	24[u]	1,701[q]	360[x]	15[t]
1986				53[6]			1[b]	73[b]		2,626[b]	9,824[b]	0[b]	4[u]	1,742[q]	434[x]	19[t]
1987							1[b]			2,835[w]	10,120[b]	2[b]	13[u]	1,524[q]	395[y]	20[t]
1988						227[8]				3,366[w]	11,035[w]	2[w]	13[u]	1,725[s]	393[y]	34[t]
1989										4,466[w]	11,079[w]	3[w]	67[u]	2,001[s]	380[y]	44[t]
1990						64[8]				4,101[w]	12,152[w]	27[w]	65[u]	1,986[s]	396[y]	52[z]
1991	391[2]	1,662+[2]	4,962[2]	68[6]					94[w]	4,082[w]	13,884[w]	73[w]	263[u]	2,248[s]	352[y]	81[z]

COLONY ABBREVIATIONS: Isla Natividad (IN), Isla Cedros (IC), Islas San Benito (ISB), Isla de Guadalupe (IG), Islas Los Coronados (ILC), San Clemente Island (SCLI), Santa Barbara Island (SBI), San Nicolas Island (SNI), San Miguel Island (SMI), Santa Rosa Island (SRI), Cape San Martin/Gorda (SMG), Año Nuevo Island and Mainland (AN), South Farallon Islands (SFAR), Point Reyes Headlands (PR).

SOURCES OF DATA: *Mexico rookeries:* [1]Rice, Kenyon, and Lluch B. 1965; [2]J. P. Gallo-Reynoso and A. Figueroa-Carranza, unpubl. data; [3]Le Boeuf, Countryman, and Hubbs 1975; [4]Antonelis, Leatherwood, and Odell 1981; [5]Le Boeuf and Mate 1978; [6]B. S. Stewart, unpubl. data; [7]K. Connally, pers. comm.; [8]W. T. Everett, pers. comm. *California rookeries:* [a]Bartholomew and Boolootian 1960; [b]Stewart 1989; [c]Orr and Poulter 1965; [d]Odell 1971; [e]Klopfer and Gilbert 1966; [f]Peterson, Gentry, and Le Boeuf 1968; [g]R. L. DeLong, unpubl. data; [h]Poulter and Jennings 1966; [i]Le Boeuf and Briggs 1977; [j]Odell 1974; [k]Antonelis, Leatherwood, and Odell 1981; [l]Le Boeuf, Ainley, and Lewis 1974; [m]Huber 1987; [n]Le Boeuf and Panken 1977; [o]Le Boeuf and Bonnell 1980; [p]Le Boeuf and Mate 1978; [q]Le Boeuf and Reiter 1991; [r]Stewart and Yochem 1984; [s]Le Boeuf and colleagues, unpubl. data; [t]Allen, Peaslee, and Huber 1989; [u]R. Jameson, unpubl. data; [v]Stewart and Yochem 1986; [w]Stewart and Yochem 1986; [x]Huber et al. 1991; [y]W. Sydeman, unpubl. data; [z]S. G. Allen, unpubl. data.

Nuevo Island and Año Nuevo mainland are combined for the same reason. From the data in table 2.1, we conclude the following.

The total elephant seal population, as reflected by births, increased 6.3% annually (= finite rate of increase, λ, where $\lambda = e^r$; see appendix 2.2) from 1965 through 1991 (r = .061; R^2, the coefficient of determination = .947; p, the significance of slope $\neq$ 0, <.001; see appendix 2.2). C. F. Cooper and B. S. Stewart (1983) calculated its increase at 8.3% from 1965 through 1977. The lower rate that we calculated here for the entire period (1965–1991) is evidently due to the lack of any apparent increase in Mexico since 1970. Growth of the total population from 1965 through 1991 was due primarily to growth at California rookeries, where births increased 14.1% annually (r = .132, R^2 = .901, p < .001), only slightly less than from 1965 through 1982 (λ = 1.145; Cooper and Stewart 1983).

Births increased slightly in Mexico between 1965 and 1970 but have not changed since then (fig. 2.2; slope of regression of births on time = 0, p = .903; see appendix 2.2). D. W. Rice, K. W. Kenyon, and D. Lluch B. (1965) suggested that carrying capacity of the Isla de Guadalupe rookery was reached by 1960. Counts made since then at the largest breeding beaches at Isla de Guadalupe support that conclusion; virtually all breeding space is now occupied and crowded during peak breeding season (J. P. Gallo-Reynoso and A. Figueroa-Carranza, unpubl. data). Because there are few recent counts at Islas San Benito, the trends on these islands are less clear (table 2.1). However, surveys of the central island (the easiest of the three to census and the site at which the data are most complete) show steady growth since 1970 (B. J. Le Boeuf, unpubl. data; B. S. Stewart, unpubl. data; J. P. Gallo-Reynoso and A. Figueroa-Carranza, unpubl. data). Births almost tripled from the early to mid-1970s (table 2.1). The central island accounted for 28.1, 37.2, and 45.1% of births on the entire island group in 1970, 1977, and 1980, respectively. If we assume that the 1,666 pups produced on the central island in 1991 accounted for 37% of the total pup production in that year, then the rookery produced 4,500 pups in 1991 and the colony is evidently still increasing. This is our tentative conclusion, but we must be guarded about the accuracy of the estimate. Some of the increase of central island numbers may have resulted from movements of seals from the west island where tourist and fishing activities have increased during the past two decades. Despite the increases at Isla Cedros and Islas San Benito, the Mexican population has not changed substantially during the past two decades, evidently because births at Isla de Guadalupe have declined, after peaking in the late 1960s (table 2.1).

The rapid increase in births at San Miguel Island, the largest colony in the species' range, accounts for most of the growth in California. Elephant seals bred only at the western tip of the island in 1968 (Le Boeuf and Bon-

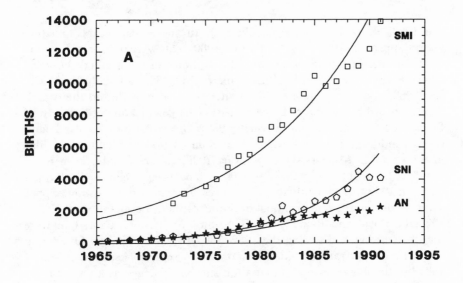

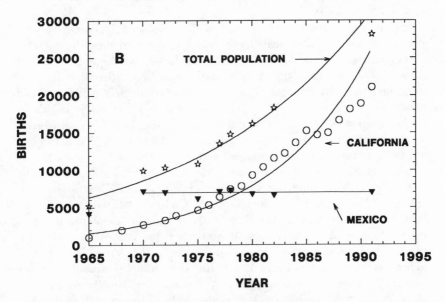

Fig. 2.2. Growth of the northern elephant seal population as reflected by births. *(A)* Increases in births at San Miguel Island (SMI), San Nicolas Island (SNI), and Año Nuevo Island and mainland combined (AN). *(B)* Growth of the entire northern elephant seal population, California segment, and Mexican segment. An intense El Niño affected North Pacific waters from late 1982 through 1983 (see text for details of immediate and delayed effects on northern elephant seals).

nell 1980; R. L. DeLong, pers. observ.). In subsequent years, breeding groups appeared farther east, so that by 1980, seals were breeding along the entire southern coast (Stewart 1989, 1992). Some of the northern beaches. however, are unused still. Births increased 9.3% annually (r = .089, R^2 = .963, p < .001) from 1965 through 1991; that growth and the coincident eastward expansion of breeding led to the colonization of Santa Rosa Island in 1985 (Stewart and Yochem 1986). Growth at San Nicolas Island (the second-largest colony), where expansion has followed patterns similar to those at San Miguel Island (Stewart 1989, 1992), has also been rapid (λ = 1.158, r = .147, R^2 = .976, p < .001). The brief decline in California births in 1985 was evidently due to poor recruitment of pups (owing to poor survival or retarded maturation or both; Huber, Beckham, and Nisbet 1991; Le Boeuf and Reiter 1991; B. S. Stewart, unpubl. data) that were born just before and during the 1982–1983 El Niño Southern Oscillation event. Pregnancy rates declined temporarily at some rookeries in 1984 and 1985 but there is no evidence that adult survival changed as a result of this intense oceanographic perturbation (Huber, Beckham, and Nisbet 1991; Le Boeuf and Reiter 1991).

Many new colonies formed in the last three decades, including at least three in Mexico. Elephant seals have clearly established breeding colonies on Isla Cedros and Islas Los Coronados. Births increased eightfold at Isla Cedros, an island that could sustain many more seals. Breeding space is limited on Islas Los Coronados, so carrying capacity has evidently been reached. Pups have been born on Isla Natividad, but monitoring of this island has been poor. At least two pups were produced on Isla San Martín (not shown in table 2.1) in 1978 (Le Boeuf and Mate 1978), but heavy human traffic on this island may preclude future growth.

California has at least six colonies that were founded since 1960. The San Clemente Island and the Santa Rosa Island colonies are in southern California. The other four colonies are in Central California; Cape San Martín/Gorda and Point Reyes Headlands are on the mainland, the Año Nuevo colony occupies both a small island and the immediate mainland, and the South Farallons colony is on an island (fig. 2.1). Año Nuevo Island reached carrying capacity in the late 1970s with annual production slightly under 1,000 pups. The colonization of Point Reyes Headlands in 1981 (Allen, Peaslee, and Huber 1989) is evidently linked to growth of the Año Nuevo and South Farallon Islands colonies. Births are still increasing at Año Nuevo and at Point Reyes Headlands. The recent explosive increase in births on beaches near Cape San Martín/Gorda is difficult to explain. Pups were first born in the area on a small, steep-backed gravel beach about 1 km north of Cape San Martín in 1981 or perhaps 1980. Breeding was restricted to that exposed site until 1989 when seals abandoned it and began using a longer gravel beach about 2 km south near Gorda. The better protection of that site against winter storms and surf was evidently more attractive to pregnant females, as indicated by the fourfold increase in births in the past two years (table 2.1).

COLONIZATION PROCESS, IMMIGRATION AND EMIGRATION

Births increased rapidly following colonization of all sites (table 2.1, fig. 2.2), and several colonies are still in this incipient growth stage. Immigrants from Isla de Guadalupe almost certainly colonized the other islands in Mexico and those in southern California. Our observation of the movement patterns of tagged seals during the past three decades (Condit and Le Boeuf 1984; B. J. Le Boeuf, unpubl. data; B. S. Stewart, unpubl. data) support that idea and also indicate the following: Año Nuevo was colonized by immigrants from San Miguel Island and to a lesser extent, immigrants from San Nicolas Island; the South Farallon Islands were colonized by immigrants from San Miguel, San Nicolas, and Año Nuevo islands (Le Boeuf, Ainley, and Lewis 1974; Huber et al. 1991). Some rookeries established in the 1980s were colonized by seals from neighboring rookeries. For example, Point Reyes Headlands was initially colonized by seals from the South Farallon Islands and Año Nuevo, and only recently have immigrants from San Miguel and San Nicolas islands been observed there (Allen, Peaslee, and Huber 1989; S. G. Allen, unpubl. data).

Some northern rookeries (e.g., Año Nuevo) in the expanding part of the range apparently still owe their growth more to a high immigration rate than to internal recruitment (which fuels most of the growth at rookeries at San Nicolas and San Miguel islands). Reproductive success of females at Año Nuevo has not been sufficient to account for the increases there (Le Boeuf and Reiter 1988). San Miguel Island seems to be the main source of immigrants. Immigration is also the primary cause of growth at the South Farallon Islands colony (Huber et al. 1991), where immigration rates from Año Nuevo, San Miguel Island, and San Nicolas Island were 3.9, 1.9, and 0.6%, respectively, between 1974 and 1986. These immigration rates were positively correlated with proximity to the South Farallon Islands.

Seals began colonizing new areas before carrying capacities were reached at most natal beaches. For example, Channel Islands colonists began breeding at Año Nuevo Island at least 20 years before San Miguel or San Nicolas Island habitats became crowded (see Orr and Poulter 1965; Stewart 1989, 1992). Similarly, Año Nuevo Island colonists began breeding at the South Farallon Islands and at Año Nuevo mainland 6 to 8 years before the island reached carrying capacity (see Le Boeuf, Ainley, and Lewis 1974; Reiter, Panken, and Le Boeuf 1981; Le Boeuf and Reiter 1991).

TOTAL POPULATION SIZE

The dynamic age structure of the northern elephant seal population (e.g., Huber et al. 1991) hinders accurate predictions of present population size from pup counts or of total seals hauled out at any time. Estimates of births are, however, useful for estimating rate of change in population size, although such calculations have problems (e.g., see Berkson and DeMaster 1985). Despite some obvious shortcomings, we use pup counts as a convenient index of population growth because superior measures of life history parameters are not available for each rookery.

Total population size may be about 3.5 to 4.5 times births (e.g., Hewer 1964; Bonner 1976; Harwood and Prime 1978, Stewart 1989). For comparison with southern elephant seals (Laws, this volume), we use T. S. McCann's formula and multiply births by 3.5 to estimate total population size at the end of the breeding season, exclusive of pups (McCann 1985). From table 2.1, we multiply 3.5 times the 28,164 pups born in 1991 to obtain the estimate of 98,574 elephant seals older than pups in the entire population in 1991. If the young of the year are added to this figure, there were approximately 127,000 elephant seals in existence in early spring 1991.

In 1991, Mexican rookeries contributed 25.5% of all births and California, 74.8%; San Miguel Island alone produced nearly half (49.3%) of all elephant seal pups. The world total of southern elephant seals in 1991 (Laws, this volume) was roughly 6.8 times larger than that of northern elephant seals.

FUTURE GROWTH

The northern elephant seal has lived in eastern North Pacific waters for at least several hundred thousand years. Their occurrence and apparent vitality in these waters today is remarkable considering their fortuitous emergence in the twentieth century after facing extinction in the nineteenth century. There seem to be few barriers to the species' continued population growth and range expansion. In the immediate future, growth of the population will probably be determined primarily by events on southern California rookeries. Growth at San Nicolas and San Miguel islands will almost certainly slow as the limited remaining habitat becomes occupied and as crowding on those islands constrains reproductive success. The new colony at Santa Rosa Island, however, has substantial breeding beach habitat that could support continued rapid growth in California. Neighboring Santa Cruz Island also offers some additional habitat, although of poorer quality than at Santa Rosa Island. The seals may also continue their northward expansion. They are now hauling out at Cape St. George in northern

California, at Cape Arago in Oregon, and on Vancouver Island in British Columbia. Recent information on seasonal movements and foraging locations of northern elephant seals (see DeLong, Stewart, and Hill 1992; Stewart and DeLong 1993; Stewart and DeLong, this volume; Le Boeuf, this volume) suggests that eventual breeding at these sites is quite plausible.

M. L. Bonnell and R. K. Selander (1974) found that northern elephant seals were homozygous at 23 loci coding for 20 blood allozymes and suggested that this was due to a loss of genetic variability when elephant seals were reduced to small numbers in the 1800s. Recent research on nuclear and mitochondrial DNA (Hoelzel et al. 1991; Lehman, Wayne, and Stewart 1993) reaffirms the earlier findings of low heterozygosity, although these studies revealed greater levels of variability than the electrophoretic analysis of blood allozymes did. Reduced genetic variability may compromise the population viability of some species of mammals (e.g., O'Brien et al. 1985; O'Brien et al. 1987), but many other species have persisted for a long time despite population bottlenecks, founder events, isolation, inbreeding, and low levels of genetic variation (e.g., Gill 1980; Nevo, Beiles, and Ben-Shlomo 1984; Gilbert et al. 1990; Benirshke and Kumamoto 1991; Wayne et al. 1991). A lack of substantial genetic variability has not limited the phenomenal population recovery of northern elephant seals. Indeed, their recovery contrasts ironically with the recent decline of some populations of the closely related southern elephant seal (Laws, this volume), which is genetically more polymorphic (McDermid, Ananthakrishna, and Agar 1972; Hoelzel et al. 1991). The consequences of low genetic variability for future population growth of northern elephant seals are unpredictable.

APPENDIX 2.1

Field Data Collection Methods

Survey methods differed slightly among rookeries as described below due to differences in colony size, dispersion, and logistical constraints. Nonetheless, our studies produced annual estimates of births and, in most cases, neonatal mortality at each colony.

San Miguel Island. Each year in late February two or three people walked among and counted weaned, suckling, and dead pups at all beaches on SMI. Observers' counts of live pups were compared after each relatively small group (<100) was counted; counts usually differed by less than 2%, but if the tallies differed by 5% or more, the group was counted again. In this way an entire cohort of pups distributed along approximately 30 km of shoreline could be surveyed in two or three days.

San Nicolas Island. Each breeding season, surveys were made every one to two days at three sites and once each week at all breeding sites along the

35 km of coastline of SNI. Weaned and suckling pups were tallied two or three times during each survey, and dead pups were marked and mapped to ensure that those that died prior to weaning were accounted for but none more than once. The number of births and pup deaths was determined for each breeding site and summed at the end of the season to determine total annual production.

Santa Rosa Island and Santa Barbara Island.   The numbers of live pups present on SRI and SBI were determined by photographing them during aerial surveys in late January. These counts were corrected to estimate each year's births according to seasonal phenology of births documented by Stewart (1989). Pup mortality was not determined.

Cape San Martín.   Nursing, weaned, and dead pups were counted periodically in January, February, or early March each year at various small beaches within 2 km of Cape San Martín beginning in 1981. Preweaning mortality could not be determined accurately.

Año Nuevo.   Most estimates of births were derived from daily or weekly counts during the breeding season of all seals present; those counts included dead, suckling, and weaned pups. In some circumstances births were estimated as follows: (1) counts of females present at ANI and ANML in late January were first adjusted to account for those that had already left the rookeries and for those that had not yet arrived to provide an estimate of the number of females that visited during the breeding season; (2) 98% of the females estimated to have hauled out were assumed to have given birth. Prior to 1980, all or most pups that died were accounted for by removing them from breeding aggregations or marking them with paint or dye. Since 1980, preweaning mortality on the island has been estimated as follows: (1) weaned and suckling pups were counted on March 1 or 2 to yield an estimate of pups weaned for the season; (2) mortality was then derived by subtracting that estimate from an estimate of the number of females that gave birth during the season (as summarized above). Some estimates of births reported here (table 2.1) are corrections of those published earlier.

South Farallon Islands.   Prior to 1987, births were determined in several ways depending on breeding location at SFAR. Pup carcasses were removed from the nine breeding sites whenever possible. At sites that were not washed by high tides, all deaths were accounted for because all females that gave birth, and their pups, were marked with hair dye. At other sites, observations of a female's appearance (i.e., blood on her hind quarters) or behavior were used to determine if newborn pups had disappeared (and presumably drowned) undetected during high tides or storms. Weaned pups were counted at the end of the breeding season and added to estimates of records of dead pups to derive an estimate of each season's births.

Beginning in 1987, births at the primary breeding site were determined as follows: (1) a peak count of females was made in late January; (2) based on reproductive characteristics of tagged females observed during the season, 93.2% of the females present during the peak count were assumed to have given birth; and (3) the estimate was adjusted to account for females that had departed already or had not yet arrived. Pup mortality was estimated indirectly by subtracting the number of weaned pups counted in late February from the estimate of births derived from the peak season female count. Births and pup deaths at the other eight breeding sites were determined by monitoring all females that were present and that were uniquely marked with hair dye.

Point Reyes Headland.   Counts of live and dead pups were made at least weekly from bluffs overlooking beaches along the Point Reyes Headland. Estimates of pup deaths are rough minima because some carcasses probably washed out to sea undetected between observations. Estimates of births were made by adding the peak count of live pups to estimates of deaths that occurred prior to that count.

Mexican Islands.   Surveys were made opportunistically on foot or from skiffs nearshore from 1968 to 1991 at IG, ISB, IC, and at other small rookeries in Mexico (fig. 2.1, table 2.1). Complete surveys of Isla de Guadalupe's west side were rarely made because of rough island terrain and heavy seas near the coastlines. We report only counts made near the end of the breeding season when nearly all births had occurred but when few pups had departed the rookeries.

APPENDIX 2.2

Analyses of Rates of Change of Elephant Seal Births

We calculated observed rates of increase (r = intrinsic rate of increase) in births by linear regression (Zar 1974). We examined the fit of an exponential model, Log_e number of births regressed on time where r is the slope of the regression line (Caughley and Birch 1971; Caughley 1977). We present the coefficient of determination (= R^2 to distinguish it from rate of increase) to describe the proportion of the total variation in births that is accounted for by time. For comparative purposes, we convert exponential rates (r) to finite rates ($e^r = \lambda$; Caughley 1977: 6). When the exponential model fit poorly, we used a linear model of births regressed on time. For both models, we present the level of significance (p) at which we did (if p < .05) or did not (if p > .05) reject the null hypothesis that the slope of regressions did not differ from zero.

REFERENCES

Allen, S. G., S. C. Peaslee, and H. R. Huber. 1989. Colonization by northern elephant seals of the Point Reyes Peninsula, California. *Marine Mammal Science* 5: 298–302.

Anthony, A. W. 1924. Notes on the present status of the northern elephant seal, *Mirounga angustirostris. Journal of Mammalogy* 5: 145–152.

Antonelis, G. A., S. Leatherwood, and D. K. Odell. 1981. Population growth and censuses of the northern elephant seal, *Mirounga angustirostris*, on the California Channel Islands. *United States Fishery Bulletin* 79: 562–567.

Barnes, L. G., and E. D. Mitchell. 1975. Late Cenozoic northeast Pacific phocidae. *Rapports et Proces-verbaux des Reunions Conseil International pour L'exploration de la Mer* 169: 34–42.

Bartholomew, G. A. 1951. Spring, summer, and fall censuses of the pinnipeds on San Nicolas Island, California. *Journal of Mammalogy* 32: 15–21.

———. 1952. Reproductive and social behavior of the northern elephant seal. *University of California Publications in Zoology* 47: 369–427.

Bartholomew, G. A., and R. A. Boolootian. 1960. Numbers and population structure of the pinnipeds on the California Channel Islands. *Journal of Mammalogy* 41: 366–375.

Bartholomew, G. A., and C. L. Hubbs. 1960. Population growth and seasonal movements of the northern elephant seal, *Mirounga angustirostris. Mammalia* 24: 313–324.

Benirshke, K., and A. T. Kumamoto. 1991. Mammalian cytogenetics and conservation of species. *Journal of Heredity* 82: 187–191.

Berkson, J. M., and D. P. DeMaster. 1985. Use of pup counts in indexing population changes in pinnipeds. *Canadian Journal of Fisheries and Aquatic Sciences* 42: 873–879.

Bleitz, D. 1993. The prehistoric exploitation of marine mammals and birds at San Nicolas Island, California. In *Third California Islands Symposium: Recent Advances in Research on the California Islands*, ed. F. G. Hochberg, 519–536. Santa Barbara, Calif.: Santa Barbara Museum of Natural History.

Bonnell, M. L., and R. K. Selander. 1974. Elephant seals: Genetic variation and near extinction. *Science* 184: 908–909.

Bonner, W. N. 1976. Stocks of grey seals and common seals in Great Britain. *Natural Environment Research Council Publications, Series C*, 16: 1–16.

Busch, B. C. 1985. *The War against the Seals*. Montreal: McGill-Queen's University Press.

Caughley, G. 1977. *Analysis of Vertebrate Populations*. London: John Wiley and Sons.

Caughley, G., and L. C. Birch. 1971. Rate of increase. *Journal of Wildlife Management* 35: 658–663.

Condit, R., and B. J. Le Boeuf. 1984. Feeding habits and feeding grounds of the northern elephant seal. *Journal of Mammalogy* 65: 281–290.

Cooper, C. F., and B. S. Stewart. 1983. Demography of northern elephant seals, 1911–1982. *Science* 210: 969–971.

DeLong, R. L., B. S. Stewart, and R. D. Hill. 1992. Documenting migrations of northern elephant seals using daylength. *Marine Mammal Science* 8: 155–159.

Gilbert, D. A., N. Lehman, S. J. O'Brien, and R. K. Wayne. 1990. Genetic fingerprinting reflects population differentiation in the California Channel Island fox. *Nature* 344: 764–767.

Gill, A. E. 1980. Evolutionary genetics of California Islands *Peromyscus*. In *The California Islands: Proceedings of a Multidisciplinary Symposium*, ed. D. M. Power, 719–744. Santa Barbara, Calif.: Santa Barbara Museum of Natural History.

Gill, T. 1866. On a new species of the genus *Macrorhinus*. *Proceedings of the Chicago Academy of Sciences* 1: 33–34.

Glassow, M. A. 1980. Recent developments in the archaeology of the Channel Islands. In *The California Islands: Proceedings of a Multidisciplinary Symposium*, ed. D. M. Power, 79–99. Santa Barbara, Calif.: Santa Barbara Museum of Natural History.

Hanna, G. D. 1925. Expedition to Guadalupe Island, Mexico—general report. *Proceedings of the California Academy of Sciences* 14: 217–275.

Harris, C. M. 1909. A cruise after sea elephants. *Pacific Monthly* 21: 331–339.

Harwood, J., and J. H. Prime. 1978. Some factors affecting the size of British grey seal populations. *Journal of Applied Ecology* 15: 401–411.

Hendey, Q. B. 1972. The evolution and dispersal of the monachinae. *Annals of the South African Museum* 59: 99–113.

Hewer, H. R. 1964. The determination of age, sexual maturity and a life-table in the grey seal (*Halichoerus grypus*). *Proceedings of the Zoological Society, London* 142: 593–624.

Hoelzel, A. R., J. Halley, C. Campagna, T. Arnbom, and B. J. Le Boeuf. 1991. Molecular genetic variation in elephant seals; the effect of population bottlenecks. In *Abstracts of the Ninth Biennial Conference on the Biology of Marine Mammals*, Chicago, Ill., 34.

Huber, H. 1987. Natality and weaning success in relation to age of first reproduction in northern elephant seals. *Canadian Journal of Zoology* 65: 1311–1316.

Huber, H. R., C. Beckham, and J. Nisbet. 1991. Effects of the 1982–83 El Niño on northern elephant seals at the South Farallon Islands, California. In *Pinnipeds and El Niño: Responses to Environmental Stress*, ed. F. Trillmich and K. Ono, 219–233. Berlin: Springer Verlag.

Huber, H. R., A. C. Rovetta, L. A. Fry, and S. Johnston. 1991. Age specific natality of northern elephant seals at the South Farallon Islands, California. *Journal of Mammalogy* 72: 525–534.

Huey, L. M. 1924. Recent observations on the northern elephant seal. *Journal of Mammalogy* 5: 237–242.

———. 1925. Late information on the Guadalupe Island elephant seal herd. *Journal of Mammalogy* 6: 126–127.

———. 1927. The latest northern elephant seal census. *Journal of Mammalogy* 8: 160–161.

———. 1930. Past and present status of the northern elephant seal with a note on the Guadalupe fur seal. *Journal of Mammalogy* 11: 188–194.

Klopfer, P. H., and B. K. Gilbert. 1966. A note on retrieval and recognition of young in the elephant seal, *Mirounga angustirostris*. *Zeitschrift für Tierpsychologie* 6: 757–760.

Le Boeuf, B. J., D. G. Ainley, and T. J. Lewis. 1974. Elephant seals on the Farallones: Population structure of an incipient breeding colony. *Journal of Mammalogy* 55: 370–385.

Le Boeuf, B. J., and M. L. Bonnell. 1980. Pinnipeds of the California Channel Islands: Abundance and distribution. In *The California Islands: Proceedings of a Multidisciplinary Symposium*, ed. D. M. Power, 475–493. Santa Barbara, Calif.: Santa Barbara Museum of Natural History.

Le Boeuf, B. J., and K. T. Briggs. 1977. The cost of living in a seal harem. *Mammalia* 41: 167–195.

Le Boeuf, B. J., D. A. Countryman, and C. L. Hubbs. 1975. Records of elephant seals, *Mirounga angustirostris*, on Los Coronados Islands, Baja California, Mexico, with recent analyses of the breeding population. *Transactions of the San Diego Society of Natural History* 18: 1–7.

Le Boeuf, B. J., and B. R. Mate. 1978. Elephant seals colonize additional Mexican and California islands. *Journal of Mammalogy* 59: 621–622.

Le Boeuf, B. J., and K. J. Panken. 1977. Elephant seals breeding on the mainland in California. *Proceedings of the California Academy of Sciences* 41: 267–280.

Le Boeuf, B. J., and J. Reiter. 1988. Lifetime reproductive success in northern elephant seals. In *Reproductive Success*, ed. T. H. Clutton-Brock, 344–362. Chicago: University of Chicago Press.

———. 1991. Biological effects associated with El Niño Southern Oscillation, 1982–83, on northern elephant seals breeding at Año Nuevo, California. In *Pinnipeds and El Niño: Responses to Environmental Stress*, ed. F. Trillmich and K. Ono, 206–218. Berlin: Springer Verlag.

Lehman, N., R. K. Wayne, and B. S. Stewart. 1993. Comparative levels of genetic variability between harbor seals and northern elephant seals using genetic fingerprinting. In *Marine Mammals: Advances in Behavioural and Population Biology*, ed. I. L. Boyd, 49–60. Symposia of the Zoological Society of London no. 66. London: Oxford University Press.

McCann, T. S. 1985. Size, status and demography of southern elephant seal (*Mirounga leonina*) populations. In *Sea Mammals of South Latitudes: Proceedings of a Symposium of the 52d ANZAAS Congress in Sydney—May 1982*, ed. L. K. Ling and M. M. Bryden, 1–17. Northfield: South Australian Museum.

McCullough, D. R., and R. H. Barrett, eds. 1992. *Wildlife 2001: Populations*. New York: Elsevier Applied Science.

McDermid, E. M., R. Ananthakrishna, and N. S. Agar. 1972. Electrophoretic investigation of plasma and red cell proteins and enzymes of Macquarie Island elephant seals. *Animal Blood Groups and Biochemical Genetics* 3: 85–94.

Meighan, C. W. 1959. The Little Harbor site, Catalina Island: An example of ecological interpretation in archaeology. *American Antiquities* 24: 383–405.

Muizon, C. de. 1982. Phocid phylogeny and dispersal. *Annals of the South African Museum* 89: 175–213.

Nevo, E., A. Beiles, and R. Ben-Shlomo. 1984. The evolutionary significance of genetic diversity. *Lecture Notes in Biomathematics* 53: 13–213.

O'Brien, S. J., M. E. Roelke, L. Marker, L. Neuman, C. A. Winkler, D. Meltzer, L. Colly, J. F. Evermann, M. Bush, and D. E. Wildt. 1985. Genetic basis for species vulnerability in the cheetah. *Science* 227: 1428–1434.

O'Brien, S. J., D. E. Wildt, M. Bush, T. M. Caro, C. Fitzgibbon, I. Aggundey, and R. E. Leakey. 1987. East African cheetahs: Evidence for two population bottlenecks? *Proceedings of the National Academy of Sciences* 84: 5083–5086.

Odell, D. K. 1971. Censuses of pinnipeds breeding on the California Channel Islands. *Journal of Mammalogy* 52: 187–190.

———. 1974. Seasonal occurrence of the northern elephant seal, *Mirounga angustirostris*, on San Nicolas Island, California. *Journal of Mammalogy* 51: 81–95.

Ogden, A. 1933. Russian sea otter and seal hunting on the California coast, 1803–1841. *California Historical Society Quarterly* 12: 217–239.

———. 1941. The California sea otter trade, 1784–1848. *University of California Publications in History* 26: 1–251.

Orr, P. C. 1967. Geochronology of Santa Rosa Island, California. In *Proceedings of the Symposium on the Biology of the California Islands*, 317–326. Santa Barbara, Calif.: Santa Barbara Botanic Garden.

Orr, R. T., and T. C. Poulter. 1965. The pinniped population of Año Nuevo Island, California. *Proceedings of the California Academy of Sciences* 32: 377–404.

Peterson, R. S., R. L. Gentry, and B. J. Le Boeuf. 1968. Investigations of pinnipeds on San Miguel and San Nicolas islands. Unpublished manuscript, University of California, Santa Cruz.

Poulter, T. C., and R. Jennings. 1966. *Annual Report to the Division of Beaches and Parks, State of California.* Menlo Park, Calif.: Stanford Research Institute.

Radford, K. W., R. T. Orr, and C. L. Hubbs. 1965. Reestablishment of the northern elephant seal (*Mirounga angustirostris*) off Central California. *Proceedings of the California Academy of Sciences* 31: 601–612.

Reeves, R. R., B. S. Stewart, and S. Leatherwood. 1992. *The Sierra Club Handbook of Seals and Sirenians.* San Francisco: Sierra Club Books.

Reinman, F. M. 1964. Maritime adaptation on San Nicolas Island, California: A preliminary and speculative evaluation. In *University of California at Los Angeles Archaeological Surveys, Annual Report* 1963–1964, 50–84.

Reiter, J., K. J. Panken, and B. J. Le Boeuf. 1981. Female competition and reproductive success in northern elephant seals. *Animal Behaviour* 29: 670–687.

Repenning, C. A., C. E. Ray, and D. Grigorescu. 1979. Pinniped biogeography. In *Historical Biogeography, Plate Tectonics, and the Changing Environment: Proceedings of the Thirty-seventh Annual Biology Colloquium, and Selected Papers*, ed. J. Gray and A. Boucot, 359–369. Corvallis: Oregon State University Press.

Rice, D. W., K. W. Kenyon, and D. Lluch B. 1965. Pinniped populations at Islas Guadalupe, San Benito, and Cedros, Baja California, in 1965. *Transactions of the San Diego Society of Natural History* 14: 73–84.

Rothschild, W. 1908. *Mirounga angustirostris* (Gill). *Novitates Zoologicae* 15(2): 393–394.

———. 1910. Notes on sea elephants (*Mirounga*). *Novitates Zoologicae* 17(3): 445–446.

Scammon, C. M. 1870. Sea-elephant hunting. *Overland Monthly* 4(2): 112–117.

———. 1874. *The Marine Mammals of the Northwestern Coast of North America.* San Francisco: J. H. Carmany and Co.

Snethkamp, P. E. 1987. Prehistoric subsistence variability on San Miguel Island. In *Third California Islands Symposium: Program and Abstracts*. Santa Barbara, Calif.

Stewart, B. S. 1989. The ecology and population biology of the northern elephant seal, *Mirounga angustirostris* (Gill) 1866, on the southern California Channel Islands. Ph.D. dissertation, University of California, Los Angeles.

———. 1992. Population recovery of northern elephant seals on the southern California Channel Islands. In *Wildlife 2001: Populations*, ed. D. R. McCullough and R. H. Barrett, 1075–1086. London: Elsevier Applied Science.

Stewart, B. S., and R. L. DeLong. 1993. Seasonal dispersion and habitat use of foraging northern elephant seals. In *Marine Mammals: Advances in Behavioural and Population Biology*, ed. I. L. Boyd, 179–194. Symposia of the Zoological Society of London no. 66. London: Oxford University Press.

Stewart, B. S., and H. R. Huber. 1993. Mammalian species: *Mirounga angustirostris*. *American Journal of Mammalogists* 449: 1–10.

Stewart, B. S., and P. K. Yochem. 1984. Seasonal abundance and distribution of pinnipeds on San Nicolas Island, California. *Bulletin of the Southern California Academy of Sciences* 83: 121–132.

———. 1986. Northern elephant seals breeding at Santa Rosa Island, California. *Journal of Mammalogy* 67: 402–403.

Stewart, B. S., P. K. Yochem, R. L. DeLong, and G. A. Antonelis. 1993. Trends in abundance and status of pinnipeds on the southern California Channel Islands. In *Third California Islands Symposium: Recent Advances in Research on the California Islands*, ed. F. G. Hochberg, 501–516. Santa Barbara, Calif.: Santa Barbara Museum of Natural History.

Townsend, C. H. 1885. An account of recent captures of the California sea-elephant and statistics relating to the present abundance of the species. *Proceedings of the U.S. National Museum* 8: 90–94.

———. 1912. The northern elephant seal *Macrorhinus angustirostris* (Gill). *Zoologica* 1: 159–173.

Vedder, J. G., and D. G. Howell. 1980. Topographic evolution of the southern California Borderland during late Cenozoic time. In *The California Islands: Proceedings of a Multidisciplinary Symposium*, ed. D. M. Power, 7–31. Santa Barbara, Calif.: Santa Barbara Museum of Natural History.

Walker, P. L., and S. Craig. 1979. Archaeological evidence concerning the prehistoric occurrence of sea mammals at Point Bennett, San Miguel Island. *California Fish and Game* 65: 50–54.

Wayne, R. K., S. B. George, D. Gilbert, P. W. Collins, S. D. Kovach, D. Girman, and N. Lehman. 1991. A morphologic and genetic study of the island fox, *Urocyon littoralis*. *Evolution* 45: 1849–1868.

Zar, J. H. 1974. *Biostatistical Analysis*. Englewood Cliffs, N.J.: Prentice-Hall.

THREE

History and Present Status of Southern Elephant Seal Populations

Richard M. Laws

ABSTRACT. The total world population of southern elephant seals in 1990 was estimated at 664,000. Of the three (or four) main stocks, South Georgia and Peninsula Valdes account for 60% of the total; Isles Kerguelen, 28%; and Macquarie Island, 12%. The species was hunted for its oil in the eighteenth and nineteenth centuries and subsequently recovered under protection. The only large-scale industry this century was a government-licensed one at South Georgia, from 1909 to 1964, which was restricted to adult males. The history and rational basis for this industry is described here, including some effects of sealing, population decline from 1931 to 1951, and subsequent recovery to sustained yield level under a research-based management plan. Pup production was estimated, population models were drawn up, and the size of the South Georgia population in 1951 was calculated. A comprehensive survey in the 1985 breeding season indicated the same annual pup production as for 1951 and a population of 357,000 in 1990. Population models indicate that some 75% of the adult male population is at sea during the breeding season, which has implications for aquatic mating of virgin females and nonpregnant mature females.

A literature review of other stocks examines various estimates of pup production by site and years between 1949 and 1990 and converts these by a raising factor to estimates of total population sizes by years. During this period, different stocks of southern elephant seals have increased (Península Valdes, Argentina, by 144% since 1975); have probably remained stable after rapid recovery from exploitation (South Georgia, 15% due to postexploitation recovery of the male stock); or have decreased dramatically at annual rates varying between 2.1% and 8% at different places and periods. Total percentage decreases since 1949 are estimated at 50% for Heard Island, 84% for Marion Island, 57% for Macquarie Island, 96% for Campbell Island, and 93% for Signy Island.

This chapter sets the scene for the more detailed discussions presented by those studying key colonies. Those discussions deal with the factors responsible for changes in population number and advance hypotheses to account for the precipitous decline of certain populations.

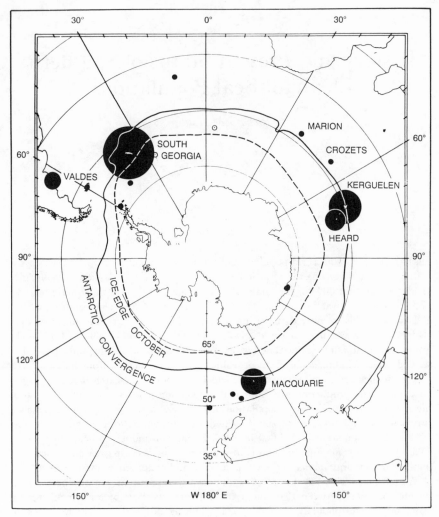

Fig. 3.1. Breeding distribution of southern elephant seals in 1990. *Closed circles:* known breeding colonies (area proportional to the estimated population sizes except for colonies of less than 5,000). *Open circles:* probable small breeding populations. Fuller data are presented in table 3.1.

In terms of numbers, the southern elephant seal, *M. leonina,* is one of the more abundant seal species in the world. There are three main stocks, defined (Laws 1960) as the South Georgia stock, the Kerguelen stock, and the Macquarie Island stock (fig. 3.1). The first is the largest numerically and includes elephant seal breeding colonies in the Scotia arc (South Georgia, South Orkney Islands, South Shetland Islands, South Sandwich Islands) and Gough and Bouvet islands, together with South America and the Falkland Islands. Although more recent information indicates movements between South American and Falkland Islands elephant seal colonies (Laws, unpubl.), up to now no movements have been reported between them and the remaining colonies; so there may be four stocks. The second stock includes Kerguelen and Heard islands, Marion and Prince Edward islands, and Iles Crozets. Single births have been recorded at Amsterdam and St. Paul islands and South Africa (Laws 1960). Finally, the Macquarie Island Stock includes Macquarie Island, Campbell Island, Auckland Islands, and Antipodes Islands, and one birth has been reported from Tasmania (Laws 1960). A very small number of births have also been reported from the Antarctic continent (Murray 1981). The centers of the three main stocks are roughly 90° and 107° longitude apart, except in the Pacific Ocean sector (163°), which is devoid of suitable islands in the right latitudes and supports no breeding colonies. Genetic studies by N. J. Gales, M. Adams, and H. R. Burton (1989) reported on elephant seal blood samples from Heard and Macquarie islands that were examined for protein variation. They concluded that the two populations "may have diverged genetically, and a very limited gene flow exists between the islands, a finding consistent with limited information from mark-recapture studies" (57). It is likely that a similar division exists between them and the South Georgia stock.

The sealers of the eighteenth and nineteenth centuries came first to hunt the more valuable fur seals for their pelts but did not ignore the larger species, which was hunted for its oil. The right whalers in southern waters also took elephant seals. As fur seals declined, the sealers concentrated more on elephant seals (Bonner 1982; Headland 1989). In his chronological list of Antarctic expeditions, R. Headland (1989) analyzed 930 voyages to the Antarctic between 1786 and 1928, the majority by U.S. (46.4%) and British (19.1%) ships, and showed the frequency of sealing voyages by year. There were progressively lower peaks, separated by periods of partial recovery, about 1820, 1842, 1855, 1875, and 1906. The early voyages primarily took fur seals, which were hunted almost to extinction; the objective of the later voyages was primarily the elephant seal. It is not known how many elephant seals were taken during this period, but it was probably in excess of one million of both sexes, assuming that the original populations totaled at least 600,000 to 750,000, as in this century (Laws 1960; McCann 1985). The sealing industry had virtually ended by 1909, and protection was subsequently conferred on most stocks, which recovered from the overhunting.

The only large-scale elephant sealing during this century was the government-licensed industry at South Georgia between 1909 and 1964 (see Laws 1953a, 1960; Bonner 1958; and Headland 1984), although small numbers were taken in other places, particularly Kerguelen, where 12,000 bulls were taken between 1958 and 1964 (van Aarde 1980). Over the course of sixty-five years, some 260,000 elephant seals, predominantly adult males, were taken at South Georgia (Headland 1984). The industry came to an end in 1964 as a result of the collapse of the whaling industry at South Georgia; the sealing industry was on a firm sustainable yield basis. As will be shown, the size of the South Georgia elephant seal population in 1951, when government-controlled sealing was at its height, was similar to the size of the population in 1985, some twenty years after sealing had ended. However, the populations in the Kerguelen and Macquarie islands stocks declined steadily over forty years from 1950, even in the absence of sealing. It seems that the elephant seal population in Argentina has been increasing in recent years.

Below I describe the recent history of twentieth-century sealing at South Georgia, because of its relevance to estimating abundance and trends, and what is known about population sizes this century in the three (or four) main stocks, which are considered separately. Estimates, updated to 1990, are given for the total world population of the species.

SOUTH GEORGIA STOCK

The Sealing Industry, 1910–1964

The biological basis underpinning rational exploitation of this species is its highly polygynous land-breeding behavior. The sexes are nearly equal in numbers at birth, and a single adult male can serve a large number of cows. In fact, the overwhelming majority of matings are carried out by a very small minority of breeding bulls; according to M. N. Bester and I. S. Wilkinson (this volume), "dominant bulls controlling the harems achieved over 98% of all matings." A large surplus of males can therefore be taken from the beaches without affecting recruitment to the population. Polygyny has resulted in the socially mature male being on average about eight times the weight of the average breeding female (Laws 1984). As the sealing industry sought oil from the blubber, it was most convenient to take the bulls in the earlier part of the breeding season in the spring, when the oil yield is greatest. Because the larger males haul out first on the beaches, the sealers automatically selected the largest animals. Oil yields from autumn sealing are lower, and this practice was discontinued in the 1950s.

To prevent recurrence of the former indiscriminate slaughter, the Falklands Islands government introduced the Seal Fishery Ordinance in 1899 (improved by further ordinances in 1904, 1909, and 1921), with the intention of placing the industry on a rational basis. The coastline of South Georgia was divided into four sealing divisions (and four reserves where no seals were taken). The divisions were generally worked in rotation, each year one of them being closed to sealing. Licenses to take a stipulated number of adult male elephant seals were issued annually to the sealing company. The first license was issued in 1909 to the oldest established whaling company at South Georgia, which held a monopoly until the 1961–1962 season. There was no sealing in 1962–1963, and licenses were issued to a Japanese company for the last two seasons, 1963–1964 and 1964–1965. With the cessation of autumn sealing in the 1950s, sealing began early in September and ended about mid-November. The sealing methods are described in detail by me (Laws 1953*a*) and by Bonner (1958). Although attempts were made to utilize carcasses, this was found to be uneconomical and the sole product of the industry continued to be blubber oil.

For many years, the annual quota under license was fixed at 6,000. But in 1948, it was raised to 7,500; in 1949 and 1950, to 9,000; and in 1951, was set at 8,000. Until 1952, the quota was divided equally among the three divisions worked each year, although seal abundances in the four divisions have probably always been unequal. The sealing regulations were altered in 1952 as a result of recommendations made by me (Laws 1960).

Effects of Sealing

Previously, I have presented a variety of evidence that indicated that from the late 1930s onward there were adverse changes in the population, despite the conservation measures in force (Laws 1960). Most significant, it was shown that the average catch of seals per catcher's day's work (CDW) in October had progressively declined by 28% from 1931 to 1951. Over these two decades, there had also been a decline in the oil yield per seal and a lengthening of the season, trends that had begun well before the higher catches in 1948–1951. Consideration of the sex ratios on the breeding beaches, harem sizes (on exploited beaches and in a reserve), and the low average age of males on the breeding beaches supported the indications that the demonstrated decline in the catch per unit effort represented a real decrease in the size of the adult male component of the population.

In Laws 1960, I reported counts of seals on beaches in Divisions I–III and one count in Division IV. These counts were corrected for date and for pup mortality, based on my research, and I concluded that approximately 20,000 pups were born in Division I and 15,000 in Division II. Supplementary information was obtained from the sealers which indicated that seals were much more abundant in Divisions III and IV, and I assumed that 32,500 pups were born in each of these divisions, making an estimated total of approximately 100,000 pups that survived to weaning. Assessing pup mortality from birth to weaning at 2% gave a figure of about 102,000 pups at birth.

I constructed population models to assess the size of the total population in each division, based on my demographic research, on the size of the commercial catches in previous years, and on the age composition of the commercial kill of males in 1951, based on growth layers in the teeth (Laws 1952, 1953*b*). Some assumptions were made, and predictions about certain future statistics were advanced, the confirmation of which later appeared to support the assumptions. The results of these calculations were estimates of the size of the population in each division and of the annual accessions to each age class (see Laws 1960 for details). Other information on breeding behavior suggested that the stock of breeding bulls in each division should be kept equal to one-twelfth of the stock of adult cows, and on this basis surplus accessions of bulls for all divisions combined were calculated at 5,842, close to the average annual catch. But the catches in Divisions I and II for this period were, respectively, about 24% and 65% higher than the estimated surpluses for those divisions, whereas the catches in Divisions III and IV were well below.

Therefore, a new system was introduced in 1952 to control sealing. Most important, divisional quotas were set for the five years 1952–1956, proportional to the estimated size of the available surplus in each division. A safeguard was introduced to monitor the age composition of the catch, by requiring canine teeth to be collected from a 5% sample of the catch which would subsequently be examined to determine age structure of the catch. A minimum length regulation was introduced.

Favorable results were immediately apparent, with a rise in the catch per CDW of 26% over the first 4-year period. In general, sealing operations were conducted earlier (midseason date two weeks earlier) and the average oil yield increased substantially. About 300 teeth were collected annually between 1952 and 1964 for age determination, and W. N. Bonner made control collections from a total of over 400 seals caught. The mean age of the catch sample in 1951 was 6.64 ± 0.34 (SE) years; in 1952, it was 7.20 ± 0.14; and it increased to level off at about 7.71 ± 0.19, 7.70 ± 0.18, and 7.69 ± 0.15 years in 1961, 1963, and 1964, respectively (Laws 1979). This leveling off confirmed that the catch size was sustainable; it corresponds to the estimated average age on attaining social maturity, and the management strategy was to monitor the age distribution of the catch and to adjust annual quotas, as necessary, so as to keep the average age of the catch at 7½ to 8 years. The success of this policy gave confidence in the general accuracy of the estimates of population abundance and life table parameters (see below).

Population Size in 1951

In formulating my recommendations, I had constructed provisional life tables for the two sexes. These life tables adjusted for 102,000 births annually indicated an average midyear population of about 310,000 elephant seals at South Georgia in 1951, that is, about 3.1 times the number of births (see Laws 1960 for details). I reviewed available information on other breeding colonies of the overall South Georgia stock and concluded that its total size was a little more than 315,100.

McCann (1985) revised the life tables, assuming higher female fecundity (88% vs. 82.5%). He found no change in the number of pups born in Cumberland Bay, no evidence for an increase or decrease in the cow population, and no significant change in the age at first pupping since 1964. He concluded that "a suitable factor by which to convert pup counts to total population of all animals older than pups of the year at the end of the breeding season is 3.5" (14). McCann (1980) also found no difference in body size in 1976–1977 compared to 1948–1951. He reported a three- to fourfold increase since exploitation ended in numbers of bulls ashore, an increase in the mean age of harem bulls, and increased average harem size; he also found that bulls hauled out earlier in the season and stayed ashore longer. There was little change in the number of breeding cows or their age structure. (Comparisons were with results in my papers, based on fieldwork in 1948–1951, listed in Laws 1960.) McCann (1985) concluded that the South Georgia population in 1980 was 350,000 (actually, it should have been 357,000 if calculated from births × 3.5).

McCann's estimate was derived from the 1951 estimate of pups born and his life tables for an unexploited population. In 1951, however, at the height of the sealing period, even assuming a similar pup production, the total population should have been somewhat lower, because the numbers of adult males were reduced in the exploited population. Comparison of Laws (1960) and McCann (1985) shows that the female life tables are really quite similar (perhaps not surprising as McCann adopted some important parameters from me, especially first- and second-year survival); the male life tables differ only in the higher first- and second-year mortality assumed by me and in the reduced survival of older males in the exploited population.

I have therefore recalculated the exploited male life table, by raising the first-year survival to the female level, as McCann did; this gives almost identical survival rates for the two models up to 6 years. Above 6 years the selective removal of older males reduced the adult male population, so that in the exploited 1951 population model there was virtually no survival beyond age 12, compared with age 20 in McCann's (1985) life table. I have calculated the number of adult males estimated by the two models, assuming 54,000 male pups at birth in both exploitation and postexploitation models; McCann reported a sex ratio of 53% males at birth. For the 1951 model there were 17,280 males aged 7 to 12 years (none were older), and for 1985 there were an estimated 36,000 aged 7 to 20 years (the upper limit). The difference is nearly 19,000, suggesting that the 1951 total South Georgia population size was actually about 338,000 (357,000 − 19,000). Sealing ended in 1964, so the recovery in adult male numbers should have been virtually complete by 1970 as full cohorts successively acceded to the adult male population.

Population Size in 1985

The elephant seal population of South Georgia was surveyed comprehensively during the 1985 breeding season (McCann and Rothery 1988). There were 87,711 females and 10,260 adult males counted on shore, and the female counts were corrected using a model of haul-out distribution over time to adjust for the date of count. The annual pup production was estimated to be about 102,000, the same as the estimate in Laws (1960). I actually counted 26,260 live pups in Divisions I and II and gave an estimate for pup production, corrected for date of count, of 31,075; I rounded up this estimate to 35,000, to allow for the beaches I did not visit. This can be compared with T. S. McCann and P. Rothery's (1988) corrected total of 34,665 in these two divisions, which is very close. McCann and Rothery obtained corrected pup counts of 42,892 and 24,676 for Divisions III and IV, respectively, and suggested that as the numbers in Divisions I and II were approximately the same in 1985 as in 1951, the same should be true of the totals in Divisions III and IV. Thus, there is a good basis for assuming that the number of pups was indeed about 102,000 in both periods, and applying the factor 3.5 gives a population other than pups of 357,000 in 1985. (A possible but small source of error lies in the adoption of my 1960 estimate of pup mortality (2%); e.g., Condy [1978] estimated preweaning mortality at 6%.)

Thus, the total pup production at South Georgia was very similar in two estimates 34 years apart. This suggests that the South Georgia population may have been almost stable since 1951, apart from an increase in adult males after sealing ended in 1964. However, McCann and Rothery (1988) show that substantial annual fluctuations occurred in the number of cows ashore on certain beaches, and their data suggest a possible decline in the 1960s. It seems likely therefore that over the last 40 years, the South Georgia population size has fluctuated around 350,000.

The 1985 count and model presented above indicate that of some 36,000 adult males (aged 7 to 20 years) estimated to be in the South Georgia population, only some 10,000 were counted on land at any one time during the breeding season, and about 26,000 were therefore in the sea, since they have not been seen elsewhere on land. This is relevant to hypotheses about the mating of the virgin females; they are very rarely seen on the breeding beaches in the breeding season, yet the 1985 model (McCann 1985) indicates that some 21,000 virgin females (equivalent to about a sixth of all mature females) mate for the first time at age 3. I observed aquatic matings and believed that newly mature females mate for the first time not on land but in the sea, probably during the usual breeding season of the adult cows (Laws 1956). (With 88% fecundity, the model also indicates that 12% of all adult females—an estimated 14,000—would be nonpregnant in the pupping season and probably also would mate aquatically). Behavior during the aquatic phase of the life cycle may have implications for the causes of the decline in other southern elephant seal populations (SCAR 1991; Bester and Wilkinson, this volume; Hindell, Slip, and Burton, this volume; Fedak et al., this volume).

Other Populations in the South Georgia Stock

Patagonia and the Falkland Islands. I estimated less than 1,000 pups born in the Falkland Islands, representing a population of less than 3,500 (Laws 1960); there is no recent information. For Patagonia, at Península Valdes, I. S. Carrara (1952) estimated a population of approximately 1,000 in 1951; J. A. Scolaro (1976) estimated 3,933 pups born in 1975, implying a population of 13,800 (3.5 × births); D. F. Vergani, M. N. Lewis, and Z. B. Stranganelli (1987) estimated 6,737 pups born there in 1982 (pop. 23,579), increasing at 5.1% annually between 1975 and 1982. C. Campagna and M. Lewis (1992) give the results of eight surveys from 1969 to 1990, corrected for date and area surveyed, concluding that 7,455 pups were born in 1982 (pop. 25,092) and 9,636 pups were born in 1990 (pop. 33,726), having increased at 3.2% annually since 1982.

South Orkney Islands. I estimated the size of the breeding population I studied at Signy Island to be 300 in 1948–1950 (Laws 1960). Subsequently, this has decreased to less than 20 (unpublished British Antarctic Survey records).

King George Island. Vergani (pers. comm.) reported 708 pups born in 1980, declining to 560 by 1990, which represents total populations of approximately 2,500 and 2,000.

Nelson Island. Vergani, Lewis, and Stranganelli (1987) reported 106 pups born in 1985, representing a population of approximately 370.

Avian Island. One birth has been reported from Avian Island (68° S.) (unpublished British Antarctic Survey record).

Gough Island. Bester (1980; pers. comm.) reported 32 pups born in 1977 and 28 in 1990, suggesting a stable population of about 105.

South Sandwich Islands and Bouvet Island. The species probably breeds in small numbers on these islands, but there is no information.

These populations represent a total of about 39,720 in 1990, about 10% of the total South Georgia stock in 1990, which is estimated at about 397,000 compared with McCann's (1985) estimate of some 369,000 based on counts spread over earlier years (table 3.1). The difference is due to a real increase in the Península Valdes population and the upward revision of the earlier South Georgia estimate to allow for the recovery of the adult male stock as discussed above.

TABLE 3.1 Estimated sizes of southern elephant seal populations in 1990 compared with earlier estimate by McCann (1985).

Stock	Area	Births[a]	Population size[b,c]
South Georgia	South Georgia	102,000	357,000
	South Orkney Islands	approx. 5	approx. 20
	South Shetland Islands	650	2,300
	Falkland Islands	approx. 1,000	3,500
	Gough Island	approx. 30	approx. 105
	Península Valdes	9,636	33,726
			397,000 (369,000)
Isles de Kerguelen	Kerguelen	41,000	143,500
	Heard Island	(11,530)	40,355
	Marion Island	(574)	2,009
	Prince Edward Island	(223)	782
	Isles Crozet	(577)	2,023
			189,000 (252,000)
Macquarie Island	Macquarie Island	(22,226)	77,791
	Campbell Island	approx. 5	approx. 20
	Antipodes Island	114	approx. 400
			78,000 (137,000)
World			664,000 (758,000)

[a] Figures in parentheses adjusted for continued decline since last estimate by applying compound % annual change to most recent estimate (see text).

[b] Population sizes—animals one year old and older (births × 3.5). Totals rounded to nearest 1,000.

[c] Totals in parentheses estimated from data relating to various years in the period 1949–1980 (McCann 1985).

THE ILES KERGUELEN STOCK

Iles Kerguelen: Courbet Peninsula

Here there are difficulties in making comparisons from year to year because of uncertainties about the proportion of the population censused. Almost all breed on the Courbet Peninsula between Point Molloy and Cap Noir, here called the study area. R. J. van Aarde (1980) summarized previous counts of breeding cows plus weaned pups (roughly equal to pups born) in the study area in 1958, 1960, and 1970 and presented his own figures for 1977.

These were, respectively, 41,970, 40,050, 55,252, and 42,420. However, the first two are underestimates because they were uncorrected and so are not comparable. On the basis of the estimates for 1970 and 1977, van Aarde calculated that there had been a decline of 4.6% annually over this period (and, raising by 3.5 times, this corresponds to a decline in total population from 193,382 to 148,470). Bester (1982) counted 36,900 pups in the study area in 1979, representing a population of 129,150; a continued decline of 4.6% from 1977 would have resulted in 38,607 pups (total 135,124). C. Guinet, P. Jouventin, and H. Weimerskirch (1992) obtained a figure of 41,000 pups in 1989 and suggest that the population has stabilized; this would indicate a population of about 143,500 in 1990.

Heard Island
R. Carrick and S. E. Ingham (1962) gave an estimate of 23,000 pups for 1949, representing a total population of 80,500. By 1985, Burton (1986) estimated pup production at 13,000, representing a population of 45,500, which had therefore declined by 2.4% annually, averaged over the period from 1949 to 1985. Continuing decline at this rate would result in a population of 40,355 in 1990, a total decline of 50%.

Marion Island
In 1973 and 1976, total numbers in various age classes were counted in the period November 13–19, when combined pup and weaner counts were at a maximum (Condy 1978); no adjustments to these were therefore necessary except for preweaning mortality (5.99%). A mean of 1,049 pups and weaners were counted, which adjusted for mortality gave an estimated 1,115 births annually in this period. Raised by 3.5, this indicates a total population of 3,903 in 1976. R. W. Rand (1962) had counted the pups, weaners, and dead pups on the island in the second half of November 1951 and obtained a figure of 3,662, corresponding to a total population of 12,817. (Condy calculated a 69.5% decline between 1951 and 1976.) P. R. Condy (1977) analyzed three comparable counts from 1951, 1965, and 1974 and estimated the rate of decrease of the Marion Island breeding population over this period at 4.8% a year.

To study the trend between 1973 and 1982 in more detail, J. D. Skinner and R. J. van Aarde (1983) analyzed results of censuses from a subset of the breeding beaches, representing about a third of the annual pup production, for the period 1973–1982 (excepting 1978); they are not directly comparable to Condy's (1977) figures. This indicated an exponential decrease in pup production over the period 1973–1982 at a mean annual rate of 8%.

The most recent data come from Bester and Wilkinson (this volume) and Wilkinson and Bester (in press). The first relate to a main study area representing about 10% of the coastline, censused from 1974 to 1989 following the methods of Condy (1978). In 1976 and 1986–1989, total pup counts of the whole island were made, which confirmed that numbers in the main study area provide a representative index of the total population. They conclude that the overall decline between 1974 and 1989 was 4.8% annually, but in 1983–1989 it had slowed to 1.9%. Wilkinson and Bester (in press) estimated the total pup production for the island in 1989 at 585; a further 1.9% decline would reduce this to 574 in 1990. Raised by 3.5, this gives an estimated total population of 2,009 in 1990, a decline of 83.7% since 1951 and of 48.5% since 1976. In percentage terms, this population has been most adversely affected of all except Signy Island and Campbell Island (see below).

Prince Edward Island
Condy (1978) also counted 386 pups and weaners on the east coast of Prince Edward Island in 1977; applying Condy's (1978) correction factor for preweaning mortality gives an estimate of 411, raised to a total population of 1,439. If this population also declined on average at 4.8% annually between 1974 and 1989 and 1.9% in 1989–1990, its 1990 population would be 782.

Iles Crozet
A. Barrat and J. L. Mougin (1978) reported about 3,000 pup births in 1976, corresponding to a total population of 10,500, having declined by 5.8% annually between 1966 and 1976 (SCAR 1991). Guinet, Jouventin, and Weimerskirch (1992) reported 612 pups in 1989, representing by inference a total population of 2,142, having declined at 5.7% annually between 1980 and 1989. Extrapolating a further year at this rate of decline implies a population of 2,019 in 1990 (SCAR 1991).

The populations comprising the Iles de Kerguelen stock represent a total of about 189,000 in 1990 (table 3.1), about 28% of the total world population, having declined dramatically over the last 40 years, one population by over 80%. This is in marked contrast to the South Georgia stock, which has increased at South Georgia since 1951 by 20,000 (15.3%) as a result of the postexploitation recovery of the male component and at Península Valdes by 20,000 (144%) since 1975, due to natural increase.

THE MACQUARIE ISLAND STOCK

Macquarie Island
M. A. Hindell and H. R. Burton (1987) have recently described the past and recent status of the elephant seal at Macquarie Island. This is one of the most continuously studied populations since observations began in

1949. Until the early 1980s, only adult females and males were counted in the isthmus study area where most of the seals haul out to breed (Carrick and Ingham 1960; Carrick et al. 1962). However, Hindell and Burton (1988) showed that the haul-out pattern of breeding females is so predictable that a census on any day within two weeks of the haul-out maximum can be used to estimate the maximum number accurately. In the 1985 breeding season, daily counts of adult females, weaned pups, dead pups, and breeding males were made on the isthmus study area, and a complete island census was carried out on a single day, a week before the peak number; an estimate of the peak number of females was derived from the haul-out curve modeled from the isthmus data.

Hindell and Burton (1987) estimated the maximum number of females ashore on the isthmus for 21 years between 1949 and 1985. They showed that there was a significant correlation between years and maximum number of females, with an average annual rate of decrease of 2.1%. There was a similar percentage decrease in the numbers of males ashore.

The estimated total island pup production in 1959 was 44,501, compared with 24,713 in 1985. Applying the 3.5 raising factor (McCann 1985) gives a total population of 156,000 in 1959 and 86,500 in 1985, that is, 44.6% less than in 1959. Hindell and Burton (1987) concluded that the population was probably declining at least from 1950, and they estimated that the whole island population in 1949 was about 183,000. Extrapolating from 1985 to 1990, assuming a continued decline at 2.1% per annum, gives a total population of 77,791 in 1990, which represents an overall decline of 57.5% since 1949.

Campbell Island

There is a small population of elephant seals at Campbell Island. J. H. Sorensen (1950) gave estimates from counts of the numbers of pups born in 1941, 1942, 1945, and 1947, increasing from 75 to 100 in 1941 to 191 in 1947. Taking these figures at face value, this suggests an average of about 140 pups born in the 1940s and, applying the raising factor used by McCann (1985), a total population of about 500 at this time. The only recent information appears to be from R. H. Taylor and G. A. Taylor (1989), who give a pup production of only 5, representing a total population numbered in tens of animals, let us say 20. This would represent a decline of 96% since the 1940s.

Antipodes Island

Taylor and Taylor (1989) also give an estimate of 113 for the pup production at Antipodes Island in 1978. There is no information on the trend in this population, but if there has been no change, applying the raising factor gives a population of about 400 in 1990.

Thus the Macquarie Island stock, which is essentially confined to Macquarie Island for breeding, had an estimated total size of about 78,000 in 1990 (table 3.1), representing only 12% of the total world population, having declined dramatically from about 183,000 in 1949, a decrease of 57% over 42 years. This is again in marked contrast to the South Georgia stock and more like the Iles Kerguelen stock.

CONCLUSIONS

As the work summarized here indicates, during the period from 1949 to the present, different populations of southern elephant seals have increased (Península Valdes), probably remained stable after rapid recovery from the exploitation that ended in 1964 (South Georgia), or decreased dramatically (Iles Kerguelen, Marion Island, Macquarie Island, Signy Island, Campbell Island).

The population sizes are derived from estimates of pups born, raised by the factor of 3.5 to obtain estimates of the size of populations, excluding pups. Different raising factors should be applied to increasing, stable, or declining populations, and in the absence of adequate demographic knowledge of the different populations, the estimates of annual pup production are more accurate than population abundance estimates. However, it is desirable to have population abundance estimates for other purposes (such as estimating population biomass and food consumption; Boyd, Arnbom, and Fedak, this volume) and since a common factor is applied, the derived population trends are the same as for pup production. Subject to this reservation, table 3.1 shows an estimated total population of the southern elephant seal in 1990 of about 664,000, which compares with an estimated total population of 758,000 derived from data obtained at different times during the period 1949–1980 (McCann 1985), an overall decline of some 12.3%. The decline is not universal, however, and 60% of the total world population is in the South Georgia (and South American) stock, which is stable or increasing.

If we look at the populations that have decreased, currently about 40% of the total world population, they have decreased on average by about 31%, but some have decreased to a much greater extent; for example, the major stock at Macquarie Island has declined by an estimated 57% since 1949. Information about the possible causes of the declines is presented elsewhere in this volume (chaps. 4, 5, and 6). A workshop meeting convened by the Scientific Committee on Antarctic Research (SCAR 1991) also addressed the possible causes of past and present population trends but did not reach any firm conclusion. It seems that the causes of the declines may differ among populations; however, the causes remain speculative and essentially unknown. In any case, the contrast between the population trends in the northern elephant seal, *M. angustirostris* (Stewart et al., this volume), and the southern elephant seal is striking and is an urgent topic for further research.

There is no evidence to suggest that the population declines are the result of factors influencing the seals while ashore, indicating that they result from factors operating while the animals are at sea (SCAR 1991). In this respect, the evidence presented here indicating that a large proportion of adult males and newly mature and fully adult females remain in the sea during the breeding season may be relevant. For successful mating, it seems unlikely that these components of the population are widely dispersed at this time. They may be in the general vicinity of the breeding islands, or relatively concentrated in other parts of the species' foraging range. Tracking studies using satellite geolocation equipment should eventually provide an answer to this question.

Much of this chapter describes the government-controlled industry at South Georgia between 1909 and 1964. This is a topic that is not covered elsewhere in this volume and has been included here because of the light it throws on the population dynamics of the species—and because this stock constitutes some 60% of the total population of the species. Despite the removal of some 260,000 bulls during a 65-year period, the South Georgia population appears to have remained fairly stable or to have fluctuated only slightly, a confirmation that the industry was on a firm sustainable yield basis after 1952. This elephant seal stock in fact provides a classic example of rational utilization of a natural renewable resource.

REFERENCES

Barrat, A., and J. L. Mougin. 1978. L'elephant de mer *Mirounga leonina* de L'île de la Possession, Archipel Crozet (46°25'S, 51°45'E). *Mammalia* 42: 143–174.

Bester, M. N. 1980. The southern elephant seal *Mirounga leonina* at Gough Island. *South African Journal of Zoology* 15: 235–239.

———. 1982. An analysis of the southern elephant seal *Mirounga leonina* breeding population at Kerguelen. *South African Journal of Antarctic Research* 12: 11–16.

Bonner, W. N. 1958. Exploitation and conservation of seals in South Georgia. *Oryx* 4: 373–380.

———. 1982. *Seals and Man*. Seattle: University of Washington Press.

Burton, H. 1986. A substantial decline in numbers of the southern elephant seal at Heard Island. *Tasmanian Naturalist* 86: 4–8.

Campagna, C., B. J. Le Boeuf, M. Lewis, and C. Bisioli. 1992. Equal investment in male and female offspring in southern elephant seals. *Journal of Zoology, London* 226: 551–561.

Campagna, C., and M. Lewis. 1992. Growth and distribution of a southern elephant seal colony. *Marine Mammal Science* 8: 387–396.

Carrara, I. S. 1952. Lobos marinos, pinguinos y guaneras de las costas del littoral maritimo e islas adyacentes de la Republica Argentina. Universidad Nacional La Plata, Facultad de Sciencias Veterinarias, Catedra de Higiene e Industrias, Publ. Espec.

Carrick, R., S. E. Csordas, S. E. Ingham, and K. Keith. 1962. Studies on the southern elephant seal, *Mirounga leonina* (L.). III. The annual cycle in relation to age and sex. *CSIRO Wildlife Research* 7: 119–160.

Carrick, R., and S. E. Ingham. 1960. Ecological studies on the southern elephant seal, *Mirounga leonina* (L.), at Macquarie Island and Heard Island. *Mammalia* 24: 325–342.

————. 1962. Studies on the southern elephant seal, *Mirounga leonina* (L.). V. Population dynamics and utilization. *CSIRO Wildlife Research* 7: 198–206.

Condy, P. R. 1977. Annual cycle of the southern elephant seal *Mirounga leonina* (Linn.) at Marion Island. *South African Journal of Zoology* 14: 95–102.

————. 1978. The distribution and abundance of southern elephant seals *Mirounga leonina* (Linn.) on the Prince Edward Islands. *South African Journal of Antarctic Research* 8: 42–48.

Gales, N. J., M. Adams, and H. R. Burton. 1989. Genetic relatedness of two populations of the southern elephant seal, *Mirounga leonina*. *Marine Mammal Science* 5: 57–67.

Guinet, C., P. Jouventin, and H. Weimerskirch. 1992. Population changes and movements of southern elephant seals on Crozet and Kerguelen archipelagos in the last decades. *Polar Biology* 12: 349–356.

Headland, R. 1984. *The Island of South Georgia.* Cambridge: Cambridge University Press.

————. 1989. *Chronological List of Antarctic Expeditions and Related Historical Events.* Cambridge: Cambridge University Press.

Hindell, M. A. 1991. Some life-history parameters of a declining population of southern elephant seals, *Mirounga leonina. Journal of Animal Ecology* 60: 119–134.

Hindell, M. A., and H. R. Burton. 1987. Past and present status of the southern elephant seal (*Mirounga leonina*) at Macquarie Island. *Journal of Zoology, London* 213: 365–380.

————. 1988. The history of the elephant sealing industry at Macquarie Island and an estimate of the pre-sealing numbers. *Papers and Proceedings of the Royal Society of Tasmania* 122: 159–176.

Laws, R. M. 1952. A new method of age determination for mammals. *Nature* 169: 172.

————. 1953a. The elephant seal industry at South Georgia. *Polar Record* 6: 746–754.

————. 1953b. A new method of age determination for mammals with special reference to the elephant seal, *Mirounga leonina* Linn. *Falkland Islands Dependencies Survey, Scientific Reports* 2: 1–11.

————. 1956. The elephant seal (*Mirounga leonina* Linn.). II. General, social, and reproductive behaviour. *Falkland Islands Dependencies Survey, Scientific Reports* 13: 1–88.

————. 1960. The southern elephant seal (*Mirounga leonina* Linn.) at South Georgia. *Norsk Hvalfangst-Tidende* 49: 446–476, 520–542.

————. 1979. Monitoring whale and seal populations. In *Monitoring the Marine Environment*, ed. D. Nichols, 115–140. New York: Praeger.

————. 1984. Seals. In *Antarctic Ecology*, vol. 2, ed. R. M. Laws, 621–715. London: Academic Press.

McCann, T. S. 1980. Population structure and social organization of southern elephant seals, *Mirounga leonina* (L.). *Biological Journal of the Linnaean Society* 14: 133–150.

————. 1985. Size, status and demography of southern elephant seal (*Mirounga leonina*) populations. In *Sea Mammals of South Latitudes: Proceedings of a Symposium of the 52d ANZAAS Congress in Sydney—May 1982*, ed. J. K. Ling and M. M. Bryden, 1–17. Northfield: South Australian Museum.

McCann, T. S., and P. Rothery. 1988. Population size and status of the southern elephant seal (*Mirounga leonina*) at South Georgia 1951–1985. *Polar Biology* 8: 305–309.

Murray, M. D. 1981. The breeding of the southern elephant seal, *Mirounga leonina* L., on the Antarctic continent. *Polar Record* 20: 370–371.

Rand, R. W. 1962. Elephant seals on Marion Island. *African Wildlife* 16: 191–198.

SCAR. 1988. Report of a meeting held at Hobart, Tasmania, August 23–25, 1988. *Biomass Report Series*, no. 59: 1–61.

————. 1991. Report of the workshop on southern elephant seals, Monterey, California, May 22–23, 1991.

Scolaro, J. A. 1976. Censo de elefantes marinos (*Mirounga leonina* L.) en el territorio continental Argentino. Informes Técnicos 1.4.2, Centro Nacional Patagonico, Puerto Madryn, R. Argentina, 1–12.

Skinner, J. D., and R. J. van Aarde. 1983. Observations on the trend of the breeding population of southern elephant seals, *Mirounga leonina*, at Marion Island. *Journal of Applied Ecology* 20: 707–712.

Sorensen, J. H. 1950. Elephant seals of Campbell Island. *Cape Expedition Series* 6: 1–31.

Taylor, R. H., and G. A. Taylor. 1989. Re-assessment of the status of southern elephant seals (*Mirounga leonina*) in New Zealand. *New Zealand Journal of Marine and Freshwater Research* 23: 201–213.

van Aarde, R. J. 1980. Fluctuations in the population of southern elephant seals *Mirounga leonina* at Kerguelen Island. *South African Journal of Zoology* 15: 99–106.

Vergani, D. F., M. N. Lewis, and Z. B. Stranganelli. 1987. Observation on haul-out patterns and trends of the breeding populations of southern elephant seals at Península Valdes (Patagonia) and Stranger Point (King George Island). *SC-CCAMLR-VI/BG/36*, Hobart, Tasmania, 1–9.

Vergani, D. F., and Z. B. Stanganelli. 1990. Fluctuations in breeding populations of elephant seals *Mirounga leonina* at Stranger Point, King George Island 1980–88. In *Antarctic Ecosystems*, ed. K. R. Kerry and G. N. Hempel, 241–245. Berlin: Springer Verlag.

Wilkinson, I. S., and M. N. Bester. 1990. Duration of postweaning fast and local dispersion in the southern elephant seal, *Mirounga leonina*, at Marion Island. *Journal of Zoology, London* 222: 591–600.

————. In press. Population parameters of the southern elephant seal population at Marion Island. *Antarctic Science*.

FOUR

Possible Causes of the Decline of Southern Elephant Seal Populations in the Southern Pacific and Southern Indian Oceans

Mark A. Hindell, David J. Slip, and Harry R. Burton

ABSTRACT. There are several characteristics of declining southern elephant seal populations that provide a basis for the formulation and testing of hypotheses designed to explain the decline. Some southern elephant seal populations seem to be declining independently of other major Southern Ocean vertebrate consumers, but only two of the three distinct stocks have exhibited a decline. At Macquarie Island, the adult male and female components of the population seem to be equally affected. Studies conducted in the 1950s and 1960s indicate that elephant seals from Macquarie Island had lower survivorship and slower growth rates and reached sexual maturity later than those from South Georgia. Survival of first-year seals decreased dramatically during the 1960s and led to the almost total failure of the 1965 cohort; the fate of later cohorts has not been determined.

Attempts to explain the recent behavior of southern elephant seal populations have so far concentrated on finding a single cause to account for the characteristics of all populations, rather than a series of independent explanations. Two main hypotheses are advanced, one involving equilibration processes after intense sealing pressure and the other concerned with fluctuations in the ocean environment. However, while there may be a single driving factor influencing the populations of southern elephant seals in the Indian and Pacific Ocean sectors, there may be additional local factors that also regulate the populations. In some cases, the interaction between these effects may obscure understanding of the main driving force. While the hypotheses have not yet been tested, they help in planning future research. Priorities should be the maintenance of censusing programs, long-term and cross-sectional demographic studies, and investigations into several aspects of the biology of first-year seals, particularly diet and foraging ranges.

Southern elephant seals, *M. leonina* Linn., have been known to science since the time of Linnaeus. For most of that time, however, their contact with man was purely as an exploitable resource, and they were ruthlessly hunted at all of their major breeding grounds as a source of high-quality oil. Only since the middle of this century have these animals been the focus of serious scientific attention.

The 1950s and 1960s saw major research programs at both the South Georgia and Macquarie Island breeding sites. The result of this research effort was a new understanding of many important aspects of the basic biology of the species, such as reproductive biology (Laws 1956a), developmental biology (Laws 1953, Bryden 1968a), behavior and social structure (Laws 1956b; Carrick, Csordas, and Ingham 1962; Carrick et al. 1962), and population biology (Laws 1960).

A major shift in research perspectives for southern elephant seals occurred in the mid-1980s, with reports of significant declines in several major populations. These reports coincided with an increasing awareness of the value of the Southern Ocean ecosystem both commercially and scientifically. The decline of a major vertebrate predator was regarded as possible cause for concern that other components of the ecosystem were also changing in as yet undetected ways.

This chapter summarizes what is known about the population declines in southern elephant seals at Macquarie Island and at other breeding sites to compare this to trends in other Southern Ocean vertebrate populations and to review the various explanations that have been advanced to explain the decline so far. A new hypothesis is also advanced, and the most fruitful directions for future research are suggested.

POPULATION TRENDS

Macquarie Island

In 1985, the population at Macquarie Island was found to have decreased at a rate of about 2% per annum (Hindell and Burton 1987), calculated on the assumption that the decline had been operating since the early 1950s. The nature of the census data was such that there was no good a priori reason for assigning the beginning of the decline to any point during the 30-year time series (fig. 4.1). More recent analysis of demographic data collected at Macquarie Island during the 1950s and 1960s suggested that the decline may have started during the early 1960s (Hindell 1991). However, this has little effect on the estimated rate of decline, which remains at about 2% per annum.

There were no apparent differences in the overall rate of decline of adult males and females at Macquarie Island. And as adult male and female elephant seals occupy largely separate regions of the Southern Ocean during their time at sea (Hindell, Burton, and Slip 1991), whatever factor is causing the decline appears to be operating on the younger age classes, possibly before sexual differences in foraging patterns develop.

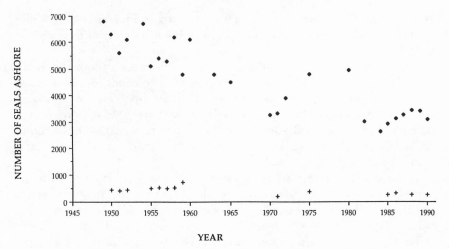

Fig. 4.1. The peak number of breeding females (diamonds) and males (crosses) on
the isthmus beaches at Macquarie Island between 1950 and 1990.

Kerguelen and South Georgia Stocks

The population decline at Macquarie Island should not be considered in
isolation, as similar declines are occurring elsewhere in the Southern
Ocean. Study of the behavior of all populations of southern elephant seals
will help to understand the situation at Macquarie Island.

Southern elephant seal populations can be grouped into three apparently
distinct stocks: Macquarie Island, Iles Kerguelen, and South Georgia. The
most recent census data on subpopulations in each of these stocks are listed
in table 4.1. The relatively inaccessible nature of many of the rookeries
combined with the brevity of the breeding season makes accurate population estimates difficult in some cases, and interpopulation comparisons
should be made with care. However, one major trend stands out. All the
populations from the Kerguelen stock have been declining at least until the
mid-1980s, while the populations from the South Georgia stock have been
generally stable or increasing. The populations within the Kerguelen stock
may be declining asynchronously with each other and with the Macquarie
Island population, but this could be partly due to artifacts of the limited
data. Nonetheless, these data suggest that there is something different
about the South Georgia seals or about some aspect of their environment.

Other Species

Until recently, elephant seals were not regarded as truly Antarctic seals and
were seen as ecologically distinct from the "ice seals." Recent studies on the

foraging ranges of elephant seals (see Slip, Hindell, and Burton, this volume) indicate that many adults spend a considerable proportion of their annual cycle in Antarctic waters. Therefore, it is relevant to compare the population trends of other Antarctic vertebrate consumers with those of elephant seals, and as yearling elephant seals may not move far from their natal islands (Carrick et al. 1962), it is also appropriate to examine population trends of other subantarctic species.

The status of other phocid species in the Southern Ocean is somewhat uncertain. Too little is known about the status of Ross seals, *Ommatophoca rossi*, and leopard seals, *Hydrurga leptonyx*, to make an assessment of any population trends at the moment (Erikson and Hanson 1990). There is no evidence of long-term changes in Weddell seal, *Leptonychotes weddelli*, populations at a number of sites around Antarctica (Testa and Siniff 1987; Green, Wong, and Burton 1992), although cycles in reproductive performance have been recorded (Testa et al. 1990). A. W. Erikson and M. B. Hanson (1990) reported a drop in the numbers of crabeater seals, *Lobodon carcinophagus*, in the Weddell Sea, but it is uncertain whether this represents a real decline in abundance, a change in distribution, or even sampling artifacts (see Green, Wong, and Burton 1992). Both Antarctic fur seal, *Arctocephalus gazella*, and sub-Antarctic fur seal, *Arctocephalus tropicalis*, populations are increasing at all of their breeding sites (e.g., Shaughnessy, Shaughnessy, and Keage 1988; Bengtson et al. 1990).

The baleen whale species that were once heavily harvested in the Southern Ocean all seem to be increasing in numbers (Bengtson 1984; Bryden, Kirkwood, and Slade 1990). Penguins constitute over 90% of the Antarctic avian biomass, but to date there have been no reports of major population declines in this important component of the Antarctic marine ecosystem (Woehler and Johnstone 1991). On the contrary, several species have increased (Laws 1985).

Of all the vertebrate consumers breeding on subantarctic islands, only the wandering albatross has declined in numbers (Hindell and Burton 1987), but this has been attributed to interactions with long-line fishing fleets (Croxall et al. 1990). Giant petrel populations have also been reported as declining in recent years (E. Woehler, pers. comm.), but as these animals rely on elephant seal carrion for a significant part of their diet (Hunter 1983), this decline could simply be a consequence of the elephant seal decline. Some rockhopper penguin, *Eudyptes chrysocome*, populations have also declined in recent years; the causes for these declines are as yet unknown (Moors 1986).

With the possible exception of crabeater seals and giant petrels and the certain exception of rockhopper penguins and wandering albatross, no other Southern Ocean consumer has exhibited large-scale population de-

TABLE 4.1 Size and status of southern elephant seal populations within the three stocks of the Southern Ocean.

Stock	Locality	Year	Pup production		Annual rate of change	Period	Status	Source
			Observed	1990				
South Georgia	South Georgia	1985	102,000	102,000	?	1951–1985	Uncertain	McCann and Rothery (1988)
	South Orkney Islands	1985	<100	—	?	1948–1985	Uncertain	McCann (1985)
	Falkland Islands	1980s	5 to 10	approx. 5	?	1970s–1980s	Declining	Boyd (pers. comm.)
		1960	approx. 1,000	approx. 1,000	?	—	Uncertain	Laws (1960)
	Gough Island	1989	28	28	0	1973–1989	Stable	Bester (pers. comm.)
	King George Island	1980	708	560	−0.05	1980–1990	Declining	Vergani (pers. comm.)
	Nelson Island	1985	106	106	?	—	Uncertain	Vergani; Lewis, and Stranganelli (1987)
	Peninsula Valdes	1982	6,737	—	5.1	1975–1982	Increasing	Vergani, Lewis, and Stranganelli (1987)
		1990	9,636	9,636	3.2	1982–1990	Increasing	Campagna and Lewis (pers. comm.)
Iles Kerguelen	Marion Island	1989	585	585	−4.8	1951–1989	Declining	Wilkinson and Bester (pers. comm.)
	Heard Island	1985	13,000	13,000	−2.4	1949–1985	Declining	Burton (1986)
	Iles Kerguelen (Courbet)	1977	45,000	—	−4.1	1970–1977	Declining	van Aarde (1980a)
		1989	41,000	41,000	0	1984–1989	Stable	Guinet, Jouventin, and Weimerskirch (1992)

Iles Crozet	(Possession)	1976	approx. 3,000	—	−5.8	1966–1976	Declining	Barrat and Mougin (1978)
		1989	612	578	−5.7	1980–1989	Declining	Guinet, Jouventin, and Weimerskirch (1992)
Macquarie Island	Macquarie Island	1985	24,000	—	−2.1	1949–1985	Declining	Hindell and Burton (1987)
		1990	22,068	22,068	−1.6	1985–1990	Declining	Slip (pers. comm.)
	Campbell Island	1986	5	4	−8.6	1947–1986	Declining	Taylor and Taylor (1989)
	Antipodes Island	1978	113	113	?	—	Uncertain	Taylor and Taylor (1989)

NOTE: Data from SCAR 1991.

clines. Several, such as the fur seals, baleen whales, and some penguin species, are actually increasing. This suggests that whatever is responsible for
the decline in some southern elephant seal populations is acting on some
aspect of their life history that is unique to these populations.

DEMOGRAPHIC DATA

There have been three studies of population biology of southern elephant
seals. Two were based at South Georgia, one in the 1950s when sealing was
still a commercial enterprise (Laws 1960) and the other during the 1970s,
some 20 years after sealing had ceased (McCann 1985). Both of these
studies were cross-sectional in nature, with life tables derived from age
structure data, which were in turn derived from age estimates made from
tooth rings. The population at South Georgia is thought to have been stable
over the time between the two studies, and the population had largely recovered from the effects of sealing by the late 1970s (McCann and Rothery
1988). Consequently, the South Georgia demographic data from McCann's
(1985) study may be regarded as representing a stable, undisturbed population and a valuable source of data for comparison with the declining
populations.

A longitudinal study based on 15 cohorts of seals branded at Macquarie
Island between 1950 and 1965 (Hindell 1991) revealed several features of
their population biology pertinent to understanding the decline in numbers.
First-year survival was essentially stable during the 1950s (fig. 4.2). The
mean first-year survival for females was 46% and for males 42%. In the
early 1960s, first-year survival declined dramatically, with the almost complete failure of the 1965 cohort (survivorship was less than 2%). However,
survival of all other age classes appeared to be stable between the 1950s
and 1960s. Unfortunately, it was not possible to analyze the success of later
cohorts.

There were several significant differences between the Macquarie Island
population parameters and those found in the stable South Georgia population. First-year survival of the Macquarie Island elephant seals during the
1950s was substantially lower than the 60% for both sexes estimated for
South Georgia (McCann 1985). Survival in all other age classes was also
lower at Macquarie Island than at South Georgia (fig. 4.3), although the
differences in methodology make quantitative comparisons difficult. Seals
from the South Georgia stock have higher growth rates, both preweaning
and postweaning, than those from Macquarie Island (Bryden 1968b). The
slower growth rates of the Macquarie Island seals may be responsible for
the 12-month delay in the average age at first breeding compared with
South Georgia (Carrick, Csordas, and Ingham 1962).

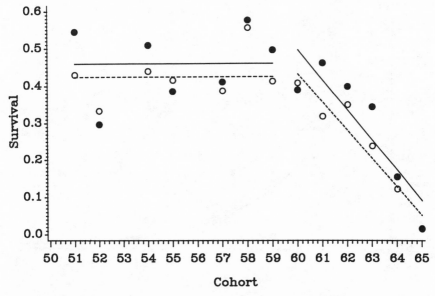

Fig. 4.2. Survival to age = 1 year for males (dashed lines, open circles) and females (solid lines, solid circles) in 13 cohorts between 1951 and 1965 (from Hindell 1991).

POSSIBLE EXPLANATIONS

An acceptable explanation of population trends in the southern elephant seal will need to account for the known characteristics of the declines. While it is possible that the declines in different populations may have independent causes, it is worthwhile to attempt to find a single explanation that will account for the decrease of all elephant seal populations.

Southern elephant seal declines are characterized by the following:

1. Both the adult male and adult female components of the Macquarie Island population appear to have declined at the same rate.
2. Only populations within the Macquarie and Kerguelen stocks have declined, while those from the South Georgia stock are probably stable. The timing and rates of the declines may also differ between subpopulations.
3. Southern elephant seal populations appear to have declined independently of other mammalian or avian species in the Antarctic or subantarctic sectors of the Southern Ocean.
4. Growth rates were lower at Macquarie Island than at South Georgia

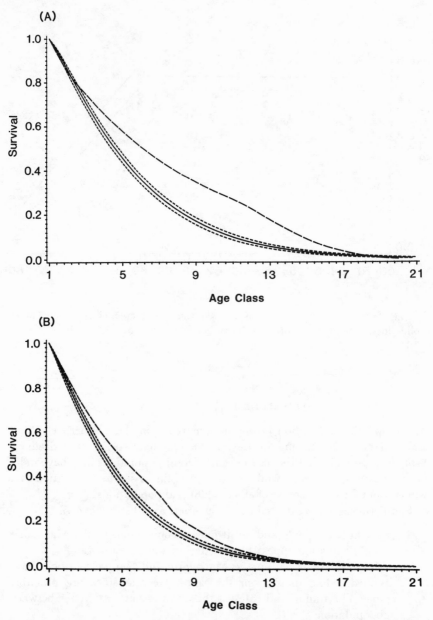

Fig. 4.3. Post-year 1 survival curves for *(A)* females from Macquarie Island (showing 95% confidence limits in dotted lines) and South Georgia, and *(B)* males from Macquarie Island (showing 95% confidence limits in dotted lines) and South Georgia (from Hindell 1991).

during the 1950s and 1960s. Age at first breeding was also one year later at Macquarie Island than at South Georgia.

5. Survivorship in all age classes was lower in the Macquarie Island population than in the South Georgia population.

6. There was a marked decline in the first-year survival of Macquarie Island animals between 1960 and 1965 which resulted in the almost complete failure of the 1965 cohort; post-year-one survival did not appear to change at this time (Hindell 1991).

A number of explanations have been advanced to explain these observations with varying degrees of success. The hypotheses can be loosely divided into three groups: intrinsic factors, predation, and the availability of food.

Intrinsic Factors

The most likely intrinsic factor that could influence elephant seal population size is some kind of density-dependent mechanism. Both northern and southern elephant seals can exhibit density-dependent pup mortality, with pup mortality increasing on very crowded beaches (Reiter, Stinson, and Le Boeuf 1978; van Aarde 1980b). Although it is possible that this mechanism might regulate elephant seal populations, it fails to explain many of the known characteristics of the decline. For example, it cannot account for both the apparent stability of the South Georgia population and the concurrent decline at Macquarie and Kergeulen. Also, as the overcrowding of harems results in increases in preweaning mortality, this mechanism does not account for the sharp decline in first-year survival during the 1960s (the animals used in that study were all branded after they were weaned).

It has been suggested that mass mortalities caused by disease, particularly parasite infestations, may be an important factor in population regulation in marine mammals (Harwood and Hall 1990). However, there is no evidence of widespread disease in southern elephant seal breeding aggregations.

Predation

Killer whales, *Orcinus orca*, are undoubtedly a major cause of juvenile mortality in southern elephant seals (Condy, van Aarde, and Bester 1978). Increases in the number of killer whales preying on newly weaned elephant seals might produce the changes in juvenile mortality described above and may also help explain the lower adult survival at Macquarie Island. However, there is no evidence for increased abundance of killer whales, and it would not satisfactorily explain the differences in growth rates and attainment of age of first breeding between the Macquarie and South Georgia populations.

Availability of Food

This category can be divided into three subgroups: competition, equilibration processes, and changes in the ocean environment.

Competition. Southern elephant seals could be competing with other Southern Ocean vertebrate consumers for basic food resources, particularly as a number of these species are currently increasing in number. However, adult elephant seals forage on deep-dwelling benthic and pelagic prey, which are generally unavailable to other mammal or bird species (Hindell, Slip, and Burton 1991), so the chances of direct competition must be limited. As we know little about the prey of first-year elephant seals, it is possible that this age class does share a prey species with one of the other increasing species, but recent data from northern elephant seals suggest that juveniles of 2 years are capable of undertaking the same dive patterns as adults (Le Boeuf, this volume). If young animals from the southern species can also make deep dives, they will equally be able to avoid direct competition with shallower-diving species. This hypothesis fails to account for the sudden decline in first-year survival observed in the 1960s, as it is difficult to understand how this could be caused by competition with other populations that were increasing at a steady rate.

It has also been suggested that elephant seals have had their resource base reduced by the recent introduction of commercial fishing operations in various regions of the Southern Ocean. M. Pascal (1986) suggested that the decline in elephant seal numbers at Iles Kerguelen may be linked to the very intense fishing activity in the area which occurred concurrently with the decline. However, there is little or no commercial fishing for deep water fishes in the areas identified as principal foraging grounds for elephant seals from Macquarie Island (Hindell, Burton, and Slip 1991; Slip, Hindell, and Burton, this volume).

Equilibration Processes. One explanation consistent with the known characteristics of the decline is the population "overshoot" hypothesis. This hypothesis states that the current decline in elephant seal numbers at Macquarie Island (and the other populations) is a direct consequence of the heavy exploitation of elephant seals during the eighteenth and early nineteenth century. The sealing industry reduced the elephant seal population at Macquarie Island to far below the original level and continued to suppress it for almost one hundred years (Hindell and Burton 1988a). During this time, prey species may have increased so that when commercial sealing ended in the early part of this century, an abundant resource was available to the elephant seals, which enabled an increase in the population such that it overshot the presealing population size. Thus, the population decline described since then may be the result of subsequent overexploitation of the food resource and a decrease toward equilibrium.

This hypothesis accounts for many of the observed characteristics of the decline. The South Georgia population may not have demonstrated an overshoot and decline because it was exploited in a managed fashion (at least for this century), while the differences in timing and rates of the declines in the other populations could be explained by different timing and extents of sealing pressure. The reduced growth rates and lower adult survival at Macquarie Island during the 1950s may have been a result of the population approaching the maximum carrying capacity of an inflated food resource. As the food resources became inadequate, the population began to decrease, principally driven by an increase in juvenile mortality as documented for the early 1960s.

The observation that both the male and female components of the population declined at a similar rate further supports the notion that the decline is a consequence of juvenile mortality. As adult male and female elephant seals use different foraging grounds, it seems most likely that the factor causing the decline has acted on the age classes in which pronounced sexual differences in morphology and behavior have not yet developed.

If the populations in the Kerguelen and Macquarie stocks declined synchronously, then the overshoot hypothesis would be substantially weakened. The hypothesis predicts that the timing and rate of the decline depends on the severity and duration of the sealing at each island (or group of islands where a common food source is used). It is unlikely that these would result in synchronous declines at a large number of populations over a considerable geographic range. Unfortunately, the time series data on abundance are inadequate to solve this problem.

Another problem with the overshoot hypothesis is that it may not account for declines in small populations, like Marion Island or Campbell Island, that may never have been heavily exploited. The hypothesis will only apply if (a) these populations share common feeding grounds with other, exploited populations and are thus exposed to the same "boom and bust" responses of the prey species or (b) such small populations are only splinter groups of the larger populations, perhaps formed when the exploited populations were approaching their maximum predecline densities.

Changes in the Ocean Environment. Another possible explanation is that the decline is due to changes in the ocean environment that have had an impact on the food species used by southern elephant seals (Burton 1986). Recent studies on global climate change have brought an increased awareness of large-scale environmental patterns and correlations between climatic fluctuations and population changes. For example, air pressure distribution over the Southern Hemisphere is influenced irregularly by the Southern Oscillation, which results in El Niño off Peru (van Loon and Shea 1988), and a scarcity of krill and a drop in breeding success of some land-based predators around South Georgia was observed in years that followed a strong El Niño Southern Oscillation (ENSO) event (Croxall et al. 1988; Priddle et al. 1988). Anomalies in air pressure at sea level affect the generation and movements of cyclones and anticyclones in the area between the subtropical ridge and the Antarctic continent (van Loon and Shea 1988), and these atmospheric processes force changes in the physical environment of the Southern Ocean, with subsequent effects on the biota (Sahrhage 1988).

The foraging grounds of adult southern elephant seals from the Macquarie stock lie off the coast of the Antarctic continent between longitude 135°E and 165°W (see Slip, Hindell, and Burton, this volume), and elephant seals from the Kerguelen stock have been regularly sighted ashore between 38°E and 110°E (Bester 1988; Gales and Burton 1989). Fluctuations in air pressure, affecting the physical environment of this area of the Southern Ocean might influence distribution and abundance of prey of these stocks. ENSO events are poorly correlated with longer-term climatic variability across the western margins of the Australian continent (Allen, Beck, and Mitchell 1990), and there appears to be no relationship between ENSO events and elephant seal numbers at Macquarie Island. However, climatic fluctuations in this area and the southern Indian Ocean appear to be a consequence of other forcings (Allen and Haylock 1993).

Analyses of mean sea level pressure data since 1957 from the wider Indian Ocean region, including coastal Antarctic stations, show that weather patterns in southwestern Australia are associated with fluctuations in the continental anticyclone and the long-wave trough to the south-southeast of Australia. This high latitude trough has shown strong persistence near the Antarctic station of Casey (110.5°W) in winter (June, July, August) and has fluctuated only in intensity, deepening to pressures below 985 hPa between 1966–1971 and 1975–1989 and becoming shallow with pressures above 985 hPa between 1957–1965 and 1972–1974 (Allan and Haylock 1993). This permanent low pressure region centered north of Casey has broadened since the early 1960s to include the Vestfold Hills on the shores of Pridz Bay, a known foraging area for seals of the Kerguelen stock (Bester 1988; Gales and Burton 1989). The deepening of this trough between 1966 and 1971 corresponds to the increase in first-year mortality at Macquarie Island.

Thus, although the causative factors are unknown, the fluctuations in this trough may have forced or have been associated with changes in the ocean environment that resulted in reduced food resources for elephant seal populations of the Macquarie and Kerguelen stocks and led to their decline. The South Georgia population may have been unaffected by such changes and remained stable because variation in environmental conditions differ from one area of the Southern Ocean to another (Sahrhage 1988). In the absence of hard data, this is very speculative.

PREDICTIONS FROM THESE HYPOTHESES

Of the possible explanations presented above, two (the equilibration and the ocean environment hypotheses) are consistent with what we know of the decline of elephant seals at Macquarie Island and provide a possible explanation of the decline of other elephant seal populations. There is an obvious need for more research into many aspects of elephant seal population biology and ecology.

One approach to deciding future research directions is to test the various predictions arising from the alternative hypotheses and perhaps eliminate them. Not all have testable predictions, but the last two can be tested. This offers a framework for further research.

Predictions from the Equilibration Hypothesis

The equilibration or overshoot hypothesis predicts that elephant seals have a very simple predator-prey relationship and that they utilize a resource not exploited by other major predators. This is partly supported by recent studies of adult foraging behavior (see above).

As first-year survival decreased rapidly in advance of adult survival, some factor limited first-year survival. This suggests that first-year animals will be fundamentally different from adults either in their diet, their foraging grounds, or their ability to find and capture prey in competition with adult seals.

The hypothesis also predicts that the population should eventually stabilize at around the presealing population size. As this was estimated to be about 90,000 to 100,000 seals for Macquarie Island (Hindell and Burton 1988a) and the present population size is approximately 90,000 seals, the decline should slow down or stop in the near future. However, the equilibration process may be expected to continue with a series of "undershoots" and "overshoots" of diminishing magnitude.

The seals would show an increase in growth rate and adult body size coincidental with the stabilization of population size. This would also be accompanied by decreases in the age of first breeding and increases in both adult and first-year survival.

As northern elephant seals are very similar in many ways to their southern congener, and they were also subjected to intense sealing pressure in the last century, they may exhibit population trends similar to southern elephant seals. At present, northern elephant seal populations are increasing rapidly (Stewart et al., this volume). If the equilibration hypothesis holds and the populations are left completely undisturbed, they should eventually level off and then decline as they also fully exploit their major food species.

Predictions from the Environmental Change Hypothesis

The central premise of this hypothesis is that the Southern Ocean presents elephant seals with an environment where their specific food resources are patchily distributed in both space and time, and this patchiness is influenced by ENSO events. Accessibility to these patches directly affects survival of first-year animals (and thus regulates population size) and can change over short periods (less than a decade).

Thus, if it is valid, first-year survivorship should change markedly over quite short time intervals, certainly within a decade and possibly at even shorter intervals; they should be nondirectional, responding directly to variations in the marine environment. In contrast, while the equilibration hypothesis also predicts changes in first-year survival, it predicts them to be directional with a trend for increasing survival as the population returns to its original level.

If changes in weather patterns and ocean circulation have a direct effect on the distribution and abundance of elephant seal food resources (particularly those of first-year seals), then the seals would be expected to exhibit changes in feeding strategies and foraging grounds over quite short time spans, as they attempt to adjust to changes in density and location of prey species. The hypothesis further predicts, as does the equilibration hypothesis, that first-year animals will have different feeding strategies and foraging grounds than older seals.

DIRECTIONS FOR FUTURE RESEARCH

There are three main lines of investigation that should be pursued to test the predictions from these hypotheses and to promote understanding of the underlying reasons for the observed changes in elephant seal numbers. With most species of large mammals, juvenile survival is a major component of the process of population regulation. The single most striking aspect of the population dynamics of the Macquarie Island elephant seals was the decline in first-year survival that occurred in the early 1960s. Changes in population size since that time may have been largely due to this major perturbation in the recruitment process, although it is unknown whether reduced first-year survival was sustained after 1965. Both the equilibration and the environmental change hypotheses rely heavily on the premise that first-year elephant seals are different from adult seals, particularly with respect to diet and foraging areas or to their ability to exploit them. First-year animals are also the component of the population that we know least about, and so they are an obvious target for research efforts. Foraging behavior and foraging areas can be studied and contrasted with those of adults, using time-depth and geolocation recorders. Although a considerable number of recorders will inevitably be lost through natural mortality, making it difficult and expensive to obtain acceptable sample sizes, this remains the best method for collecting the relevant data. It may also be possible to study diet directly as these young seals often haul out on sub-Antarctic islands during the winter months (Hindell and Burton 1988*b*).

Another priority is the establishment of demographic studies. There are two possible approaches. The first is to renew long-term mark-resighting programs similar to the one at Macquarie Island in the 1950s or the one that recently began at Marion Island (Bester and Wilkinson, this volume). Although these studies may not produce valuable data for a number of years, long-term demographic information is fundamental to testing several of the predictions outlined above and is also central to management decisions that may need to be made in the future. The second approach is to conduct cross-sectional age structure studies similar to those from South Georgia. These have the advantage of supplying some types of demographic data, such as estimates of age-specific survival, quickly. It may be possible to obtain age estimates from incisors painlessly removed under anesthetic (Arnbom et al. 1992), which will greatly increase the practicality of such an approach.

Another research priority is the maintenance of simple censusing programs. In many ways, this should be seen as the primary research priority, as it can be performed cheaply and easily at a large number of locations. Data on any future population fluctuations will be required to test predictions from both the equilibration and environmental change hypotheses and provide ongoing information on the status of the species throughout its range.

REFERENCES

Allen, R. J., K. Beck, and W. M. Mitchell. 1990. Sea level and rainfall correlations in Australia: Tropical links. *Journal of Climate* 3: 838–846.

Allen, R. J., and M. Haylock. 1993. Circulation features associated with the winter rainfall decrease in southwestern Australia: Implications for greenhouse and natural variability studies. *Journal of Climate* 6: 1356–1367.

Arnbom, T. A., N. J. Lunn, I. L. Boyd, and T. Barton. 1992. Ageing live Antarctic fur seals and southern elephant seals. *Marine Mammal Science* 8: 37–43.

Barrat, A., and J. L. Mougin. 1978. L'elephant de mer *Mirounga leonina* de L'ile de la Possession, archipel Crozet (46°25′S, 51°45′E). *Mammalia* 42: 143–174.

Bengtson, J. L. 1984. Review of *Antarctic Marine Fauna. Selected Papers Presented to the Scientific Committee of CCAMLR 1982–1984*, 1–226.

Bengtson, J. L., L. M. Ferm, T. J. Harkonen, and B. S. Stewart. 1990. Abundance of Antarctic fur seals in the South Shetland Islands, Antarctica, during the 1986–87 austral summer. In *Antarctic Ecosystems: Ecological Change and Conservation*, ed. K. R. Kerry and G. Hempel, 265–270. Berlin and Heidelberg: Springer Verlag.

Bester, M. N. 1988. Marking and monitoring studies of the Kerguelen stock of southern elephant seals *Mirounga leonina* and their bearing on biological research in the Vestfold Hills. *Hydrobiologica* 165: 269–277.

Bryden, M. M. 1968a. Development and growth of the southern elephant seal (*Mirounga leonina* Linn.). *Papers and Proceedings of the Royal Society of Tasmania* 102: 25–30.

———. 1968b. Control of growth in two populations of elephant seals. *Nature* 217: 1106–1108.

Bryden, M. M., G. P. Kirkwood, and R. W. Slade. 1990. Humpback whales, Area V. An increase in numbers off Australia's east coast. In *Antarctic Ecosystems: Ecological Change and Conservation*, ed. K. R. Kerry and G. Hempel, 271–277. Berlin and Heidelberg: Springer Verlag.

Burton, H. R. 1986. A substantial decline in numbers of the southern elephant seal at Heard Island. *Tasmanian Naturalist* 86: 4–8.

Carrick, R., S. E. Csordas, and S. E. Ingham. 1962. Studies on the southern elephant seal, *Mirounga leonina* (L.). IV. Breeding and development. *CSIRO Wildlife Research* 7: 161–197.

Carrick, R., S. E. Csordas, S. E. Ingham, and K. Keith. 1962. Studies on the southern elephant seal, *Mirounga leonina* (L.). III. The annual cycle in relation to age and sex. *CSIRO Wildlife Research* 7: 119–160.

Condy, P. R., R. J. van Aarde, and M. N. Bester. 1978. The seasonal occurrence and behaviour of killer whales *Orcinus orca*, at Marion Island. *Journal of Zoology, London* 184: 449–464.

Croxall, J. P., T. S. McCann, P. A. Prince, and P. Rothery. 1988. Reproductive performance of seabirds and seals at South Georgia and Signy Island, South Orkney Islands, 1976–1987: Implications for Southern Ocean monitoring studies. In *Antarctic Ocean and Resources Variability*, ed. D. Sahrhage, 261–285. Berlin and Heidelberg: Springer Verlag.

Croxall, J. P., P. Rothery, S. P. C. Pickering, and P. A. Prince. 1990. Reproductive performance, recruitment and survival of wandering albatrosses *Diomedea exulans* at Bird Island, South Georgia. *Journal of Animal Ecology* 59: 775–796.

Erickson, A. W., and M. B. Hanson. 1990. Continental estimates and population trends of Antarctic seals. In *Antarctic Ecosystems: Ecological Change and Conservation*, ed. K. R. Kerry and G. Hempel, 253–264. Berlin and Heidelberg: Springer Verlag.

Gales, N. J., and H. R. Burton. 1989. The past and present status of the southern elephant seal (*Mirounga leonina* Linn.) in greater Antarctica. *Mammalia* 53: 35–47.

Green, K., V. Wong, and H. R. Burton. 1992. A population decline in Weddell seals: Real or sampling artifact? *CSIRO Wildlife Research* 19: 59–64.

Guinet, C., P. Jouventin, and H. Weimerskirch. 1992. Population changes, movements of southern elephant seals on Crozet and Kerguelen archipelagos in the last decades. *Polar Biology* 12: 349–356.

Harwood, J., and A. Hall. 1990. Mass mortality in marine mammals: Its implications for population dynamics and genetics. *Trends in Evolution and Ecology* 5: 254–257.

Hindell, M. A. 1991. Some life history parameters of a declining population of southern elephant seals, *Mirounga leonina*. *Journal of Animal Ecology* 60: 119–134.

Hindell, M. A., and H. R. Burton. 1987. Past and present status of the southern elephant seal (*Mirounga leonina*) at Macquarie Island. *Journal of Zoology, London* 213: 365–380.

———. 1988*a*. The history of the elephant seal industry at Macquarie Island and estimates of the pre-sealing numbers. *Papers and Proceedings of the Royal Society of Tasmania* 122: 159–176.

———. 1988*b*. Seasonal haul-out patterns of the southern elephant seal (*Mirounga leonina* L.) at Macquarie Island. *Journal of Mammalogy* 69: 81–88.

Hindell, M. A., H. R. Burton, and D. J. Slip. 1991. Foraging areas of southern elephant seals, *Mirounga leonina*, as inferred from water temperature data. *Australian Journal of Marine and Freshwater Research* 42: 115–128.

Hindell, M. A., D. J. Slip, and H. R. Burton. 1991. Diving behaviour of adult male and female southern elephant seals. *Australian Journal of Zoology* 39: 595–619.

Hunter, S. 1983. The food and feeding ecology of the giant petrels *Macronectes halli* and *M. giganteus* at South Georgia. *Journal of Zoology, London* 200: 521–538.

Laws, R. M. 1953. The elephant seal (*Mirounga leonina* Linn.). I. Growth and age. *Falkland Islands Dependencies Survey, Scientific Reports* 8: 1–61.

———. 1956*a*. The elephant seal (*Mirounga leonina* Linn.). II. General, social and reproductive behaviour. *Falkland Islands Dependencies Survey, Scientific Reports* 13: 1–88.

———. 1956*b*. The elephant seal (*Mirounga leonina* Linn.). III. The physiology of reproduction. *Falkland Islands Dependencies Survey, Scientific Reports* 15: 1–66.

———. 1960. The southern elephant seal (*Mirounga leonina* Linn.) at South Georgia. *Norsk Hvalfangst-Tidende* 49: 466–476, 520–542.

———. 1985. Ecology of the Southern Ocean. *American Scientist* 73: 26–40.

McCann, T. S. 1985. Size, status and demography of southern elephant seal *Mirounga leonina* populations. In *Sea Mammals of South Latitudes: Proceedings of a Symposium of the 52d ANZAAS Congress in Sydney—May 1982*, ed. L. K. Ling and M. M. Bryden, 1–17. Northfield: South Australian Museum.

McCann, T. S., and P. Rothery. 1988. Population size and status of the southern elephant seal (*Mirounga leonina*) at South Georgia, 1951–1985. *Polar Biology* 8: 305–309.

Moors, P. J. 1986. Decline in numbers of Rockhopper penguins at Campbell Island. *Polar Record* 23: 69–73.

Pascal, M. 1986. Numerical changes in the population of elephant seals (*Mirounga leonina* L.) in the Kerguelen Archipelago during the past 30 years. In *Marine Mammal Fishery Interactions*, ed. J. R. Beddington, R. J. H. Beverton, and D. M. Lavigne, 170–186. London: George Allen and Unwin.

Priddle, J., J. P. Croxall, I. Everson, R. B. Heywood, E. J. Murphy, P. A. Prince, and C. B. Sear. 1988. Large-scale fluctuations in distribution and abundance of krill—A discussion of possible causes. In *Antarctic Ocean and Resources Variability*, ed. D. Sahrhage, 261–285. Berlin and Heidelberg: Springer Verlag.

Reiter, J., N. L. Stinson, and B. J. Le Boeuf. 1978. Northern elephant seal development: The transition from weaning to nutritional independence. *Behavioral Ecology and Sociobiology* 3: 337–367.

Sahrhage, D. 1988. Some indications for environmental and krill resources variability in the Southern Ocean. In *Antarctic Ocean and Resources Variability*, ed. D. Sahrhage, 33–40. Berlin and Heidelberg: Springer Verlag.

SCAR. 1991. Report of the workshop on southern elephant seals, Monterey, California, May 22–23, 1991.

Shaughnessy, P. D., G. L. Shaughnessy, and P. L. Keage. 1988. Fur-seals at Heard Island: Recovery from past exploitation? In *Marine Mammals of Australasia—Field Biology and Captive Management*, ed. M. L. Augee, 71–77. Sydney: Royal Zoological Society of New South Wales.

Taylor, R. H., and G. A. Taylor. 1989. Re-assessment of the status of southern elephant seals (*Mirounga leonina*) in New Zealand. *New Zealand Journal of Marine and Freshwater Research* 23: 201–213.

Testa, J. W., G. Oehlert, D. G. Ainley, J. L. Bengtson, D. B. Siniff, R. M. Laws, and D. Rounsevell. 1990. Temporal variability in Antarctic marine ecosystems: Periodic fluctuations in the phocid seals. *Canadian Journal of Fisheries and Aquatic Science* 48: 631–639.

Testa, J. W., and D. B. Siniff. 1987. Population dynamics of Weddell seals (*Leptonychotes weddelli*) in McMurdo Sound, Antarctica. *Ecological Monographs* 57: 149–165.

van Aarde, R. J. 1980a. Fluctuations in the population of southern elephant seals *Mirounga leonina* at Kerguelen Island. *South African Journal of Zoology* 15: 99–106.

———. 1980b. Harem structure of the southern elephant seal *Mirounga leonina* at Kerguelen Island. *Review Ecologie (Terre Vie)* 34: 31–44.

van Loon, H., and D. J. Shea. 1988. A survey of atmospheric elements at the ocean's surface south of 40°S. In *Antarctic Ocean and Resources Variability*, ed. D. Sahrhage, 3–20. Berlin and Heidelberg: Springer Verlag.

Vergani, D. F., M. N. Lewis, and Z. B. Stranganelli. 1987. Observation on haul-out patterns and trends of the breeding populations of southern elephant seals at Península Valdes (Patagonia) and Stranger Point (King George Island). *SC-CAMLR-VI/BG/* 36: 1–9.

Woehler, E. J., and Johnstone, G. W. 1991. Status and conservation of the seabirds of the Australian Antarctic Territory. In *Seabird Status and Conservation: A Supplement*, ed. J. P. Croxall, 279–308. Technical Publication No. 11, International Council for Bird Preservation, Cambridge, England.

FIVE

Population Ecology of Southern Elephant Seals at Marion Island

Marthan N. Bester and Ian S. Wilkinson

ABSTRACT. Research has highlighted the continued decline in the southern elephant seal population at Marion Island, and life table data implicate recently matured cows as the most vulnerable part of the population. Observation of onshore behavior suggests that the factors producing the elevated mortality rates of these cows operate at sea. Current research is concentrated on the investigation of at-sea behavior of this component of the population.

Declines in the southern elephant seal populations in the southern Indian Ocean, including the population at Marion Island, have been evident since the 1970s (Condy 1978; van Aarde 1980; Skinner and van Aarde 1983; Burton 1985). Notwithstanding a number of hypotheses that have been offered to explain the observed declines (ibid.), the driving forces behind them are, as yet, poorly understood.

Comprehensive tagging studies commenced at Marion Island in 1983 to permit the first assessment of life history parameters such as age-specific survival, age-specific fecundity, and age at maturity for females in the Marion Island population based on data collected at Marion Island. Previous comments on population characteristics for this population (Condy 1977) were based largely on statistics drawn from the populations at Macquarie Island (Carrick and Ingham 1962) and South Georgia (Laws 1960). Such population parameters cannot tell us what the cause of the decline is, but they can highlight weak links within the population at this site, which allows the focusing of future research efforts.

This chapter describes the present status of the Marion Island population and analyzes the population based on parameters derived from the 1983 cohort. In addition, we present the first data on movements of cows from Marion Island during the pelagic phase of their life cycle.

METHODS

Counts of pups were conducted in a main study area (MSA) representing ±10% of the coastline at Marion Island (46°54′S, 37°45′E) during the breeding seasons from 1974 to 1989 following the methods of P. R. Condy (1978). In 1976 and 1986–1989, the number of births for the whole of Marion Island was determined from counts made at all the breeding beaches at the end of October (Wilkinson and Bester, in press). The number of pups born in the MSA expressed as a percentage (range: 24–31%) of the total island pup production was calculated for 1976 and the years 1986–1989. Given that the proportion of births occurring in the MSA did not change significantly in these years (Wilkinson 1991), the pup population in the MSA is assumed to be representative of the entire population as an index of population size. Rates of population change were calculated using the exponential equation (Caughley 1977)

$$N_t = N_o e^{rt}$$

for the periods 1974–1989 and 1983–1989. The years 1983–1989 coincide with the period for which data on population parameters are available. From 1983 onward, weaned pups were tagged and beaches around the whole of Marion Island searched at 7- to 10-day intervals to resight these individuals (Bester 1989; Wilkinson 1991). Age-specific survival rates incorporating tag losses were calculated using Jolly-Seber mark-recapture analysis (see Wilkinson 1991). Rates of loss for individual tags ranged from 0.5% in the first year to 4.5% in the fifth year, which yielded combined loss rates (two tags) of 0.0 to 0.6%. Intercohort differences in first-year survival were calculated according to I. S. Wilkinson (1991). Calculation of age at first reproduction followed the method of A. E. York (1983), and fecundity followed the method of Wilkinson (1991). Sex ratio at birth was taken to be the same as the ratio of male:female weanlings that were tagged between 1983 and 1989 (Wilkinson 1991). The net reproductive rate (R_o), calculated using the equation

$$R_o = \Sigma l_x m_x$$

where $\Sigma l_x m_x$ = the sum of the product of the age-specific survival and fecundity values of females of a cohort (Caughley 1977), of the population at Marion Island and other components of the life table are described in Wilkinson (1991). In calculating the life table, empirical data were used up to age 5, after which the mortality rate was assumed to be constant (Harwood and Prime 1978) until a year or two before death (McCann 1985). Fecundity rates were assumed to remain constant after the age of full recruitment (age 6 in this population) to the breeding population until death, with no reproductive senescence (Hindell and Little 1988). All cows were

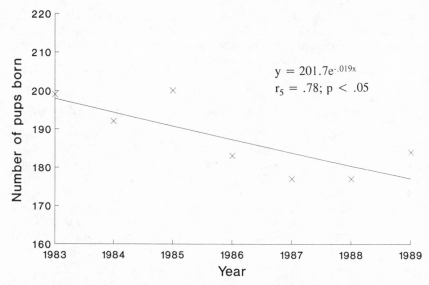

Fig. 5.1. Annual pup production for southern elephant seals within the MSA, Marion Island, for the years 1983 to 1989. The equation refers to the line fitted through the data points using least squares regression analysis (Zar 1984).

assumed to produce only one pup (Laws 1956), at a sex ratio found in the present study, and female longevity was 23 years (Hindell and Little 1988).

Data on individual movement of postbreeding cows (n = 3, aged 4, 7, and 7 years) at sea were collected with microprocessor-controlled Time-Depth Recorders (Wildlife Computers, Woodinville, Wash.) using the geolocation option, which included measurement of surface water temperature. The recorders were deployed on cows that were immobilized chemically using a remote injection method (Bester 1988b) and following procedures detailed in M. N. Bester and H. M. Pansegrouw (1992). The Geolocation analysis software package of Wildlife Computers was used to calculate the daily longitude and latitude from light-level data, the theory and precision of which are discussed by R. D. Hill (1991).

RESULTS AND DISCUSSION

Trends in Population Size and Population Parameters

The Marion Island population declined an average of 4.8% per annum between 1974 and 1989; it had slowed to 1.9% per annum between 1983 and 1989 (fig. 5.1). The total island pup production for 1989 was 585 individuals. As immigration and emigration are virtually nonexistent (Burton

TABLE 5.1   Life table of female southern elephant seals of the 1983 cohort at Marion Island. Values after age 5 are calculated according to the assumptions mentioned in the methods section (from Wilkinson 1991). Net reproductive rate $R_o = 0.661$.

Age (x)	Probability of survival (l_x)	Probability of dying (d_x)	Mortality rate (q_x)	Pregnancy rate	Fecundity (m_x)	$l_x m_x$
0	1.000	0.382	0.382	0.000	0.000	0.000
1	0.618	0.168	0.272	0.000	0.000	0.000
2	0.450	0.098	0.218	0.000	0.000	0.000
3	0.352	0.096	0.273	0.262	0.128	0.045
4	0.256	0.045	0.176	0.565	0.276	0.071
5	0.211	0.037	0.176	0.763	0.373	0.079
6	0.174	0.031	0.176	1.000	0.489	0.085
7	0.143	0.025	0.176	1.000	0.489	0.070
8	0.118	0.021	0.176	1.000	0.489	0.058
9	0.097	0.017	0.176	1.000	0.489	0.047
10	0.080	0.014	0.176	1.000	0.489	0.039
11	0.066	0.012	0.176	1.000	0.489	0.032
12	0.054	0.009	0.176	1.000	0.489	0.026
13	0.045	0.008	0.176	1.000	0.489	0.022
14	0.037	0.007	0.176	1.000	0.489	0.018
15	0.030	0.005	0.176	1.000	0.489	0.015
16	0.025	0.004	0.176	1.000	0.489	0.012
17	0.021	0.004	0.176	1.000	0.489	0.010
18	0.017	0.003	0.176	1.000	0.489	0.008
19	0.014	0.002	0.176	1.000	0.489	0.007
20	0.012	0.002	0.176	1.000	0.489	0.006
21	0.010	0.003	0.250	1.000	0.489	0.005
22	0.007	0.004	0.500	1.000	0.489	0.003
23	0.003	0.003	1.000	1.000	0.489	0.001

1985; Bester 1989; Gales, Adams, and Burton 1989), the observed decline must be a consequence of an imbalance between births and deaths within the population.

Cows of the 1983 cohort produced pups for the first time at age 3. An estimated 26.2% of cows had produced pups at age 3, 56.5% at age 4, 76.3% at age 5 and 100% of those cows known to still be alive at age 6. The sex ratio of 3,856 pups tagged between 1983 and 1989 was 1.04:1 (male:female), which does not differ significantly from unity ($\chi^2(1) = 1.82$, $p > .05$). The mean age of these females at first pupping was 4.41 years.

Survival of females was 61.8% to age 1, 45% to age 2, 35.2% to age 3, 25.6% to age 4, and 21.1% to age 5 (table 5.1). The mortality rate decreased up to age 3, then increased in the fourth year, and further decreased from age 4 to 5 (table 5.1). Intercohort comparison of first-year survival showed no differences for the years 1983 to 1988.

An indication of the viability of a population is shown by its net reproductive rate (R_o). The R_o represents the number of daughters born to a female during her lifetime, or the rate of increase in the population with each passing generation (Caughley 1977). Therefore, to maintain itself, a population must have an R_o of 1. In the case of the population at Marion Island, the R_o is 0.661, resulting in a loss of 34% of the population with each passing generation. As R_o is a composite of both survival and fecundity, it can be affected by age-specific survival, pregnancy rate, and sex ratio of offspring.

The mean age at first pupping and hence recruitment to the adult component of the population is similar for Marion Island and the stable South Georgia population (McCann 1985; McCann and Rothery 1988) and occurs a year earlier than at Macquarie Island (Hindell 1991). The apparent 100% pupping rates of females age 6 and older are higher than the 85% reported at South Georgia (McCann 1985), and only reports for hooded seals (98%; Øritsland 1964, in Riedman 1990) and northern elephant seals (97.8%; Le Boeuf and Reiter 1988) are similar in magnitude.

Observations of sexual behavior at a single beach at Marion Island over three consecutive summers showed that dominant bulls controlling the harems in these years achieved over 98% of all matings and that the success of these matings was not affected by their timing in the season, or by prior (during the same season) levels of sexual activity of the bull (Wilkinson 1991). In view of the high pupping rates observed and the observed effectiveness of the few bulls that did breed, the "paucity of males" hypothesis (Skinner and van Aarde 1983) can be dismissed as a possible factor in the decline.

Manipulation of Life Table Data

The proportion of female pups at Marion Island (48.9%) is higher than the figure of 47% quoted for South Georgia, which resulted in an R_o of greater than 1 (McCann 1985). The proportion of female pups born would need to increase to 75% to allow the R_o to reach a value of 1 if pupping rates and age-specific survival remained constant (table 5.2).

When considering the age at primiparity, maturity would have to be advanced by two years, with breeding beginning at age 1 (26.2% pupping) and recruitment to the adult population complete by age 4 (100% pupping), to realize an R_o of 1. The present mean age at maturity (4.41) is already lower than previously reported at this site (Condy 1977) and similar to the stable South Georgia population (McCann 1985). Given that the cows are reproducing at an age that is already early for this species and

TABLE 5.2   The effect on net reproductive rate, R_o, of changing the proportion of female pups born to southern elephant seal cows at Marion Island, while maintaining age-specific pupping and survival rates at the levels in table 5.1 (from Wilkinson 1991).

Percentage of female pups born	New reproductive rate (R_o)
48.9	0.661
50.0	0.675
55.0	0.743
60.0	0.811
65.0	0.878
70.0	0.946
75.0	1.013

TABLE 5.3   The effect on net reproductive rate, R_o, of changing the first-year survival rate of female southern elephant seals at Marion Island, while maintaining age-specific fecundity and survival rates (after age 1) at the levels in table 5.1 (from Wilkinson 1991).

Percentage of female pups surviving to age 1	Net reproductive rate (R_o)
61.8	0.661
65.0	0.693
70.0	0.750
75.0	0.802
80.0	0.863
85.0	0.933
90.0	0.985
95.0	1.033
100.0	1.087

that they are producing offspring at the maximum rate, it would seem that age-specific survival is the key to the decline. Higher survival rates would improve age-specific survival/fecundity values and the resultant R_o value.

Assuming all other parameters are held constant, first-year survival would have to increase to 95% to produce a stable population (table 5.3). The present observed survival to age 1 is already the highest ever reported for this species, and given the major improvement that is required to stabilize the population, it seems unlikely that this is the vulnerable age group. Studies on the reproductive success of cows at Marion Island also showed low preweaning mortality rates (Wilkinson 1991).

The annual adult survival rate of females (82.4%) is too low to maintain

TABLE 5.4   The effect on net reproductive rate, R_o, of changing the annual adult survival rate of female southern elephant seals at Marion Island, while maintaining age-specific fecundity and survival rates (up to age 5) at the levels in table 5.1 (from Wilkinson 1991).

Annual survival rate of adult females	Net reproductive rate (R_o)
82.4	0.661
83.0	0.676
84.0	0.709
85.0	0.762
86.0	0.817
87.0	0.866
88.0	0.917
89.0	0.981
90.0	1.043

the population and needs to be increased to 90% to stabilize the population (table 5.4). The assumption made that survival rate of adults remains constant from age 5 until a year or two before death at age 23 (Hindell and Little 1988) contrasts with the view of T. S. McCann (1985) and R. M. Laws (1960) who assumed survival to remain constant from maturity to age 10 and then decline annually until death at age 20. However, the assumption of constant adult survival agrees with a study on gray seals, *Halichoerus grypus*, by J. Harwood and J. H. Prime (1978) and provides the highest R_o that is possible. If survival declined after age 10, the R_o would be lower still.

This manipulation of life table data shows the relative importance of the first year and adult components of the population. A 1% change in annual adult survival results in an approximately 5% change in R_o (table 5.4) but only a 1% change in R_o when first-year survival is changed by a similar amount (table 5.3). As mentioned, survival to age 1 is higher than ever reported in this species, and if it is then assumed that adult annual survival rate is not abnormally low, then it may be juvenile survival rates that are the problem, resulting in lowered recruitment to the adult age class.

If the figures for first-year survival and adult annual survival (from age 5 onward) described above are assumed correct and maintained, while replacing juvenile survival ($l_2 - l_4$; see table 5.1) with the values calculated for South Georgia (McCann 1985), the R_o value exceeds one (table 5.5).

Mortality Rates of Recently Matured Cows
The data for the 1983 cohort show that there is an increase in the mortality of 3-year-old females. This increase in mortality comes at a crucial time for the population. Although this age group is not the most reproductively

TABLE 5.5   Life table of female southern elephant seals of the 1983 cohort at Marion Island. Mortality rates from age 1 to 5 years ($q_1 - q_4$) have been replaced by those reported for South Georgia (McCann 1985), while maintaining the reported first-year mortality rate and those after age 5 (from Wilkinson 1991). Net reproductive rate $R_o = 1.072$.

Age (x)	Probability of survival (l_x)	Probability of dying (d_x)	Mortality rate (q_x)	Pregnancy rate	Fecundity (m_x)	$l_x m_x$
0	1.000	0.382	0.382	0.000	0.000	0.000
1	0.618	0.108	0.175	0.000	0.000	0.000
2	0.510	0.061	0.120	0.000	0.000	0.000
3	0.449	0.054	0.120	0.262	0.128	0.058
4	0.395	0.047	0.119	0.565	0.276	0.109
5	0.348	0.061	0.176	0.763	0.373	0.130
6	0.287	0.050	0.176	1.000	0.489	0.140
7	0.237	0.042	0.176	1.000	0.489	0.116
8	0.195	0.034	0.176	1.000	0.489	0.095
9	0.161	0.028	0.176	1.000	0.489	0.079
10	0.133	0.023	0.176	1.000	0.489	0.065
11	0.110	0.019	0.176	1.000	0.489	0.054
12	0.091	0.016	0.176	1.000	0.489	0.044
13	0.075	0.013	0.176	1.000	0.489	0.037
14	0.062	0.011	0.176	1.000	0.489	0.030
15	0.051	0.009	0.176	1.000	0.489	0.025
16	0.042	0.007	0.176	1.000	0.489	0.021
17	0.035	0.006	0.176	1.000	0.489	0.017
18	0.029	0.005	0.176	1.000	0.489	0.014
19	0.024	0.004	0.176	1.000	0.489	0.012
20	0.020	0.004	0.176	1.000	0.489	0.010
21	0.016	0.004	0.250	1.000	0.489	0.008
22	0.012	0.006	0.500	1.000	0.489	0.006
23	0.006	0.006	1.000	1.000	0.489	0.003

valuable (evidenced by the $l_x m_x$ values in table 5.1), a high mortality rate at this age will, combined with high juvenile mortality, reduce the level of recruitment to the adult population and thus affect the survival/fecundity schedule, lowering R_o. The sharp increase in mortality among 3-year-old females comes at a time when animals at Marion Island are maturing sexually and are exposed to greater physiological stress levels resulting from gestation and lactation.

Gestation and the postpartum lactation period impose increased energetic demands on the female, and these costs are relatively higher among young females (Reiter and Le Boeuf 1991) that are still in a more rapid phase of growth and development than their older counterparts (Laws 1953; Reiter, Panken, and Le Boeuf 1981). During the lactation period the cows may also lose up to 43% of their initial prepartum mass (Costa et al., 1986). Cows at Marion Island that were observed leaving the breeding beaches at the end of lactation were noticeably emaciated (Wilkinson 1991), implying a severe drain on their body reserves. In contrast, some cows at South Georgia were observed to leave the beaches with large blubber stores intact (McCann, Fedak, and Harwood 1989), possibly indicating that the nutritional status of the cows at the two sites differs.

Given current knowledge of the terrestrial component of the life cycle at Marion Island, it would appear that factors operating during the pelagic phase of the annual cycle, of subadult and adult cows in particular, hold the key to the decline process.

Cow Movements at Sea

The three cows that were tracked ranged over entirely different areas (fig. 5.2). None were followed for the total time at sea (67, 69, and 78 days, respectively), but two cows appeared to be returning to the island when recordings ceased. The cows spent a large proportion of their time in reasonably well-defined areas at, or near, the limit of their feeding range (between 1,100 and 1,400 km) (Bester and Pansegrouw 1992). The most circumscribed foraging area (that of the 4-year-old cow) lay south of the Antarctic Polar Front at 54°–57°S latitude and 25°–29°E longitude in cold surface water (minimum temperature of −1.7°C). The two other cows moved north to widely separated (by 15°–20° longitude) areas between approximately 40°–45°S latitude (fig. 5.2). The northwest-bound cow encountered warmer surface water (maximum temperature of 14.1°C) consistent with the mean position of the Subtropical Convergence at 41°40′S latitude which is recognizable at the sea surface by a mean decrease in temperature from 17.9°C to 10.6°C (Lutjeharms, Walters, and Allanson 1985). The northeast-moving cow appeared to remain in water between 5.3°C to 7.3°C at the Subantarctic Front (central surface temperature 7°C; Lutjeharms 1985). The cows therefore ranged widely without overlap in their feeding areas, the choice of which might have been influenced by biological enhancement at oceanic frontal systems (see Lutjeharms, Walters, and Allanson 1985). It is likely, however, that the small sample size has identified only a small part of the actual feeding range of elephant seal cows from Marion Island during the postbreeding period (Bester and Pansegrouw 1992), since three of five postbreeding cows from Macquarie Island foraged in deep oceanic waters off the Antarctic coast (Hindell, Burton, and Slip 1991).

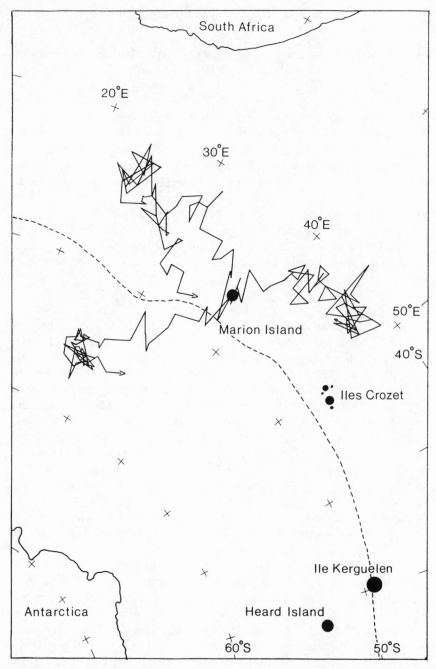

Fig. 5.2. At-sea movements of 3 southern elephant seal cows from Marion Island, during their postbreeding pelagic period. The mean position of the Antarctic Polar Front is shown (dashed line).

REFERENCES

Bester, M. N. 1988a. Marking and monitoring studies of the Kerguelen stock of southern elephant seals, *Mirounga leonina*, and their bearing on biological research in the Vestfold Hills. *Hydrobiologia* 165: 269–277.

———. 1988b. Chemical restraint of Antarctic fur seals and southern elephant seals. *South African Journal of Wildlife Research* 18: 57–60.

———. 1989. Movements of southern elephant seals and subantarctic fur seals in relation to Marion Island. *Marine Mammal Science* 5: 257–265.

Bester, M. N., and H. M. Pansegrouw. 1992. Ranging behaviour of southern elephant seal cows from Marion Island. *South African Journal of Science* 88: 574–577.

Burton, H. R. 1985. Tagging studies of male southern elephant seals (*Mirounga leonina* L.) in the Vestfold Hills area, Antarctica, and some aspects of their behaviour. In *Sea Mammals of South Latitudes: Proceedings of a Symposium of the 52d ANZAAS Congress in Sydney—May 1982*, ed. L. K. Ling and M. M. Bryden, 19–30. Northfield: South Australian Museum.

Carrick, R., and S. E. Ingham. 1962. Studies of the southern elephant seal *Mirounga leonina* (L.). V. Population dynamics and utilization. *CSIRO Wildlife Research* 7: 198–206.

Caughley, G. 1977. *Analysis of Vertebrate Populations*. New York: John Wiley and Sons.

Condy, P. R. 1977. The ecology of the southern elephant seal, *Mirounga leonina* (Linnaeus 1758), at Marion Island. D.Sc. dissertation, University of Pretoria, South Africa.

———. 1978. The distribution and abundance of southern elephant seals *Mirounga leonina* (Linn.) on the Prince Edward Islands. *South African Journal of Antarctic Research* 8: 42–48.

Costa, D. P., B. J. Le Boeuf, A. C. Huntley, and C. L. Ortiz. 1986. The energetics of lactation in the northern elephant seal, *Mirounga angustirostris*. *Journal of Zoology, London* 209: 21–33.

Gales, N. J., M. Adams, and H. R. Burton. 1989. Genetic relatedness of two populations of the southern elephant seal, *M. leonina*. *Marine Mammal Science* 5: 57–67.

Harwood, J., and J. H. Prime. 1978. Some factors affecting the size of British grey seal populations. *Journal of Applied Ecology* 15: 401–411.

Hill, R. D. 1991. *Geolocation by Light-Level Readings*. Woodinville, Wash.: Wildlife Computers Geolocation Instruction Manual.

Hindell, M. A. 1991. Some life-history parameters of a declining population of southern elephant seals, *Mirounga leonina*. *Journal of Animal Ecology* 60: 119–134.

Hindell, M. A., and H. R. Burton. 1987. Past and present status of the southern elephant seal (*Mirounga leonina*) at Macquarie Island. *Journal of Zoology, London* 213: 365–380.

Hindell, M. A., H. R. Burton, and D. J. Slip. 1991. Foraging areas of southern elephant seals, *Mirounga leonina*, as inferred from water temperature data. *Australian Journal of Marine and Freshwater Research* 42: 115–128.

Hindell, M. A., and G. J. Little. 1988. Longevity, fertility and philopatry of two female southern elephant seals (*Mirounga leonina*) at Macquarie Island. *Marine Mammal Science* 4: 168–171.

Krebs, C. J. 1985. *Ecology: The Experimental Analysis of Distribution and Abundance.* New York: Harper and Row.

Laws, R. M. 1953. The elephant seal (*Mirounga leonina* Linn.). I. Growth and age. *Falkland Islands Dependencies Survey, Scientific Reports* 8: 1–62.

———. 1956. The elephant seal (*Mirounga leonina* Linn.). II. General, social and reproductive behaviour. *Falkland Islands Dependencies Survey, Scientific Reports* 13: 1–88.

———. 1960. The southern elephant seal (*Mirounga leonina* Linn.) at South Georgia. *Norsk Hvalfangst-Tidende* 49: 466–476, 520–542.

Le Boeuf, B. J., and J. Reiter. 1988. Lifetime reproductive success in northern elephant seals. In *Reproductive Success,* ed. T. H. Clutton-Brock, 344–362. Chicago: University of Chicago Press.

Lutjeharms, J. R. E. 1985. Location of frontal systems between Africa and Antarctica. *Deep-Sea Research* 32: 1499–1509.

Lutjeharms, J. R. E., N. M. Walters, and B. R. Allanson. 1985. Oceanic frontal systems and biological enhancement. In *Antarctic Nutrient Cycles and Food Webs,* ed. W. R. Siegfried, R. M. Laws, and P. R. Condy, 11–31. Berlin: Springer Verlag.

McCann, T. S. 1985. Size, status and demography of southern elephant seal *Mirounga leonina* populations. In *Sea Mammals of South Latitudes: Proceedings of a Symposium of the 52d ANZAAS Congress in Sydney—May 1982,* ed. L. K. Ling and M. M. Bryden, 1–17. Northfield: South Australian Museum.

McCann, T. S., M. A. Fedak, and J. Harwood. 1989. Parental investment in southern elephant seals, *Mirounga leonina. Behavioral Ecology and Sociobiology* 25: 81–87.

McCann, T. S., and P. Rothery. 1988. Population size and status of the southern elephant seal (*Mirounga leonina*) at South Georgia, 1951–1985. *Polar Biology* 8: 305–309.

Reiter, J., and B. J. Le Boeuf. 1991. Life history consequences of variation in age at primiparity in northern elephant seals. *Behavioral Ecology and Sociobiology* 28: 153–160.

Reiter, J., K. J. Panken, and B. J. Le Boeuf. 1981. Female competition and reproductive success in northern elephant seals. *Animal Behavior* 29: 670–687.

Riedman, M. L. 1990. *The Pinnipeds: Seals, Sea Lions, and Walruses.* Berkeley and Los Angeles: University of California Press.

Skinner, J. D., and R. J. van Aarde. 1983. Observations on the trend of the breeding population of southern elephant seals, *Mirounga leonina,* at Marion Island. *Journal of Applied Ecology* 20: 707–712.

Taylor, R. H., and G. A. Taylor. 1989. Reassessment of the status of southern elephant seals (*Mirounga leonina*) in New Zealand. *New Zealand Journal of Marine and Freshwater Research* 23: 201–213.

van Arade, R. J. 1980. Fluctuations in the population of southern elephant seals *Mirounga leonina* at Kerguelen Island. *South African Journal of Zoology* 15: 99–106.

Wilkinson, I. S. 1991. Factors affecting reproductive success of southern elephant seals, *Mirounga leonina,* at Marion Island. Ph.D. dissertation, University of Pretoria, South Africa.

Wilkinson, I. S., and M. N. Bester. In press. Population parameters of the southern elephant seal population at Marion Island. *Antarctic Science*.

York, A. E. 1983. Average age at first reproduction of the northern fur seal (*Callorhinus ursinus*). *Canadian Journal of Fisheries and Aquatic Science* 40: 121–127.

Zar, J. H. 1984. *Biostatistical Analysis*. 2d ed. Englewood Cliffs, N.J.: Prentice-Hall.

SIX

Biomass and Energy Consumption of the South Georgia Population of Southern Elephant Seals

Ian L. Boyd, Tom A. Arnbom, and Michael A. Fedak

ABSTRACT. The total annual energy expenditure was estimated for different age and sex classes of southern elephant seals, *Mirounga leonina*, that breed at South Georgia. The estimated energy costs of reproduction, growth, foraging, and molt were used to calculate an annual energy budget for individuals in each age and sex class. This was combined with population size and age structure to estimate population energy requirements. The estimated average metabolic cost of maintenance for adult males and females was 0.17 and 0.39 MJ/year, respectively. Male biomass accounted for 63% of the total population biomass (222,903 metric tonnes), and the metabolic power for the whole population averaged over one year was 190 MWatts. Total energy expenditure of each age class declined during the first two years but then began to increase because of the onset of reproduction in females and because of increased energy costs of foraging and growth in males. Foraging accounted for 63.2% and 68.2% of the annual energy budget in males and females, respectively. The total annual energy expenditure was 6.01×10^9 MJ/year, and 59% of this was accounted for by males. The gross energy requirement was 7.89×10^9 MJ/year. The production efficiency was 8.2% Average daily gross energy intake during potential foraging periods was 77.3 and 43.2 MJ/day for males and females, respectively. This suggested a capture rate of 0.26 and 0.15 kg of fish or muscular squid per dive for males and females, respectively. Biomass of food consumed depended on assumptions about diet composition. If southern elephant seals at South Georgia fed exclusively on squid, the consumption biomass was 2.28×10^6 tonnes/year.

Total food or energy requirements of pinnipeds are important for assessing the impact of their populations on marine resources and the potential effects of changes in the abundance of prey on population size and distribution. Within a pinniped population, demand for food can vary between different age and sex classes because the abundance of individuals is a declining function of age and there are differing nutritional requirements for growth or reproduction within each age and sex class. Although an estimate of the total energy requirements of a population can be useful, it is often more instructive to estimate energy demand for different age and sex classes because this can help to identify which parts of a population are most vulnerable to food shortage or which have the greatest potential impact on food resources.

This study is concerned exclusively with southern elephant seals, *Mirounga leonina*, that breed at South Georgia. This population has remained stable since the 1950s (Laws 1960; McCann and Rothery 1988), but this study has clear implications for interpretations of factors leading to declining numbers of southern elephant seals in other populations (van Aarde 1980; Skinner and van Aarde 1983; Hindell and Burton 1987; Hindell 1991). In addition, calculation of the total energy requirements of a population involves the synthesis of data about age structure, growth, and energy costs of reproduction, molt, locomotion, and foraging. Therefore, it is a means of highlighting the strengths and weaknesses of current data sets within a structured model.

Estimates of annual food consumption among phocids have been made for harp, *Phoca groenlandica* (Lavigne et al. 1985), gray, *Halichoerus grypus* (Fedak, Anderson, and Harwood 1981), and southern elephant seals (Laws 1977; McCann 1985). The estimates for harp and gray seals were based on empirical and best estimates of the energy budgets of individuals (derived from analyses of carcass composition), energy investment in reproduction, and measured metabolic costs of maintenance, growth, and age-dependent survivorship, combined with diet composition and assimilation efficiency to estimate food consumption. In contrast, estimates for southern elephant seals were largely based on crude estimates of food consumption per unit of biomass, which gives little scope for estimating the effects of changes in prey abundance on the population or of changes in the population parameters on food consumption. More recently, studies using time-depth-temperature recorders have provided significant information about the foraging of southern elephant seals (Boyd and Arnbom 1991; Hindell 1991) in terms of both swimming behavior and the location of prey in the water column. This has been augmented by detailed studies of the diet (Rodhouse et al. 1992) and population size (McCann and Rothery 1988).

The main objectives of this paper were to: (1) bring together recent information about southern elephant seal energetics (Fedak et al., this volume) by considering energy consumption in the broad context of the South Georgia population; (2) show how energy demand changes in relation to age both for individuals and for whole age classes; (3) contrast patterns of energy demand in the two sexes; and (4) provide an indication of what interaction there is likely to be between elephant seals and commercial fisheries in the Atlantic sector of the Southern Ocean.

TABLE 6.1   Population parameters used to estimate energy requirements for the South Georgia population of elephant seals.

	Females				*Males*			
Age	Survival rate[a]	N–T length (m)	Mass (kg)	Girth (m)	Survival rate[a]	N–T length (m)	Mass (kg)	Girth (m)
0	1.000	1.50	112	1.28	1.000	1.50	119	1.48
1	0.600	1.71	171	1.38	0.600	1.74	176	1.40
2	0.510	2.00	247	1.56	0.510	2.10	271	1.58
3	0.449	2.28	323	1.68	0.422	2.56	486	2.08
4	0.395	2.40	373	1.78	0.350	2.98	771	2.51
5	0.348	2.45	395	1.83	0.290	3.33	1,086	2.84
6	0.306	2.50	418	1.87	0.241	3.62	1,402	3.12
7	0.269	2.55	443	1.91	0.193	3.82	1,652	3.31
8	0.237	2.58	458	1.94	0.145	3.95	1,829	3.43
9	0.209	2.60	468	1.96	0.101	4.04	1,959	3.51
10	0.184	2.62	478	1.97	0.071	4.10	2,048	3.57
11	0.162	2.64	489	2.00	0.050	4.14	2,110	3.60
12	0.136	2.65	494	2.00	0.035	4.16	2,141	3.62
13	0.106	2.65	494	2.00	0.024	4.18	2,172	3.64
14	0.079	2.66	499	2.00	0.017	4.19	2,188	3.65
15	0.055	2.66	499	2.00	0.012	4.19	2,188	3.65
16	0.036	2.66	499	2.00	0.008	4.20	2,204	3.66
17	0.022	2.66	499	2.00	0.005	4.20	2,204	3.66
18	0.011	2.66	499	2.00	0.003	4.20	2,204	3.66
19	0.004	2.66	499	2.00	0.002	4.20	2,204	3.66
20	0.001	2.66	499	2.00	0.001	4.20	2,204	3.66

[a] From McCann 1985.

METHODS

Population Parameters, Growth, and Biomass

Annual Survival Rate.   T. S. McCann's (1985) revision of R. M. Laws's (1960) life tables for male and female southern elephant seals was used to provide statistics on survival rates from birth to 20 years of age (table 6.1). These were based on cross-sectional age distributions from material collected by Laws and from a collection made in 1978. Comparison was made between these survival rates and those for the Macquarie Island stock (Hindell 1991), where longitudinal data have been collected on survival rates of individuals. Estimates of survivorship differed for the two stocks, especially survival rates up to 1 year and in the 5- to 12-year age classes (ibid.). Given that the Macquarie Island stock has been in decline while the South Georgia stock has been stable in recent years (Hindell and Burton

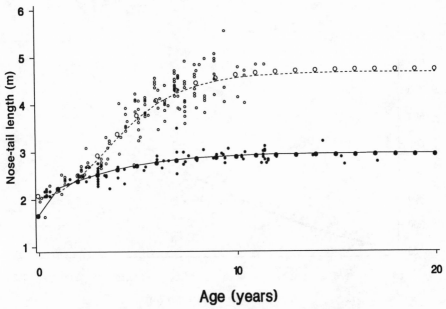

Fig. 6.1. The relationship between nose-tail (N–T) straight length and age for female (dots) and male (open circles) southern elephant seals. Data were from Laws (1953), and Gompertz growth models (large dots and solid line for females; large open circles and dashed line for males) were fitted by least squares regression. The regression equations were:

$$\text{female length} = 168 + 128 \times e(-e - 0.028[\text{age} - 4.5])$$
$$\text{male length} = 202.4 + 264.9 \times e(-e - 0.039[\text{age} - 39.73])$$

1987; McCann and Rothery 1988), reduced survival at Macquarie Island would be sufficient to account for this difference. The maximum longevity for a female elephant seal from South Georgia is 21 years (Arnbom et al. 1992), which is close to the 20 years predicted by McCann's life table.

Survival rate estimates for males are less certain than those for females. Males were assumed to become sexually mature at 4 years of age but did not breed until they were > 6 years of age (Laws 1953).

Growth. For the purposes of this analysis, we define the mean size of seals in any age class as the size at the end of the period of reproductive investment each year (the end of the breeding season). For seals < 1 year old, this means from the time of nutritional independence or weaning. Laws (1953) gave the nose-tail curved lengths of male and female elephant seals in relation to age. Curved lengths are about 10% greater than nose-tail straight length, so Laws's data were corrected to the equivalent nose-tail

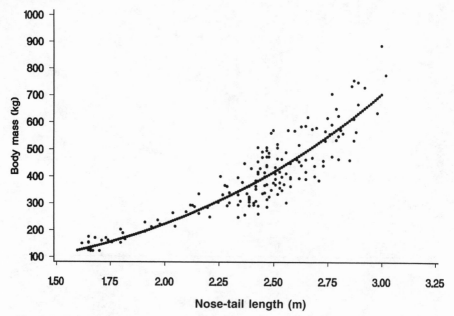

Fig. 6.2. The relationship between nose-tail straight length and mass for 162 female elephant seals. The line was fitted by least squares regression (length = 22.61 + 25.33 mass^3).

straight length and asymptotic growth curves were fitted to the data by least squares regression (fig. 6.1). The predicted values for each age were used as the mean length for a given age (table 6.1). The model overestimated weaning mass of both sexes, so the values given by T. S. McCann, M. A. Fedak, and J. Harwood (1989) were used for this age class.

Recent studies in which females were chemically immobilized (Baker et al. 1990; McCann, Fedak, and Harwood 1989; Boyd, Arnbom, and Fedak 1993) have yielded a relationship between mass and nose-tail length for females (fig. 6.2). This was used to estimate mean mass for each age class of elephant seal (table 6.1). The relationship between mass and length in males described by Bryden (1969b, 1972) was used to calculate the equivalent mass-age relationship for males. The pectoral girth of seals was also estimated using the equations in M. M. Bryden (1969b) which were subsequently corrected (Bryden 1972; see table 6.1).

Biomass. The number of individuals in each age class was calculated from a pup production of 102,000 (McCann and Rothery 1988), assuming a 1:1 sex ratio at birth (McCann 1985) and a stable population age structure. The biomass of each age class was derived from the product of the number of individuals in each age class and the mean mass of those individuals (table 6.1). The total body gross energy (TBGE) for that biomass was calculated using the equation provided by J. J. Reilly and M. A. Fedak (1990) derived from body composition analyses of gray seals. We assumed a mean water content of 55% derived from estimates of northern elephant seal, *M. angustirostris*, water content during lactation and molt (Costa et al. 1986; Worthy et al. 1992). The validity of using the Reilly and Fedak equation on different species was discussed in relation to other work on ringed seals, *Phoca hispida* (Stirling and McEwan 1975), and was broadly confirmed by I. L. Boyd and C. D. Duck (1991). Estimates of biomass are calculated in metric tonnes; 1 tonne equals 1,000 kg.

Energy Expenditure

The annual cycle of southern elephant seals involves three distinct activities: (1) reproduction in adult seals, (2) molt, and (3) foraging at sea (Laws 1960). Each of these will have a characteristic energy cost depending on the mass, reproductive status, and sex of the seal. An additional cost will be incurred in terms of body growth, which will vary with age, although reversible growth, such as seasonal fattening before breeding or molt, is considered as one of the costs of reproduction or molt in this analysis. The total energy expenditure (E) for an individual southern elephant seal can be expressed as

$$E = ER + EG + EF + EM, \tag{1}$$

where ER, EG, EF, and EM are the energy costs of reproduction, growth, foraging, and molt, respectively. ER includes the heat increments of lactation, gestation, and, in the case of males, harem defense and competition for mates. EF includes the heat increments of locomotion and feeding the estimates of assimilation efficiency and duration of foraging. EG includes the energy value of tissue growth but excludes seasonal changes in body reserves.

Energy of Reproduction. The metabolic rate, measured by metabolic water production, of female southern elephant seals during lactation was 88 MJ/day (Fedak et al., this volume) or 3.1 times the predicted standard metabolic rate (SMR). The general relationship ($SMR = 0.293 \ W^{0.75}$ MJ/d) between SMR and body mass of mammals has been discussed and confirmed for seals by Lavigne et al. (1986). Energy expenditure on metabolism was 46% of total energy expenditure. This shows that the total energy expenditure of lactation is 6 times SMR, which is in close agreement with measurements made in northern elephant seals (Costa et al. 1986). The mean duration of suckling used to calculate the heat increment of lactation was 23 days for mothers with either male or female pups (McCann, Fedak, and

Harwood 1989; Fedak et al., this volume). Thus, the energy of reproduction for an individual female was obtained from $6 \times SMR \times 23$.

The mean mass of elephant seals at birth was 42.5 kg (McCann, Fedak, and Harwood 1989). The energy density of fetal seals at term has been calculated for ringed seals (9.04 MJ/kg; Stirling and McEwan 1975), harp seals (7.24 MJ/kg; Worthy and Lavigne 1983) and gray seals (5.76 MJ/kg; Anderson and Fedak 1987). The variation in these figures probably reflects species differences in the fat content of pups at birth. We do not know which value is the most appropriate for southern elephant seals, but because some of the ringed seal pups in I. Stirling and E. H. McEwan's study (1975) could have been suckled and this figure may be an overestimate of energy density, we used a conservative value of 6.5 MJ/kg in this analysis. The mass of southern elephant seal placentas is 5 kg, and an energy density of 0.46 MJ/kg (Lavigne and Stewart 1979) was used to calculate total energy content.

There is less information about the energy costs of reproduction in males. The only pinniped studies that have measured energy expenditure showed that breeding male northern elephant seals and territorial male Antarctic fur seals, *Arctocephalus gazella*, both have metabolic rates 3.3 times SMR (Deutsch, Haley, and Le Boeuf 1990; Boyd and Duck 1991). The competitive mating behavior of these species is similar to the southern elephant seal (McCann 1981). Therefore, we used the same value to calculate the metabolic costs of reproduction in males > 6 years of age. According to McCann (1980), the number of bulls ashore at the breeding grounds increases from September 1 to October 1 and remains roughly stable thereafter until November 18 when the number declines rapidly. We have assumed that they are present from September 14 until November 20 (58 days). The duration of the postweaning fast for pups was 45 days.

Energy of Molt. The energy cost of molt in southern elephant seals was estimated to be 2.4 times SMR (Boyd, Arnbom, and Fedak 1993), which is similar to that measured for adult female northern elephant seals (Worthy et al. 1992). Molting lasts 28 and 40 days on average in adult female and male southern elephant seals, respectively (Ling and Bryden 1981). These values were used to calculate the heat increment of molt in both sexes.

Energy of Growth. The body composition of southern elephant seals shows minor changes at different stages of life (Bryden 1969a). However, because the scale of these changes is small, we have assumed that growth of body components is proportional to growth in total mass. The annual growth increment of southern elephant seals was calculated from the data on body mass (table 6.1), and the energy increment it represents was calculated using the equation for TBGE provided by Reilly and Fedak (1990).

Energy of Foraging. Activity budgets of seals during the aquatic phases of the annual cycle are based on time depth recorder outputs for southern elephant seals (Boyd and Arnbom 1991; Slip, Hindell, and Burton, this volume). While at sea, adult elephant seals spend 88% of their time diving. Velocities of descent and ascent were 1.64 m/s and 1/44 m/s, which makes sense in terms of empirical measurements of the cost of transport in phocids (Davis, Williams, and Kooyman 1985; Williams and Kooyman 1985). Seventy-four percent of dives had a profile showing that the seals remained at a specific depth for 54% of the duration of the dive. A further 10% of dives had a period when the seals remained at a constant depth before the resumption of a descent phase, and 16% of dives had no period spent at constant depth. Mean dive duration was 17 minutes for each type of dive.

The work required to transport a body through seawater is given by

$$E = 0.5 \times r \times CD \times A \times V^2 \times L, \tag{2}$$

where CD is the coefficient of drag, V is the velocity, r is the average water density (1.027×103 kg/m^3 for the Southern Ocean, data from British Antarctic Survey), A is the maximum cross-sectional area, and L is the distance. This is an extended form of the equation of drag (Williams 1987). Estimates of CD will depend on the shape and size of the seal. Values of CD measured for seals range from 0.06 to 0.09 (Williams and Kooyman 1985; Feldkamp 1987), but these are for gliding seals, and it is well known that the drag associated with active swimming is greater. Furthermore, an average adult female elephant seal (length 3.0 m) has a Reynolds number (Re) of 4.3×10^6, which is close to $Re = 5 \times 10^6$ for turbulent flow at a velocity of 1.5 m/s. This suggests that most elephant seals are close to the boundary of turbulent flow. Put together, this suggests that the CD for southern elephant seals is greater than measured values for other species would suggest. Therefore, a value of 0.12 was used in the calculation.

T. M. Williams, G. L. Kooyman, and D. A. Croll (1990) showed that harbor seals, *Phoca vitulina*, had a swimming efficiency (power output as a percentage of power input) that varied with velocity but was approximately 10% at 1.5 m/s.

Using the parameters given in table 6.1, we calculated the average cost of transport for the South Georgia elephant seal population. This assumed that most elephant seals forage in a way similar to the small sample for which detailed measurements are available. It also assumed that nonbreeders continued to forage during the breeding season but that all seals molted on land.

For harp seals, the heat increment of digestion was estimated to be 17% of the gross energy intake (Gallivan and Ronald 1981), and for harp seals and northern fur seals, the assimilation efficiency was estimated to be 90%

of gross energy intake (Keiver, Ronald, and Beamish 1984; Fadely, Worthy, and Costa 1990) on a diet of fish. There is no information for elephant seals, so we have used these figures in our calculations of the heat increment of foraging.

Diet Composition and Population Energy Consumption

It has been difficult to obtain an accurate measure of the diet of southern elephant seals at South Georgia mainly because it has not yet proved possible to measure diet at the foraging grounds. We have to rely on stomach samples obtained from elephant seals at the breeding and molting areas, which can be many hundreds of kilometers from the foraging areas (Fedak et al, this volume). Laws (1956) examined 139 stomachs, of which 108 contained no food; of the remainder, 24 contained squid and 9 fish. Laws (1977) therefore suggested that the diet of southern elephant seals consisted of 75% squid and 25% fish by weight. P. G. Rodhouse et al. (1992) have described the cephalopod prey, from samples obtained by stomach flushing, and confirmed that squid made up the greatest proportion of the diet. No fish remains were discovered, but fish are probably underrepresented in stomach flushes because of their rapid rate of digestion. Therefore, we have little justification for departing from Laws's overall figure for diet composition. In addition, we have to assume for the present that diet measured from stomach lavaging is broadly indicative of overall diet, although the analysis presented here also provides an indication of what amounts of different types of prey would be taken if the proportions of squid and fish departed from those suggested by Laws (1977).

Energy densities vary considerably between different species of squid, so the squid diet was divided into three parts: (1) muscular squid, (2) gelatinous squid, and (3) cranchiid squid, which have a leathery mantle. The classification of each species plus its representation by percentage mass of squid in the diet is given in table 6.2. Energy values were assigned following A. Clarke et al. (1985). We assumed that fish taken were a mixture of myctophids and notothenids. Although there is no evidence that southern elephant seals feed on myctophids, they are dominant among species of shoaling fish in the parts of the water column in which elephants seals apparently feed (Hindell 1991; Boyd and Arnbom 1991), which suggests that they are potential prey. Energy densities of myctophids are in the range of 7–8 MJ/kg (Cherel and Ridoux 1992), and we have used a value of 7.5 MJ/kg. The energy density of other fish species was assumed to be 4 MJ/kg (ibid.).

The total biomass of each item consumed in the diet was calculated from the gross energy requirement of the population (GE), so that

$$GE = I1_m I1_e + I2_m I2_e + I3_m I3_e + \ldots + In_m In_e, \tag{3}$$

TABLE 6.2   Cephalopods in the diet of southern elephant seals.

	Species	% Total squid biomass in diet Females	% Total squid biomass in diet Males	Type	Energy density (MJ/kg)
A	Gonatus antarcticus	3.9	3.2	muscular	4.0
B	Kondakovia longimanna	10.6	41.6	cranchiid	1.7
C	Moroteuthis knipovitchi	43.7	15.4	muscular	4.0
D	Brachioteuthis sp.	0.5	1.0	muscular	4.0
E	Batoteuthis skolops	<0.1	<0.1	muscular	4.0
F	Histioteuthis sp.	1.9	0.9	gelatinous	2.0
G	Psychroteuthis glacialis	12.1	19.4	muscular	4.0
H	Alluroteuthis glacialis	9.6	12.2	muscular	4.0
I	Martalia hyadesi	17.5	3.1	muscular	4.0
J	Chiroteuthis	0.2	0.2	gelatinous	2.0
K	Mastigoteuthis psychrophila	—	0.3	gelatinous	2.0
L	Caliteuthis glacialis	0.2	0.3	cranchiid	1.7
M	Mesonychoteuthis hamiltoni	0.1	0.2	cranchiid	1.7
N	Pareledone charcoti	—	1.7	muscular	4.0
O	Pareledone polymorpha	—	0.5	muscular	4.0

NOTE: Modified from Rodhouse et al. 1992 and Clarke et al. 1985.

where $I1_m$, $I2_m$, ..., In_m denote the mass of each item in the diet and $I1_e$, $I2_e$, ..., In_e denote the energy value (MJ/kg) of each item. This equation was solved for the mass of each item from knowledge of the proportion by mass of each item in the diet. Therefore,

$$In_m = E/(r1I1_e + r2I2_e + r3I3_e + ... + In_e),$$ (4)

where r1, r2, r3, and so on were the ratios of In_m to the mass of the other items in the diet.

RESULTS

Annual Energy Budget

The pattern of energy demand by each age class reflected the changing number, mass, and reproductive condition with age (fig. 6.3). Total energy expenditure in both sexes declined initially, because of the high rate of juvenile mortality, and then began to increase. Among females, this was due to the introduction of the additional burden of reproduction, while among males, the increase was the result of growth and increased costs of locomotion associated with larger body size.

The energy expenditure of individual males increased to 55×10^3 MJ up to age 10 with a sharp increase in energy expenditure associated with the

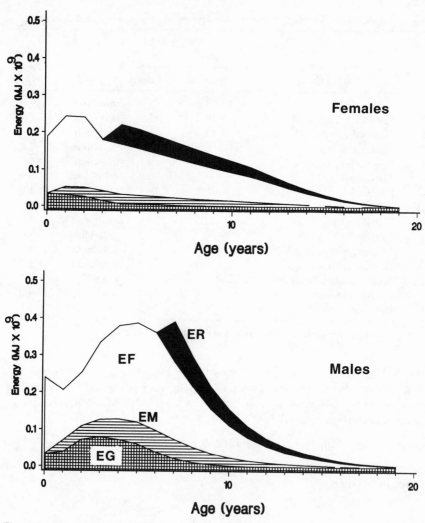

Fig. 6.3. Changes in the energy expenditure of different age classes of the stable southern elephant seal stock at South Georgia.

beginning of reproduction at age 7 (fig. 6.4). In females, the asymptote of annual energy expenditure was reached at a similar age but at about one quarter of the energy expenditure of males.

The total population size was 469,000, with a sex ratio of 1.25 females for each male. The total biomass was 222,903 tonnes, of which male biomass made up 63% and the mean mass was 205 kg. The estimated total annual energy expenditure for the population was 6.01×10^9 MJ, which was equivalent to 190 Mw averaged over the whole year.

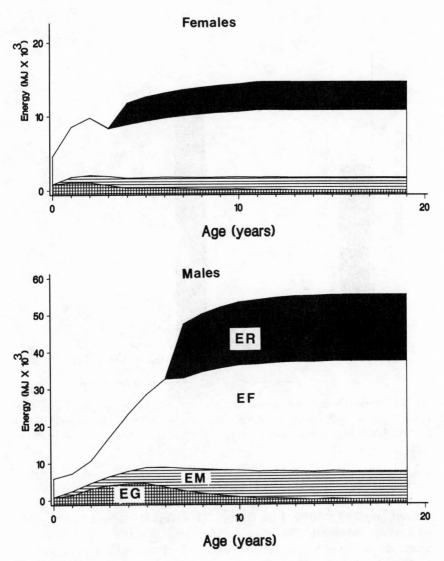

Fig. 6.4. Changes in the per capita energy expenditure of female and male southern elephant seals in relation to age.

The annual energy expenditure associated with foraging (EF) was 63.2% and 68.2% of the total energy costs in males and females, respectively (fig. 6.5), and EF was the largest single form of energy expenditure. Overall, 65.2% of the energy expenditure of the population was spent on foraging. Among males, the energy expenditure of molt (EM) was the next largest energy demand (13.7%), and this was greater than the total energy

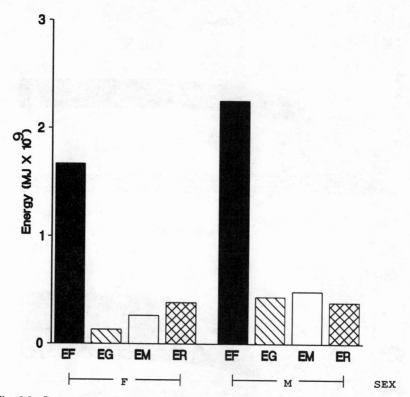

Fig. 6.5. Components of the annual energy budget of southern elephant seals from South Georgia. EF = foraging energy expenditure; EG = energy expenditure of growth; EM = energy expenditure of molt; ER = energy expenditure of reproduction.

expenditure of molt in females (10.8% of female total). Energy expenditure on reproduction (ER) was similar in the two sexes (3.87×10^8 MJ for females and 3.88×10^8 MJ for males). This was 10.9% and 15.8% of the total energy expenditure for males and females, respectively. The energy expenditure on growth was 12.3% and 5.3% of the total expenditure for males and females, respectively.

Food Consumption

The total biomass of squid and fish (assuming 100% myctophids) consumed was 1.71 and 0.26 million metric tonnes per year, respectively (table 6.3). The female section of the population consumed 0.63 and 0.11 million tonnes of squid and fish, while equivalent values for males were 1.08 and 0.15 million tonnes, respectively. The biomass of each squid species consumed suggested that both males and females specialized in the large,

TABLE 6.3   Expected biomass of different food items consumed by the South Georgia elephant seal population given different diet compositions by weight. This takes account of different energy values of prey items including those given for squid in table 6.2.

Diet composition (%)			Biomass ($\times$ 10^6 tonnes)		
Squid	Myctophids	Other fish	Squid	Myctophids	Other fish
0	100	0	0	1.04	0
25	75	0	0.30	0.90	0
50	50	0	0.70	0.70	0
75	25	0	1.28	0.43	0
100	0	0	2.17	0	0
0	0	100	0	0	1.95
0	50	50	0	0.68	0.68
50	25	25	0.84	0.42	0.42
33	33	33	0.52	0.52	0.52
60	30	10	0.97	0.49	0.16

muscular species of high energy density (table 6.2) compared with the species of lower energy density (fig. 6.6).

Production Efficiency

The gross energy requirement (GE) of the population was 7.82×10^9 MJ (3.18×109 MJ for females and 4.64×10^9 MJ for males), and the production energy was 0.65×10^9 MJ. The production efficiency was therefore 8.2%. On average, gross energy intake during potential foraging time was 77.3 and 43.2 MJ/day for males and females, respectively. This suggests a capture rate of 0.26 and 0.15 kg of muscular squid or fish per dive for males and females, respectively, based on dive rates obtained from Boyd and Arnbom (1991) and D. Slip, M. Hindell, and H. Burton (this volume).

DISCUSSION

It was not possible to provide statistical confidence intervals on estimates of energy expenditure because of the very variable quality of the data used to produce these estimates. Therefore, the results should be interpreted with care, although there is evidence supporting their accuracy. For example, our estimate of the gross energy requirements of the South Georgia elephant seal population compares well with the estimate for harp seals given by D. M. Lavigne et al. (1985). They estimated a per capita energy requirement of 11.3×10^3 MJ/year, while for the larger elephant seal, the value was 16.8×10^3 MJ/year. In addition, W. Sakamoto et al. (1989) esti-

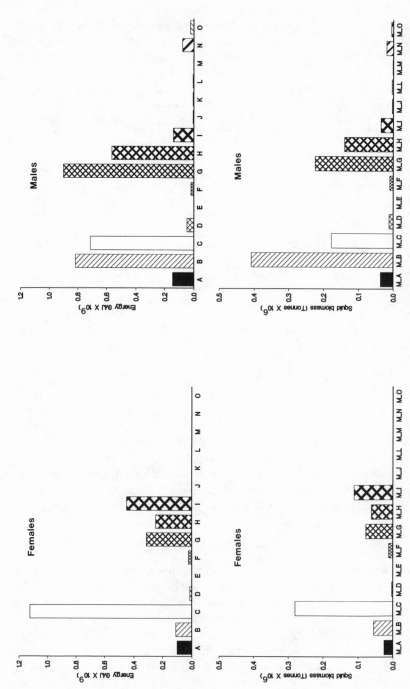

Fig. 6.6. Composition of the squid portion of the diet of southern elephant seals from the South Georgia stock expressed in terms of energy value and biomass. The squid species are those labeled in table 6.2.

mated the daily gross energy requirement for an average adult female northern elephant seal when foraging as about 40 MJ/day. In this study we estimate a value of 41 MJ/day for the daily energy expenditure of adult females averaged over the whole year.

We have estimated population biomass to be 1.21 times that estimated by McCann (1985). This is because we calculated biomass at the end of the breeding season, whereas McCann did not include the biomass of the 0+ age class. His estimates of increasing mass with age were also more conservative, but our experience of weighing both male and female elephant seals has suggested that the estimates of mass given in table 6.1 are realistic. There are also significant differences between the food consumption of the population estimated in this study compared with the estimate made by McCann (1985). We found a total biomass consumption of 2.1 million tonnes (given the 3:1 numerical ratio of squid to fish in the diet), whereas McCann (1985) estimated this to be 3.68 million tonnes. The difference is largely caused by the figures used by McCann to translate population biomass into an estimate of food consumption. Following Laws (1977), he assumed that the annual food consumption was 20 times body mass. Our estimates suggest that for elephant seals the multiplier should be closer to 10. This is similar to the estimate made for harp seals on a fish diet by Lavigne et al. (1985).

A further reason for the reduced size of the multiplier is the apparently high production efficiency shown by southern elephant seals. W. F. Humphreys (1979) showed that production efficiency for mainly small herbivorous mammals rarely exceeded 5%. Respiratory costs are reduced in elephant seals because of the large body size (Lavigne et al. 1986) and because the cost of transport per unit of mass declines with increased mass. Since the cost of foraging, which is largely associated with the cost of transport, was the major form of energy expenditure, then the respiratory costs of a population of large swimming mammals will be lower than a population with an identical biomass but made up of a greater number of small individuals. Thus, the production efficiency of a harp seal population is about half that of the elephant seal. The long delay in reproductive maturity of males also contributes greatly to the production efficiency because energy normally expended on reproduction is available to be routed into growth.

Apart from the suggestion of Laws (1977), there is no evidence that elephant seals feed with a ratio of three parts squid to one part fish by weight. This highlights one of the greatest deficiencies in our knowledge of southern elephant seal foraging ecology. The diet may contain more or less fish, and it may also contain more benthic and demersal species that will have energy densities more similar to the muscular squid (4 MJ/kg, table 6.2, Cheral and Ridoux 1992) than to myctophids, which are particularly oily. Table 6.3 compares the biomass of fish and squid consumed with different proportions of fish and squid in the diet. Estimates of total food consumed also depend on the energy values applied to each item of the diet. In this study these are only crude estimates that, in the case of squid, are based on estimates of the energy density of related temperate squid species (Clarke et al. 1985). Little is known about some of the species found in the diet of southern elephant seals, and we cannot be certain that the general classification into "muscular" or "gelatinous" forms is correct. Estimates of total consumption must also be interpreted with care because lavaging animals soon after they arrive on land probably only indicates diet in the immediate surroundings of the landing site or is biased by items, such as squid beaks, that persist in the stomach longer than other food remains.

One of the most significant potential threats to the stability of the South Georgia elephant seal stock is competition for food with commercial fisheries. This study is a synthesis of knowledge about the biology of southern elephant seals at South Georgia that provides an important, albeit imprecise, insight into the potential interactions between the South Georgia southern elephant seal population and its food supply. The metabolic costs of maintenance for fully grown adult female southern elephant seals was 0.39 MJ/kg$^{.75}$, and for adult males it was 0.17 MJ/kg$^{.75}$. The differences between the sexes is probably caused by differences in the foraging component of energy expenditure because the cost of locomotion increases in direct proportion to cross-sectional area. Large changes in mass are accompanied by relatively small changes in cross-sectional area, resulting in reduced costs of locomotion per unit mass. The total annual energy expenditure for the southern elephant seal population at South Georgia was 6.01×10^9 MJ/year, and 59% of this was accounted for by males. Despite contributing nothing to the energy costs of raising young and having only a small chance of reproducing (Deutsch, Haley, and Le Boeuf 1990), males demand a greater proportion of the total resources than females. They may also require richer foraging grounds because during potential periods of foraging, the rate of energy intake of males has to be almost twice that of females. These differences suggest that male survival, growth, and condition may be more responsive to changes in environmental conditions than that of females. However, the high production efficiency of southern elephant seals also suggests that they may be able to exploit highly dispersed prey resources.

ACKNOWLEDGMENTS

This chapter is the result of a collaborative research project between the British Antarctic Survey (Natural Environment Research Council) and the Sea Mammal Research Unit (NERC). We thank T. Barton, C. Chambers, B. J. McConnell, A. Morton, and A. Taylor for assistance in the field.

REFERENCES

Anderson, S. S., and M. A. Fedak. 1987. Gray seal, *Halichoerus grypus*, energetics: Females invest more in male offspring. *Journal of Zoology, London* 211: 667–679.

Arnbom, T. A., N. J. Lunn, I. L. Boyd, and T. Barton. 1992. Ageing live Antarctic fur seals and southern elephant seals. *Marine Mammal Science* 8: 37–43.

Baker, J. R., M. A. Fedak, S. S. Anderson, T. Arnbom, and J. R. Baker. 1990. Use of tiletamine-zolazepam mixture to immobilise wild gray seals and southern elephant seals. *Veterinary Record* 126: 75–77.

Boyd, I. L., and T. Arnbom. 1991. Diving behaviour in relation to water temperature in the southern elephant seal: Foraging implications. *Polar Biology* 11: 259–266.

Boyd, I. L., T. Arnbom, and M. A. Fedak. 1993. Water flux, body composition and metabolic rate during molt in female southern elephant seals (*Mirounga leonina*). *Physiology Zoology* 66: 43–60.

Boyd, I. L., and C. D. Duck. 1991. Mass changes and metabolism in territorial male Antarctic fur seals (*Arctocephalus gazella*). *Physiological Zoology* 64: 375–392.

Bryden, M. M. 1969a. Relative growth of the major body components of the southern elephant seal, *Mirounga leonina* (L.). *Australian Journal of Zoology* 17: 153–177.

———. 1969b. Growth of the southern elephant seal, *Mirounga leonina* (Linn.). *Growth* 33: 69–82.

———. 1972. Body size and composition of elephant seals (*Mirounga leonina*): Absolute measurements and estimates from bone dimensions. *Journal of Zoology, London* 167: 265–276.

Cherel, Y., and V. Ridoux. 1992. Prey species and nutritive value of food fed during summer to king penguin, *Aptenodytes patagonica*, chicks at Possession Island, Crozet Archipelago. *Ibis* 134: 118–127.

Clarke, A., M. R. Clarke, L. J. Holmes, and T. D. Waters. 1985. Calorific values and elemental analysis of eleven species of oceanic squids (Mollusca: Cephalopoda). *Journal of the Marine Biological Association* 65: 983–986.

Costa, D. P., B. J. Le Boeuf, A. C. Huntley, and C. L. Ortiz. 1986. The energetics of lactation in the northern elephant seal, *Mirounga angustirostris*. *Journal of Zoology, London* 209: 21–33.

Davis, R. W., T. M. Williams, and G. L. Kooyman. 1985. Swimming metabolism of yearling and adult harbor seals, *Phoca vitulina*. *Physiological Zoology* 58: 590–596.

Deutsch, C. J., M. P. Haley, and B. J. Le Boeuf. 1990. Reproductive effort of male northern elephant seals: Estimates from mass loss. *Canadian Journal of Zoology* 68: 2580–2593.

Fadely, B. S., G. A. J. Worthy, and D. P. Costa. 1990. Assimilation efficiency of northern fur seals determined using dietary manganese. *Journal of Wildlife Management* 54: 246–251.

Fedak, M. A., S. S. Anderson, and J. Harwood. 1981. The energetics of the gray seal (*Halichoerus grypus*) in European waters: Energy flow and management implications. Final Report to the European Communities on Contract ENV 405-80-UK (B).

Feldkamp, S. D. 1987. Swimming in the California sea lion: Morphometrics, drag, and energetics. *Journal of Experimental Biology* 131: 117–135.

Gallivan, G. J., and K. Ronald. 1981. Apparent specific dynamics action in the harp seal (*Phoca groenlandica*). *Comparative Biochemistry and Physiology* 69A: 579–581.

Hindell, M. A. 1991. Some life-history parameters of a declining population of southern elephant seals, *Mirounga leonina*. *Journal of Animal Ecology* 60: 119–134.

Hindell, M. A., and H. Burton. 1987. Past and present status of the southern elephant seal (*Mirounga leonina*) at Macquarie Island. *Journal of Zoology, London* 213: 365–380.

Humphreys, W. F. 1979. Production and respiration in animal populations. *Journal of Animal Ecology* 48: 427–453.

Keiver, K. M., K. Ronald, and F. W. H. Beamish. 1984. Metabolizable energy requirements for maintenance and faecal and urinary losses of juvenile harp seals (*Phoca groenlandica*). *Canadian Journal of Zoology* 62: 769–776.

Lavigne, D. M., S. Innes, R. E. A. Stewart, and G. A. J. Worthy. 1985. An annual energy budget for northwest Atlantic harp seals. In *Marine Mammals and Fisheries* ed. J. R. Beddington, R. J. H. Beverton, and D. M. Lavigne, 319–336. London: George Allen and Unwin.

Lavigne, D. M., S. Innes, G. A. J. Worthy, K. M. Kovacs, O. J. Schmitz, and J. P. Hickie. 1986. Metabolic rates of seals and whales. *Canadian Journal of Zoology* 64: 279–284.

Lavigne, D. M., and R. E. A. Stewart. 1979. Energy content of harp seal placentas. *Journal of Mammalogy* 60: 854–856.

Laws, R. M. 1953. The elephant seal (*Mirounga leonina* Linn.). I. Growth and age. *Falkland Islands Dependencies Survey, Scientific Reports* 8: 1–62.

———. 1956. The elephant seal (*Mirounga leonina* Linn.). I. General, social and reproductive behavior. *Falkland Islands Dependencies Survey, Scientific Reports* 13: 1–88.

———. 1960. The elephant seal (*Mirounga leonina* Linn.) at South Georgia. *Norsk Hvalfangst-Tidende* 49: 466–476, 520–542.

———. 1977. Seals and whales of the Southern Ocean. *Philosophical Transactions of the Royal Society of London, Series B*, 279: 81–96.

Ling, J. K., and M. M. Bryden. 1981. Southern elephant seal, *Mirounga leonina* Linnaeus, 1758. In *Handbook of Marine Mammals. 2. Seals*, ed. S. H. Ridgway and R. J. Harrison, 297–327. London: Academic Press.

McCann, T. S. 1980. Population structure and social organization of southern elephant seals, *Mirounga leonina* (L.). *Biological Journal of the Linnaean Society* 14: 133–150.

———. 1981. Aggression and sexual activity of male southern elephant seals, *Mirounga leonina*. *Journal of Zoology, London* 195: 295–310.

———. 1985. Size, status and demography of southern elephant seal (*Mirounga leonina*) populations. In *Sea Mammals in South Latitudes: Proceedings of a Symposium of the 52d ANZAAS Congress in Sydney—May 1982*, ed. J. K. Ling and M. M. Bryden, 1–17. Northfield: South Australian Museum.

McCann, T. S., M. A. Fedak, and J. Harwood. 1989. Parental investment in southern elephant seals, *Mirounga leonina*. *Behavioral Ecology and Sociobiology* 25: 81–87.

McCann, T. S., and P. Rothery. 1988. Population size and status of the southern elephant seal (*Mirounga leonina*) at South Georgia, 1951–1985. *Polar Biology* 8: 305–309.

Reilly, J. J., and M. A. Fedak. 1990. Measurement of the body composition of living gray seals by hydrogen isotope dilution. *Journal of Applied Physiology* 69: 885–891.

Rodhouse, P. G., T. A. Arnbom, M. A. Fedak, J. Yeatman, and A. W. A. Murray. 1992. Cephalopod prey of the southern elephant seal *Mirounga leonina* L. *Canadian Journal of Zoology* 70: 1007–1015.

Sakamoto, W., Y. Naito, A. C. Huntley, and B. J. Le Boeuf. 1989. Daily gross energy requirements of female northern elephant seal, *Mirounga angustirostris*, at sea. *Nippon Suisan Gakkaishi* 55: 2057–2063.

Skinner, J. D., and R. J. van Aarde. 1983. Observations on the trend of the population of southern elephant seals, *Mirounga leonina*, at Marion Island. *Journal of Applied Ecology* 20: 707–712.

Stirling, I., and E. H. McEwan. 1975. The caloric value of whole ringed seals (*Phoca hispida*) in relation to polar bear (*Ursus maritimus*) ecology and hunting behavior. *Canadian Journal of Zoology* 53: 1021–1027.

van Aarde, R. J. 1980. Fluctuations in the population of southern elephant seals, *Mirounga leonina*, at Kerguelen Island. *South African Journal of Zoology* 15: 99–106.

Williams, T. M. 1987. Approaches to the study of exercise physiology and hydrodynamics in marine mammals. In *Approaches to Marine Mammal Energetics*, ed. A. C. Huntley, D. P. Costa, G. A. J. Worthy, and M. A. Castellini, 127–146. Special Publication no. 1. Lawrence, Kan.: Society for Marine Mammalogy.

Williams, T. M., and G. L. Kooyman. 1985. Swimming performance and hydrodynamic characteristics of harbor seals, *Phoca vitulina*. *Physiological Zoology* 58: 576–589.

Williams, T. M., G. L. Kooyman, and D. A. Croll. 1990. The effect of submergence on heart rate and oxygen consumption of swimming seals and sea lions. *Journal of Comparative Physiology* 160B: 637–644.

Worthy, G. A. J., and D. M. Lavigne. 1983. Changes in the energy stores during postnatal development of the harp seal, *Phoca groenlandica*. *Journal of Mammalogy* 64: 89–96.

Worthy, G. A. J., P. A. Morris, D. P. Costa, and B. J. Le Boeuf. 1992. Molt energetics of the northern elephant seal. *Journal of Zoology, London* 227: 257–265.

PART II

Behavior and Life History

SEVEN

Juvenile Survivorship of Northern Elephant Seals

Burney J. Le Boeuf, Patricia Morris, and Joanne Reiter

ABSTRACT. The aim of this study was to determine the juvenile survivorship rate of northern elephant seals, *Mirounga angustirostris*, throughout the first four years of life and to assess the role of year, cohort, and condition at weaning on survival.

The study was conducted at Año Nuevo, California, during the years 1971–1978, a time when colony size was increasing. Pup mortality on the rookery prior to weaning was estimated from daily censuses during the breeding season. Juvenile survivorship was determined from resights of 8,362 individuals tagged on the rookery at weaning (about 30 days of age); systematic searches were conducted on the natal rookery as well as on neighboring rookeries. The effect of mass and length at weaning on juvenile survivorship to 1 and 2 years of age was determined from 734 weaned pups weighed and measured during the years 1978 and 1984–1988.

Mean percentage survival to age 1 was 36.8 ± 8.5; to age 2, 26.3 ± 6.3; to age 3, 19.4 ± 5.1; and to age 4, 16.3 ± 5.2. Most of the first-year mortality occurred at sea; on average, $31.5 \pm 12.4\%$ of the first-year mortality was due to neonate death on the rookery. Juvenile survivorship rates were lowest in El Niño years (1978, 1983, and 1986). As colony size increased fivefold over the study period, survivorship to age 1 did not change significantly, but survivorship to age 4 decreased significantly. No significant relationship was found between weanling mass and survival to 1 and 2 years of age. Survivorship to 1 year of age was positively correlated with standard length, but this relationship did not hold for survivorship to age 2.

The juvenile survivorship rate of seals from the Año Nuevo colony is too low to support the observed growth rate of the colony and of the population as a whole. Other California rookeries, such as San Miguel Island, must have significantly higher juvenile survivorship rates to account for the recent population increase. The causes of high juvenile mortality at sea are unknown; they do not appear to be related to condition at weaning, as reflected by weaning weight.

Survivorship and fertility schedules shape life history tactics (Stearns 1976, 1980) and provide vital demographic data for estimating population growth (Wilson and Bossert 1971). This chapter addresses juvenile survivorship in northern elephant seals, the percentage of individuals born that survive to each of the first four years of life. Our aim is to describe age-specific survival rates and age-specific mortality rates of seals born at Año Nuevo, California, over the last two decades when colony size was increasing and to examine the role of year, cohort, and condition at weaning—as reflected by weight, length, or an index of the two measures—on survival. Because juvenile survivorship is an important determinant of the growth or decline of a population, data presented here may elucidate the rapid growth of the northern elephant seal population over the last few decades as well as provide an instructive comparison with southern elephant seals, whose numbers are declining at several rookeries (Hindell, Slip, and Burton, this volume).

This chapter summarizes and augments data on juvenile survivorship of Año Nuevo-born seals presented in J. Reiter, N. L. Stinson, and B. J. Le Boeuf (1978), J. Reiter (1984), B. J. Le Boeuf and J. Reiter (1988), and B. J. Le Boeuf and J. Reiter (1991).

BACKGROUND

Año Nuevo is a peripheral colony in the northern elephant seal range. Since breeding began here in 1961 (Radford, Orr, and Hubbs 1965), it has received immigrants from larger southern rookeries in southern California, San Miguel and San Nicolas islands. Throughout this period, the entire population has grown steadily (Stewart et al., this volume). Births at the Año Nuevo colony have increased at the rate of 14% per year, and annual pup production is now on the order of 2,000 pups. The growth, however, is due mainly to immigration from southern rookeries, for internal recruitment is too low for the colony to sustain itself (Le Boeuf and Reiter 1988).

Most females give birth for the first time at age 4 (range 3–6 years of age) and then give birth annually until death. Single pups are produced, nursed 25 to 28 days, and weaned abruptly when the mother returns to sea. The weaned pup fasts for 2½ months on the rookery while learning to swim and dive before going off on its first foraging trip (Reiter, Stinson, and Le Boeuf 1978).

Juveniles make two foraging trips per year, each lasting about five months (see fig. 13.1 in chap. 13). As a result, they appear on the rookery twice a year, in the spring and in the fall, each haul-out lasting about one month. At this time, they are identified and survival is estimated. The pattern changes when females begin giving birth. Consequently, sex differences in survival begin to appear in year 3, partially a result of a sex difference in time spent at sea.

METHODS

Pups born in the years 1971 to 1988 at Año Nuevo, California, were tagged shortly after weaning with one or two cattle ear tags in the interdigital webbing of the hind flippers (Le Boeuf and Peterson 1969). The number of seals tagged per year varied from 100 to 900. Survivors were identified when their tags were read at approximately weekly intervals on the island and the mainland resting and molting sites at Año Nuevo. Seals dispersing to other rookeries were identified and reported to us by H. Huber and W. Sydeman for Southeast Farallon Island, S. Allen for Point Reyes Headlands, and B. Stewart and R. DeLong for San Miguel and San Nicolas islands. Seals that stranded along the central California coast were reported to us by researchers at the California Marine Mammal Center.

From resights of tagged seals, we calculated life tables to age 4. The criterion for survival to age 1 was resighting the seal after the first trip to sea, when it was 9 to 10 months old in the fall or 15 to 17 months old in the spring. Survival to age 2, 3, and 4 was recorded similarly. In our experience, only slightly more juveniles are observed in the spring than in the preceding fall haul-out.

The survivorship data for each cohort and age class were adjusted to account for unrecognizable survivors that lost their tag identification. The proportion of animals that lost their tags varied from year to year because cohorts varied with respect to the proportion of single- and double-tagged individuals. Nearly all seals in the cohorts during the interval 1985–1988 were double tagged. For single-tagged seals, we assumed a tag loss rate of 11% per annum for the first two years of life and 6% per annum thereafter. The tag loss rate of double-tagged seals was assumed to be the rate of single tag loss squared, or 1.21% per annum for the first two years and 0.36% per annum thereafter. Tag loss estimates were based on the loss rate of single tags determined from double-tagged individuals (Reiter 1984)

The influence of weight, length, and a condition index on the probability of first-year survival was investigated. During 1978 and the years 1984–1988, 734 weaned pups were weighed and measured within a month of weaning. Weaning mass was estimated by back calculation based on known rates of mass lost per day (see equation in Appendix 10.1 of Deutsch et al., this volume). All pups that weighed less than 50 kg were excluded from the analysis. These were orphaned pups that did not suckle normally; most of them died on the rookery or stranded nearby shortly after going to sea. Standard length was measured in a straight line from tip of nose to tip of

TABLE 7.1 Partial life table for the northern elephant seal, *M. angustirostris*, constructed from resightings of 8,362 individuals tagged at weaning at Año Nuevo, California, during the years 1971–1988. This table includes a correction for tag loss (see text).

Age interval	Number dying during age interval	Number surviving at beginning of age interval	Number surviving as a fraction of newborn (l_x)	Mortality rate as fraction of number surviving at beginning of age interval (m_x)
0–1	5,281	8,362	1.000	0.632
1–2	907	3,081	0.368	0.294
2–3	551	2,174	0.263	0.253
3–4	260	1,623	0.194	0.160
4–5		1,363	0.163	

tail above the dorsal surface. A condition index, ostensibly reflecting a pup's stored energy reserves, was calculated as mass divided by length.

RESULTS

Age-specific Survival

Of the pups born during the years 1971–1988, the mean percentage that survived to age 1 was 36.8 ± 8.5; to age 2, 26.3 ± 6.3; to age 3, 19.4 ± 5.1; and to age 4, 16.3 ± 5.2. These data are presented as partial life tables in table 7.1 and figure 7.1. Age-specific survival varied widely over the years, with the following range of values being observed: 19.9–48.7% to age 1; 11.1–37.4% to age 2; 7.1–28.9% to age 3; and 5.0–26.5% to age 4. Survival rates were highest for the year 1971 and lowest for the year 1983 (fig. 7.2).

Age-specific Mortality

Age-specific mortality for the entire sample was highest during the first year of life, 63.2%, and then dropped steadily until reaching a low of 16% between 3 and 4 years of age (table 7.1, fig. 7.1).

Mortality on the Rookery and at Sea during the First Year

Over the study period, a mean of $31.5 \pm 12.4\%$ of the first-year mortality was due to neonate death on the rookery; the majority of the first-year mortality occurred at sea. The proportion of first-year mortality occurring on the rookery reached a high of 61% of pups born in 1983 due to the rookery being inundated by storm-whipped high surf at high tide during the peak

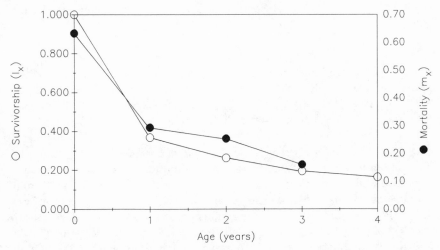

Fig. 7.1. Survivorship (l_x) and mortality (m_x) curves for the northern elephant seal at Año Nuevo, California, during the years 1971–1988. Based on data in table 7.1.

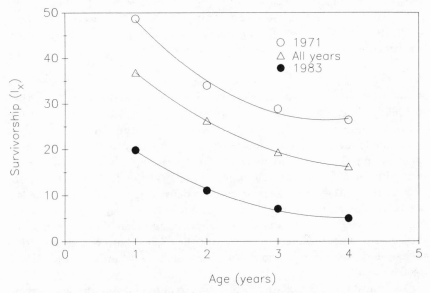

Fig. 7.2. Survivorship (l_x) curves for northern elephant seals from Año Nuevo, California, during the best (1971) and worst (1983) years in the study period, 1971–1988, and for all years combined.

pupping period (see also Le Boeuf and Condit 1983; Le Boeuf and Reiter 1991).

Survival Rates of Weanlings

For all years combined, the mean survival rate of weanlings over the periods at sea was $46.0 \pm 7.7\%$ (range = 35.0%–61.1%) to age 1; $32.8 \pm 5.9\%$ (range = 21.6–44.5%) to age 2; $24.2 \pm 5.5\%$ (range = 13.8–33.8%) to 3; and $20.3 \pm 5.8\%$ (range = 9.8–30.9%) to age 4.

The Effect of Year and Associated Conditions

The role of a specific year and associated conditions, such as weather and prey availability, is reflected by the annual survival rates of 1-, 2-, 3- and 4-year-olds in that year. For example, in 1981, survival to age 1 was $42.5 \pm 4\%$; the survival rate of the 1980 cohort from age 1 to 2 during 1981 was 91.9%; the 1979 cohort survival from age 2 to 3 in 1981 was 96.0%; and the 1978 cohort survival from age 3 to 4 in 1981 was 97.5%, giving 1981 a mean score of 82.0%.

Calculated in this way, the mean score of all cohorts was $79.1 \pm 3.6\%$. There were three years with mean scores one standard deviation or more above the mean, an indication that they were exceptionally good years: 1974 (84.6%), 1985 (83.3%), and 1980 (82.6%). There were three poor years: 1983 (72.7%), 1986 (74.0%), and 1978 (76.0%). All three poor years are categorized as El Niño years by oceanographers.

Cohort Variation

As the size of the Año Nuevo colony increased more than fivefold from 1971 to 1988 (an increase similar to that of the entire population; see Stewart et al., this volume), one might expect lower survivorship values in the later years due to increased competition for resources either on the rookery or on the foraging grounds. Pup mortality on the island prior to weaning increased from a low of 14.5% of pups born in 1971 to a high of 70% of pups born in 1983; preweaning pup mortality is density dependent, and there is a significant interaction with weather (Le Boeuf and Briggs 1977; Le Boeuf and Reiter 1991). There was no tendency for survivorship to age 1 to decrease with time and increasing density (fig. 7.3a); however, survivorship to age 4 decreased with time (the regression of y on $x = 90.1 - 0.94x$; $r = 0.81$).

An indication of the relative long-term strength of a cohort is the percentage decrease in the juvenile survival rate from year 1 to year 4. The percent decrease in survivorship from age 1 to age 4 increased over the study period (fig. 7.3b); that is, the mortality rate over the juvenile years increased with time.

The mean percentage decrease in survivorship from age 1 to 4 (fig. 7.3b)

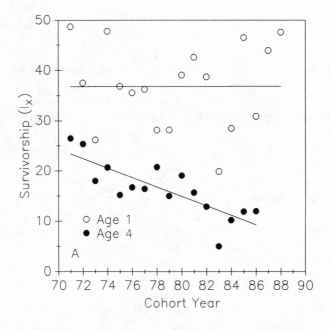

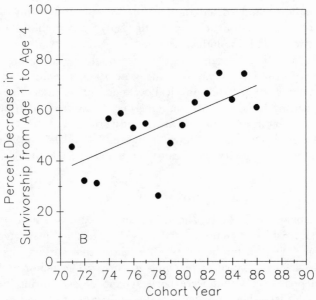

Fig. 7.3. *(A)* Survivorship to age 1 (open circles) and to age 4 (closed circles) as a function of cohort year. The regression equation for the latter is $y = 90.1 - 0.94x$; $r = .81$. *(B)* The percent decrease in survivorship from age 1 to age 4 as a function of cohort year. The regression equation is $y = -111 + 2.1x$; $r = .69$.

for all 18 cohorts was $-55.2 \pm 14.1\%$. The three strongest cohorts, those with the lowest percentage decline over the three-year period, were 1978 $(-26.2 \pm 3\%)$, 1973 (-31.3%), and 1972 (-32.3%). The three weakest cohorts were 1983 (-74.4%), 1985 (-74.4%), and 1988 (-66.9%).

The Effect of Weaning Mass, Size, and Condition on Age-specific Survival
Although the weight of weanlings varied greatly, with animals in the highest weight category being almost twice as large as those in the lowest weight category (fig. 7.4a), there was no significant relationship between mass at weaning and survival to 1 year of age (chi-square = 8.17, df = 8, p > .05) or to 2 years of age (chi-square = 6.83, df = 8, p > .05) (fig. 7.4b). Seals with the most common weights at weaning, in the range of 120 to 150 kg, had the lowest survivorship to year 1, 34.8 to 41.7%). However, these rates did not differ significantly from that of other weight categories. The extreme low and high weight categories incurred the greatest decline in survivorship from year 1 to year 2, but these differences, too, were not significantly different from the declines in other weight categories (chi-square = 8.53, df = 8, p > .05).

Survivorship to 1 year of age varied significantly as a function of standard length at weaning (chi-square = 12.9, df = 8, p < .05)—weanlings with the smallest standard lengths had the lowest survivorship—but this effect did not hold for survival to age 2 (fig. 7.5). Survivorship did not vary significantly with condition index.

Reasoning that high weight or great size might be advantageous in a poor year, we examined separately the year 1986, the only year in the weighed weanling sample that physical oceanographers categorize as an El Niño year. During El Niño years, foraging may be more difficult because of lower prey availability (Arntz, Pearcy, and Trillmich 1991). Using the same weight and length classes shown in figures 7.4 and 7.5, survivorship of the 1986 cohort did not vary significantly with any measure of condition.

DISCUSSION

Northern elephant seals exhibit the most common survivorship curve in nature, Type III, which is characterized by a steep decline in survivorship at an early age (Wilson and Bossert 1971). Those that survive the juvenile period have a good chance of reaching maturity. The majority of young northern elephant seals that were born at Año Nuevo during this study (nearly two-thirds of them, on average) did not survive to 1 year of age, and only 20%, on average, lived to age 4. Most of the juvenile mortality to age 1 occurs at sea, and all mortality to age 2, 3, and 4 occurs at sea. White sharks, *Carcharodon carcharias*, and killer whales, *Orcinus orca*, are known predators on northern elephant seals (Ainley et al. 1981; Le Boeuf, Riedman,

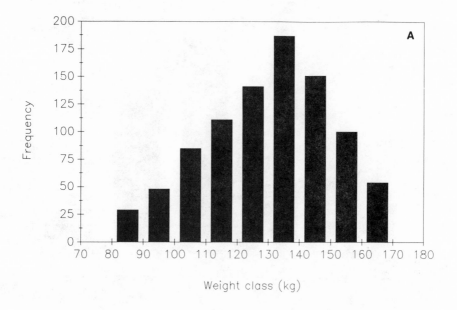

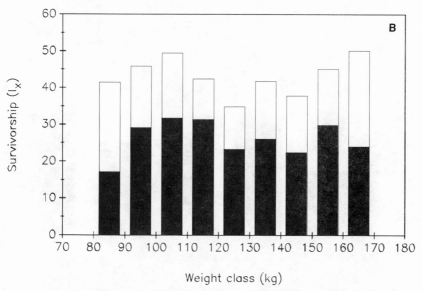

Fig. 7.4. *(A)* Frequency distribution of 734 northern elephant seal pups by weight class at weaning. *(B)* Survivorship of northern elephant seal juveniles to age 1 (open bars) and to age 2 (closed bars) as a function of their weight class at weaning. The sample is taken from Año Nuevo, California, during the years 1978 and 1984–1988.

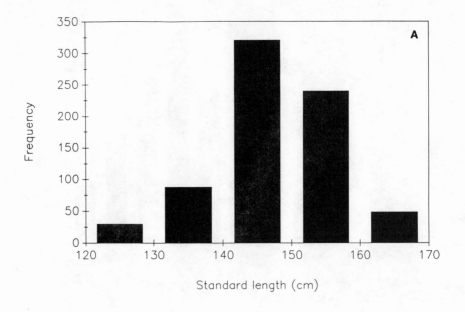

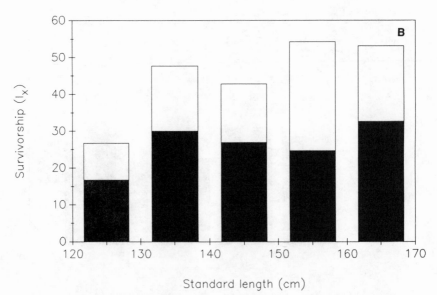

Fig. 7.5. *(A)* Frequency distribution of 734 northern elephant seal pups by standard length class at weaning. *(B)* Survivorship of juveniles to age 1 (open bars) and to age 2 (closed bars) as a function of standard length at weaning.

and Keyes 1982; M. Pierson, pers. comm.), but the degree to which they account for the observed mortality rates is not clear.

In any case, the high juvenile mortality rate at Año Nuevo does not appear to be due to condition at weaning, insofar as condition is reflected by mass at weaning. Weaning weight (above 80 kg) was not correlated with survival to age 1 or to age 2. The fattest were not the fittest (see Sinervo et al. 1992). This is surprising given the large body of information on the importance of parental investment in enhancing individual reproductive success through the production of progeny (e.g., Clutton-Brock 1991). It remains to be determined whether mass at weaning is correlated with reproductive success.

Like other northern and southern elephant seal colonies (Huber, Beckham, and Nisbet 1991; Hindell 1990), cohort variation in juvenile survivorship at Año Nuevo was great, especially to age 1. Much of this variability was due to the effect of storms that occurred during the peak pupping period and caused high pup mortality on the rookery. A decrease in survivorship from age 1 to age 4 was evident over the course of the study period. This may have been due to increasing competition as the population grew or to interactions with fisheries, for example, high seas drift net fisheries in foraging areas. The cause of this decline may become clear as we begin to accumulate knowledge of the migratory paths and foraging areas of juveniles (Le Boeuf, this volume).

The juvenile mortality rate of Año Nuevo-born elephant seals is high relative to other well-studied large mammals such as Dall mountain sheep, *Ovis dalli* (Deevey 1947), and red deer, *Cervus elephus* (Clutton-Brock, Albon, and Guinness 1988), but strikingly similar to the low rates of male, relative to female, African lions, *Panther leo*, from the Serengeti population (Packer et al. 1988) and male vervet monkeys, *Cercopithecus aethiops* (Cheney et al. 1988).

Juvenile survivorship of northern elephant seals from the Año Nuevo colony is significantly lower than that of southern elephant seals from South Georgia, a large colony that was stable in numbers during the period 1951–1985 (Laws, this volume). T. S. McCann (1985) revised the life tables of R. M. Laws (1960) and estimated survivorship to age 1 as 60%, to age 2 as 51%, to age 3 as 43.5%, and to age 4 as 37.2% (we combined the sexes in his life tables for comparability). In comparison with this Southern Hemisphere rookery, juvenile survivorship at Año Nuevo was 42% lower to age 1 and 52% lower to age 4. Estimates of juvenile survivorship at Marion Island are similar to those at South Georgia, despite declines in number at the rate of 4.5% per year during the period 1974–1989 (Bester and Wilkinson, this volume).

Juvenile survivorship rates from Año Nuevo more closely resemble those of the declining southern elephant seal colony at Macquarie Island, studied by R. Carrick and S. E. Ingham (1962) and M. A. Hindell (1990). Hindell (1991, this volume) estimated first-year survival at Macquarie Island during the 1950s as 44% (both sexes combined); however, from 1960 to 1965, first-year survival declined dramatically to 2%. Life table estimates of survival to each of the first four years of life (1_x), based on the entire study period, are 0.350, 0.298, 0.232, and 0.178, respectively. These values are similar to those reported in table 7.1 for Año Nuevo: 0.368, 0.263, 0.194, and 0.163.

These results are paradoxical, or at least do not appear to have a unitary explanation. Despite equally high juvenile survivorship rates, the Año Nuevo colony is increasing in number and the Macquarie Island colony is declining in number. Why is Marion Island declining in number despite a high juvenile survivorship rate? This state of affairs is not easily explained given current information. However, the following information is important for sorting out these incongruencies.

1. Año Nuevo may not be the ideal representative of an expanding colony for comparison with stable or declining colonies. The increase in the Año Nuevo colony, and other colonies like the Farallons at the northern boundary of the species' breeding range, is due mainly to dispersion and immigration from large rookeries in southern California, especially San Miguel Island; internal recruitment, alone, would lead to a decline in colony numbers (Le Boeuf and Reiter 1988; Huber, Beckham, and Nisbet 1991). San Miguel Island accounts for most of the growth of the entire northern elephant seal population (Stewart et al., this volume) and hence may best represent an expanding population. Presumably, juvenile survival rates at San Miguel Island are higher than those at Año Nuevo and perhaps even higher than those at South Georgia. Unfortunately, there are no data on juvenile survivorship for this rookery.

2. Considerable mixing occurs between colonies in the Northern Hemisphere. Consequently, the factors that make for growth of northern elephant seal colonies are more difficult to assess than those of southern elephant seal colonies, where immigration is rare or nonexistent (Burton 1985; Bester 1989; Gales, Adams, and Burton 1989; Hindell 1990) and growth depends ultimately on internal recruitment. For example, during the period 1969–1976, there was considerable dispersion of northern elephant seals among the seven extant rookeries (Bonnell et al. 1979). Movement was primarily in the northward direction and most prevalent during the first year of life. The majority of the movements represented permanent immigration. Southern California rookeries received immigrants from Mexican rookeries and sent out immigrants to northern California rookeries. There was

bidirectional exchange between rookeries separated by short distances, such as San Miguel and San Nicolas islands in southern California and Año Nuevo and the Farallons in central California.

3. Differences in adult female mortality might partially explain the different population trajectories of the expanding northern elephant seal population as compared to stable or declining colonies in the Southern Hemisphere. Adult female mortality is apparently lower at northern elephant seal rookeries such as Año Nuevo and the Farallons (Le Boeuf and Reiter 1988; Huber et al. 1991) than at the southern elephant seal rookery at Marion Island, where it is concluded that the colony decline is due mainly to high adult female mortality (Bester and Wilkinson, this volume). Natality of adult females does not appear to account for differences in growth of colonies since these rates are uniformly high at colonies in both hemispheres; that is, natality exceeds 85% of adult females (McCann 1985; Le Boeuf and Reiter 1988; Huber et al. 1991; Bester and Wilkinson, this volume). Longevity of females seems to be substantially greater at Año Nuevo than at Macquarie, but the data are from different eras and are not a fair comparison.

4. Differences in methodology may cause substantial variation in estimates of juvenile survivorship rates. For example, survivorship rates at Macquarie Island were determined by monitoring animals branded at weaning, while those at Año Nuevo and Marion Island were based on recovery of animals marked with cattle ear tags. Branding yields a permanent mark; tags are impermanent, and some of them are lost. When tags are used, the estimate of juvenile survivorship is affected by the estimate of tag loss. Additionally, tags are harder to see and read than brand marks. Consequently, tag loss is usually greater than estimated, leading to an underestimate of survivorship. That is, the survivorship rates based on tag resight data we have presented for the Año Nuevo colony are probably minimum estimates.

The pup mortality rate on the rookery before weaning, when the animals are marked, is another variable that affects estimates of juvenile survivorship. Deaths on the rookery are a component of the initial sample size. Neonate deaths on small rookeries like Año Nuevo can be counted directly or calculated with reasonable confidence from censuses of suckling and weaned pups. The mean preweaning pup mortality rate at Año Nuevo during the present study was 24.4 ± 10.7% of pups born. On large rookeries such as Macquarie Island, the preweaning pup mortality rate is assumed to be the rate observed in selected harems amenable to censusing. This rate was assumed to be 4.5% of pups born for South Georgia (McCann 1985) and 5% of pups born for Macquarie Island (Hindell and Burton 1987). To what extent these different methodologies explain the wide disparity in pre-weaning pup mortality rates is not clear. A similar statement could be made about search effort, which, necessarily, varies with the size, terrain, and location of rookeries.

In summary, one can obtain reasonably accurate estimates of juvenile survivorship from small, expanding northern elephant seal colonies, such as Año Nuevo, but the extent to which they elucidate the role of juvenile survivorship in declining colonies is unclear. Indeed, Año Nuevo would be declining at a similar rate as Macquarie Island were it not for the influx of animals from San Miguel Island. It may be more important to document juvenile survivorship at San Miguel Island because its growth rate drives the growth of the population. But the task is made difficult by the sheer size of the colony and the need to estimate immigration and emigration rates.

Indeed, the most appropriate comparison, if not the easiest, is to compare the entire northern elephant seal population with that of either of the three main southern elephant seal populations defined by Laws (1960) as the South Georgia stock, the Kerguelen stock, and the Macquarie Island stock (see fig. 3.1, chap. 3). Like the northern elephant seal population, each southern stock is geographically isolated; animal movements between stocks are rare, and gene flow is limited (Gales, Adams, and Burton 1989). Animal movements within each population have an important effect on juvenile survivorship and female reproductive success and, ultimately, on population regulation.

REFERENCES

Ainley, D. G., C. S. Strong, H. P. Huber, T. J. Lewis, and S. H. Morrell. 1981. Predation by sharks on pinnipeds at the Farallon Islands. *Fishery Bulletin, U.S.* 78: 941–945.

Arntz, W., W. G. Pearcy, and F. Trillmich. 1991. Biological consequences of the 1982–83 El Niño in the Eastern Pacific. In *Pinnipeds and El Niño: Responses to Environmental Stress*, ed. F. Trillmich and K. Ono, 22–24. Berlin: Springer Verlag.

Bester, M. N. 1989. Movements of southern elephant seals and subantarctic fur seals in relation to Marion Island. *Marine Mammal Science* 5: 257–265.

Bonnell, M. L., B. J. Le Boeuf, M. O. Pierson, D. H. Dettman, and G. D. Farrens. 1979. *Summary Report, 1975–1978. Marine Mammal and Seabird Surveys of the Southern California Bight Area. III. Pinnipeds.* Bureau of Land Management, Department of the Interior, Contract AA550-CT7-36.

Burton, H. R. 1985. Tagging studies of male southern elephant seals (*Mirounga leonina* L.) in the Vestfold Hills area, Antarctica, and some aspects of their behavior. In *Sea Mammals of South Latitudes: Proceedings of a Symposium of the 52d ANZAAS Congress in Sydney—May 1982*, ed. L. K. Ling and M. M. Bryden, 19–30. Northfield: South Australian Museum.

Carrick, R., and S. E. Ingham. 1962. Studies on the southern elephant seal, *Mirounga leonina* (L.). V. Population dynamics and utilization. *CSIRO Wildlife Research* 7: 198–206.

Cheney, D. L., R. M. Seyfarth, S. J. Andelman, and P. C. Lee. 1988. Reproductive success in vervet monkeys. In *Reproductive Success*, ed. T. H. Clutton-Brock, 384–402. Chicago: University of Chicago Press.

Clutton-Brock, T. H. 1991. *The Evolution of Parental Care*. Princeton: Princeton University Press.

Clutton-Brock, T. H., S. D. Albon, and F. E. Guinness. 1988. Reproductive success in male and female red deer. In *Reproductive Success*, ed. T. H. Clutton-Brock, 325–343. Chicago: University of Chicago Press.

Deevey, E. S., Jr. 1947. Life tables for natural populations of animals. *Quarterly Review of Biology* 22: 263–314.

Gales, N. J., M. Adams, and H. R. Burton. 1989. Genetic relatedness of two populations of the southern elephant seal, *M. leonina*. *Marine Mammal Science* 5: 57–67.

Hindell, M. A. 1990. Population dynamics and diving-behaviour of a declining population of southern elephant seals. Ph.D. dissertation, University of Queensland, Australia.

———. 1991. Some life history parameters of a declining population of southern elephant seals, *Mirounga leonina*. *Journal of Animal Ecology* 60: 119–134.

Hindell, M. A., and H. R. Burton. 1987. Seasonal haul-out patterns of the southern elephant seal (*Mirounga leonina*) at Macquarie Island. *Journal of Zoology, London* 213: 365–380.

Huber, H. R., C. Beckham, and J. Nisbet. 1991. Effects of the 1982–83 El Niño on northern elephant seals on the South Farallon Islands, California. In *Pinnipeds and El Niño: Responses to Environmental Stress*, ed. F. Trillmich and K.A. Ono, 219–233. Berlin: Spring Verlag.

Huber, H. R., A. C. Rovetta, L. A. Fry, and S. Johnston. 1991. Age-specific natality of northern elephant seals at the South Farallon Islands, California. *Journal of Mammalogy* 72: 525–534.

Laws, R. M. 1960. The southern elephant seal (*Mirounga leonina* Linn.) at South Georgia. *Norsk Hvalfangst-Tidende* 49: 466–476, 520–542.

Le Boeuf, B. J., and K. T. Briggs. 1977. The cost of living in a seal harem. *Mammalia* 41: 167–195.

Le Boeuf, B. J., and R. S. Condit. 1983. The high cost of living on the beach. *Pacific Discovery* 36: 12–14.

Le Boeuf, B. J., and R. S. Peterson. 1969. Social status and mating activity in elephant seals. *Science* 163: 91–93.

Le Boeuf, B. J., and J. Reiter. 1988. Lifetime reproductive success in northern elephant seals. In *Reproductive Success*, ed. T. H. Clutton-Brock, 344–362. Chicago: University of Chicago Press.

———. 1991. Biological effects associated with El Niño Southern Oscillation, 1982–83, on northern elephant seals breeding at Año Nuevo, California. In *Pinnipeds and El Niño: Responses to Environmental Stress*, ed. F. Trillmich and K. Ono, 206–218. Berlin: Springer Verlag.

Le Boeuf, B. J., M. Riedman, and R. S. Keyes. 1982. White shark predation on pinnipeds in California coastal waters. *Fishery Bulletin* 80: 891–895.

McCann, T. S. 1985. Size, status and demography of southern elephant seal (*Mirounga leonina*) populations. In *Sea Mammals of South Latitudes: Proceedings of a Symposium of the 52d ANZAAS Congress in Sydney—May 1982*, ed. J. K. Ling and M. M. Bryden, 1–17. Northfield: South Australian Museum.

Packer, C., L. Herbst, A. E. Pusey, J. D. Bygott, J. P. Hanby, S. J. Cairns, and M. B. Mulder. 1988. Reproductive success of lions. In *Reproductive Success*, ed. T. H. Clutton-Brock, 363–383. Chicago: University of Chicago Press.

Radford, K. W., R. T. Orr, and C. L. Hubbs. 1965. Reestablishment of the northern elephant seal, *Mirounga angustirostris*, off central California. *Proceedings of the California Academy of Sciences* 31: 601–612.

Reiter, J. 1984. Studies of female competition and reproductive success in the northern elephant seal. Ph.D. dissertation, University of California, Santa Cruz.

Reiter, J., N. L. Stinson, and B. J. Le Boeuf. 1978. Northern elephant seal development: The transition from weaning to nutritional independence. *Behavioral Ecology and Sociobiology* 3: 337–367.

Sinervo, B., P. Doughty, R. B. Huey, and K. Zamudio. 1992. Allometric engineering: A causal analysis of natural selection on offspring size. *Science* 258: 1927–1930.

Stearns, S. C. 1976. Life history tactics: A review of ideas. *Quarterly Review of Biology* 51: 3–47.

———. 1980. A new view of life-history evolution. *Oikos* 35: 266–281.

Wilson, E. O., and W. H. Bossert. 1971. *A Primer of Population Biology*. Stamford, Conn.: Sinauer Associates.

EIGHT

Life History Strategies of Female Northern Elephant Seals

William J. Sydeman and Nadav Nur

ABSTRACT. We review the literature on variation in life history traits, evaluate evidence on the cost of reproduction, and investigate the significance and extent of individual differences in reproductive success for female northern elephant seals in California. Studies have demonstrated considerable variation in age of primiparity and natality rates and the consequences of this variation for individual fitness. Age at first breeding varies from 3 to 6, with a mode at age 4, and is influenced by socioecological factors. Natality is related to ecological factors and other life history traits. Studies generally support the concept of the cost of reproduction for female northern elephant seals. Costs of reproduction were expressed as a reduction in survivorship associated with age at first breeding and decreased future fecundity associated with either age at first breeding or natality; ultimate consequences for fitness have yet to be addressed. Future fecundity was negatively influenced by the number of prior reproductive bouts (an estimate of breeding intensity) for females that had attempted reproduction often for their ages. Individual variation in reproductive success was apparent. Some individual females had a history of being more successful than others at weaning their pups. The interactions among female quality and causes and consequences of variation for life history traits are considered. Additional study on natality in relation to fitness will serve to illuminate life history strategies that likely vary with individual quality.

Life history theory assumes that organisms cannot maximize survival and fecundity at the same time, because of constraints that may be physiological or ecological. As a result, higher reproductive effort will entail a cost in terms of subsequent survival or future reproductive success or both (Williams 1966; Stearns 1976; Schaffer and Rosenzweig 1977). The concept of a cost of reproduction has led to many studies on the apparent trade-off between current reproductive effort and the probability of future reproductive success (see reviews in Clutton-Brock 1988 and Nur 1990). Studies of marine mammal life history traits in relation to the cost of reproduction are few because marking and tracking individuals through time is difficult. However, the northern elephant seal, *Mirounga angustirostris*, provides an excellent opportunity to examine life history parameters for pinnipeds because it is possible to track and monitor individuals throughout most of their reproductive lifetimes and to estimate fecundity and survivorship.

Our objectives here are (1) to review the literature on variation in reproductive traits that affect fitness for female northern elephant seals, (2) to evaluate evidence for a cost of reproduction, and (3) to examine variation in the pattern of successes and failures for individual females to determine the significance of factors acting between years within the same individual. The latter analysis is new and based on a long-term study conducted on the Farallon Islands, California; methods and results of this analysis will be detailed later. We focus our review on the adaptive significance of two life history traits: age at first breeding and natality rates. Natality is measured by the proportion of animals giving birth each year and varies with the frequency of intermittent reproduction or "skipping" by individual females. We evaluate life history traits, where possible, in regard to a cost of reproduction, expressed as either a reduction in survivorship or future fecundity; only three papers directly address the issue of a cost of reproduction in female northern elephant seals (Huber 1987; Reiter and Le Boeuf 1991; Sydeman et al. 1991). In reviewing the literature and the cost of reproduction, we reanalyze and reinterpret these original papers.

Life history traits and demographic parameters for northern elephant seals have been studied in detail on the Farallon Islands (38°N, 123°W) and Año Nuevo State Reserve (37°N, 121°W) in central California. Both of these colonies, located 90 km apart, are at the northern extreme of this species' distribution (see Stewart et al., this volume, for a range map). The Farallon Islands (Southeast Farallon Island, or SEFI) support 9 breeding groups or harems. At Año Nuevo (AN), there are about 15 harems on the island and the adjacent mainland. Colony numbers at the Farallones increased through 1983, before reaching a plateau (Huber 1987). The colony numbers continue to increase at AN (Stewart et al., this volume). Field methods were similar at each site and are detailed elsewhere. In our review, we consider both cross-sectional (Reiter, Panken, and Le Boeuf 1981; Huber et al. 1991; Sydeman et al. 1991) and longitudinal (Huber 1987; Le Boeuf and Reiter 1988, Le Boeuf, Condit, and Reiter 1989; Reiter and Le Boeuf 1991) studies from both the Farallon Islands and Año Nuevo.

AGE AT FIRST BREEDING

There is considerable variation in age at first breeding for female northern elephant seals. Modal age at primiparity was 4 years on SEFI (62.6% of

275 females; Sydeman et al. 1991) and AN (54% of 67 females; Le Boeuf and Reiter 1988). Age of primiparity varied from 3 to 8 for SEFI females and from 2 to 6 at AN. At AN, B. J. Le Boeuf and J. Reiter (1988) reported that 36% of the females first gave birth at 3 years of age, while at SEFI only 13% of first-time breeders were 3 years old (Sydeman et al. 1991). These values likely differ because Le Boeuf and Reiter studied the 1973 and 1974 cohorts from ANI, whereas Sydeman et al. reported data on females born from 1974 through 1985. Over the past two decades, each cohort has experienced differing social and ecological conditions that influence immature survival and recruitment probabilities (Huber et al. 1991; Le Boeuf, Morris, and Reiter, this volume). Colony size and competition for breeding space have changed dramatically during this period. H. R. Huber et al. (1991) showed that when SEFI colony growth slowed in the late 1970s and early 1980s, the proportion of 3-year-olds giving birth also decreased. However, Huber et al. also indicated that age at first breeding was delayed for cohorts affected by the 1982–1983 El Niño Southern Oscillation (ENSO). Thus, it is presently unclear whether changes in colony size or the ENSO explains changes in age at primiparity. Nevertheless, in comparison to the data set of Le Boeuf and Reiter, the SEFI data set analyzed by Sydeman et al. (1991) included many cohorts that were potentially influenced by the 1982–1983 ENSO and increased colony size. Consequently, average and modal age at primiparity for females first breeding in the 1980s was later than for females in the 1970s.

INTERMITTENT BREEDING AND NATALITY

We define intermittent breeding, or "skipping," as forgoing pupping in any year during an animal's breeding lifetime, for example, after a female has become parturient but before she dies or ceases to reproduce altogether due to senescence. It is possible that a "skip" could be confused with an animal giving birth at another colony, or giving birth and losing the pup before being observed. Although skipping may be easily confused with poor coverage, we believe that the intensive research programs on northern elephant seals make this scenario unlikely. At present, all seal rookeries in central and southern California are under study. At AN and SEFI, weather permitting, individual subcolonies are visited daily by researchers, tagged cows are recorded, and animals are temporarily marked with hair dye to facilitate individual identification. At Point Reyes, only 25 km from SEFI, coverage is less frequent (about once per week), but for the past five years most tagged animals have been identified (S. G. Allen, pers. comm.). Through a cooperative exchange of information, movement of females between AN and SEFI and between SEFI and Point Reyes has been documented, but in general, most females are extremely site tenacious once they begin pupping at a particular site. Emigration from a site followed by immigration back to the original location has not been observed. Consequently, we believe that case records of intermittent reproduction are real, not errors associated with poor coverage of breeding areas.

On SEFI, animals that forgo breeding in one year and are confirmed alive and breeding in subsequent years (i.e., "skippers") are seen during the winter breeding season (6% of all those observed to breed intermittently) or fall haul-out (15%) period, but most are seen during spring molt or in a subsequent breeding season (79%; Huber et al. 1991). Huber et al. (1991) demonstrated that skipping is fairly regular among multiparous females, although it is apparently more common for young, 4- to 5-year-old, animals (Huber 1987). Natality of females, 5 to 10 years old, averaged 80% among years, indicating that approximately 20% of the SEFI females do not pup in any given year. On AN, however, Le Boeuf and Reiter (1988) reported that age-specific natality varied little after first reproduction and that 97% of the females present on the rookery during the breeding season each year gave birth (compared with 94% on SEFI).

Whether skipping is associated with a life history strategy or a constraint imposed by ecological or social conditions is open to speculation. Intermittent breeding (natality) also varies with ecological effects, such as ENSO, and life history traits such as age at first breeding. Huber et al. (1991) documented that natality rates on SEFI were lowest in the years following the 1982–1983 ENSO. Similar to Huber et al.'s findings on age at first breeding in relation to ENSO, reduced food resources in 1983 may have had a negative effect on female body condition, which, in turn, affected natality rates in 1984 and 1985. Overall natality rates in 1986 recovered to pre-ENSO levels, indicating a short-term effect. Huber (1987) found that females who began breeding at an early age were more likely to skip breeding in subsequent years than females that deferred reproduction.

THE COST OF REPRODUCTION

Survival and Age at First Breeding

Reiter and Le Boeuf (1991) studied survivorship to age 8 of female northern elephant seals that first bred at age 3 (P3 group) versus females that deferred breeding until age 4 (P4 group). Females who bred at age 3 had lower survival to age 4 than females that did not breed at age 3. This result provides strong evidence for the cost of reproduction among the youngest parous animals. Reiter and Le Boeuf attributed this result to the immediate energetic cost of reproduction, which is greater in young primiparous females who are still growing.

Survival of P3 females between the ages of 4 and 8 was lower than the survival of P4 females to age 8 (26.8% vs. 37.0%). Using a linear regression analysis on grouped data, Reiter and Le Boeuf (1991) reported a significant difference in the slopes of these survival curves. For a variety of reasons, however, we believe this approach was questionable. On statistical grounds, the linear regression used did not take into account differences in sample size for each age, heteroscedasticity in survival variance, and lack of independence between observations and assumes that residual errors are normally distributed. On biological grounds, female survival to the year after first breeding was included for the P4 group but not for the P3 group. This is an important distinction because the survival of P3 females to the year following first breeding (78.4%) was actually greater than the survival of P4 females to the subsequent year (74.3%). Omitting these data biases the results in favor of higher survival for the P4 group. Notwithstanding these difficulties, the data presented by Reiter and Le Boeuf are critical to understanding the evolution of life history patterns.

We have reanalyzed the survival data presented by Reiter and Le Boeuf (1991) using a form of survival analysis, the Cox Proportional Hazards Model (CPHM; Kalbfleisch and Prentice 1980). This technique has become very popular among biostatisticians, in part because it does not assume a normal distribution in the response variable (survival), nor is it unduly sensitive to small sample sizes and influential data points. Using the CPHM, we considered survival for P3 females from age 3 to 7 and P4 females from age 4 to 8, a four-year comparison for both groups including the year after first breeding. For these data, the CPHM did not indicate a significant difference (LRS = 2.84, p = .092), although perhaps with additional data the trend would be significant. Other possible comparisons (e.g., age 4 to 8 for both groups) yielded higher p-values (p > .10). We believe that the difference between our conclusions about statistical significance and those of Reiter and Le Boeuf (1991) reflect the fact that soon after first breeding, when sample sizes were large, survivorship to the first year after breeding was actually higher for the P3 group (78.4% for P3 vs. 74.3% for P4). Later in life, when the sample sizes were smaller (and thus standard errors were greater), the pattern reversed; survival of P3 females from age 7 to age 8 was 64%, whereas survival of P4 females from age 7 to age 8 was 70%. Unlike Reiter and Le Boeuf's linear regression on grouped data, the CPHM takes into account variation in sample size that could potentially bias results. We find Reiter and Le Boeuf's results compelling but feel that a larger data set (or different data set) should be analyzed to confirm the trends they report.

Future Fecundity and Breeding Intensity

W. J. Sydeman et al. (1991) investigated the effects of prior reproductive effort on current and future reproductive success for northern elephant seal females. To index previous reproductive effort, the number of prior parturitions in a female's reproductive history was tallied. The number of prior parturitions for each female is equivalent to a female's "experience." It is important to note that experience not only serves as an estimate of yearly accrual of breeding skill but also reflects physical acts of reproduction. As such, experience, as measured, is an index to potentially costly reproductive activities such as gestation and parturition. The effects of "experience" on weaning success were considered, with the potentially confounding effects of female age, using multiple logistic regression.

Weaning success increased asymptotically with increasing female age (fig. 8.1a), but this relationship was clouded by the relationship between the number of times a female had previously pupped, that is, experience, and weaning success (fig. 8.1c). By statistically adjusting for the experience effect, W. J. Sydeman et al. demonstrated that the relationship between female age and weaning success was actually log-linear (fig. 8.1b), indicating that weaning success improved with each increase in female age. The relationship between female age and weaning success indicated that the benefits of increasing age were steadily realized throughout a female's lifetime, a conclusion also reached by Reiter, Panken, and Le Boeuf (1981), although these authors did not control for the confounding effects of experience.

Weaning success also improved with increasing female experience, but this relationship appeared to plateau and decline for extremely experienced animals (fig. 8.1c). Reiter, Panken, and Le Boeuf (1981) also found that weaning success improved with increasing experience for young animals. By statistically correcting for the effect of female age (as discussed above), the downturning in the relationship between experience and weaning success became more pronounced (fig. 8.1d). Thus, the benefit of increasing experience diminished throughout an animal's lifetime, and effects became negative after five prior parturitions. The interaction between female age and female experience, that the effect of experience was dependent on age, was also significant. Another way to view this result can be seen when considering experience effects within age groups. For example, among animals aged 4 and 5, the effect of experience on weaning success was positive (regression coefficient = .714), whereas for females aged 11 through 15, the effect of experience was negative (regression coefficient = −.279).

Results suggest that when females are young, experience is beneficial, whereas late in life (after five previous reproductive bouts), additional experience is detrimental. This result is not counterintuitive if one considers that experience may be an index to costly reproductive activities such as gestation, parturition, and, in some cases, lactation. Thus, the relationship between experience and weaning success demonstrates a cost of prior reproductive effort to future fecundity. However, the negative relationship between experience and weaning success was found only after adjusting for the effect of female age; that manipulation deserves clarification. Biologi-

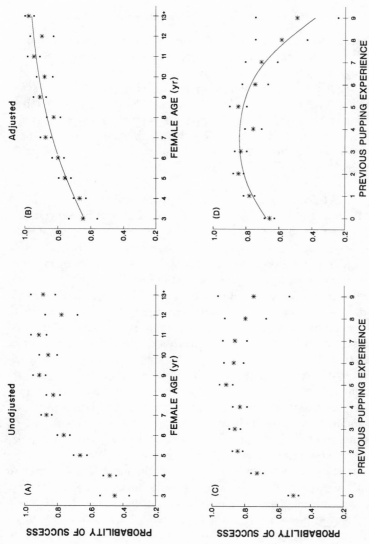

Fig. 8.1. Age-specific weaning success (A) before and (B) after adjustment for previous pupping experience. Experience-specific weaning success (C) before and (D) after adjustment for female age. Data are means ±1 SE. Results of logistic regression analyses are shown (linear in B, quadratic in D). Reprinted, by permission, from Sydeman et al. 1991.

cally, age and experience would be the same measurement if all females began to reproduce at the same age and gave birth yearly. Due to the aforementioned variation in both age at first breeding and natality/intermittent pupping, this was not the case. Experience is actually equal to:

age − age at first breeding − the number of breeding seasons skipped.

Therefore, experience adjusted for female age reflects the effects of variation in age at first breeding and intermittent reproduction. Because experience adjusted for age reflects these life history traits, results are supportive of a trade-off in the life history strategies of female northern elephant seals; females that first breed at too young an age, or too intensely (continuously rather than intermittently), pay a cost in terms of reproductive success late in life. Studies on red deer (Clutton-Brock, Albon, and Guinness 1989) and birds (reviewed by Nur 1990) also demonstrate costs of reproduction expressed as reductions in future fecundity.

Population Dynamics and the Cost of Reproduction

Reiter and Le Boeuf (1991) investigated age-specific weaning success and optimal age at first breeding in relation to rookery density and juvenile survivorship. These data allowed them to examine whether the benefit of breeding at age 3 outweighs the cost, expressed as the decrement in survival. Age-specific weaning success was greatest in low-density rookeries, and density effects were most influential on young animals. Reiter and Le Boeuf used these data, combined with information on age-specific survivorship, to calculate the net reproductive rate and intrinsic rate of increase for populations composed of either P3 or P4 females and to estimate reproductive value of P3 versus P4 females. To model population growth, juvenile survivorship to the age of recruitment was assumed to be either 40% or 80%; results of the model depend heavily on which juvenile survivorship value is used. Model projections indicate that at either low or high density, P4 females maintain greater reproductive value at each age class than P3 females. Estimates of the intrinsic rate of increase for P3 versus P4 populations indicated higher growth rates for the P4 group.

We converted the intrinsic rate of increase to the finite (annual) growth rate (λ) using the formula $\lambda = e^r$, where r = the intrinsic rate of increase (table 8.1). Because elephant seal populations do not grow continuously through time but grow in a discrete manner, we believe that λ provides a more useful and easily interpretable metric than the intrinsic rate of increase (r). The population generated by P3 females in a high-density rookery with low (40%) juvenile survival was decreasing at a rate of 14.8% per annum compared to the P4 population, which declined at 11.4% per annum. The population generated by P3 females in a low-density rookery with 40% juvenile survival was decreasing at 8.1% per annum, compared

TABLE 8.1   The finite growth rate (λ) for hypothetical northern elephant seal populations generated by females primiparous at age 3 (P3) versus age 4 (P4). Growth rates are presented for high- and low-density breeding conditions and in relation to low and high levels of juvenile survival.

Juvenile survival	Breeding density	P3	P4	% Difference
Low (40%)	high	−0.852	−0.886	3.4
	low	−0.919	−0.941	2.2
High (80%)	high	−0.941	−0.966	2.5
	low	1.038	1.040	0.2

NOTE: Values are calculated from table 2 of Reiter and Le Boeuf 1991.

to a decline of 5.9% per annum for the P4 population. Except for conditions of low density and extremely high (80%) juvenile survival, model populations were declining. Declining model populations are consistent with field data from AN and SEFI indicating that immigration maintains current populations and that internal recruitment is very low (Le Boeuf and Reiter 1988; Huber et al. 1991). For example, Huber et al. estimate that only 17% of the females born on SEFI return there to breed.

Reiter and Le Boeuf (1991) conclude that at very low densities, for example, for an incipient colony where weaning success and juvenile survival is high, breeding at age 3 may be optimal, but under other circumstances, P4 is optimal. Northern elephant seals at AN have very low success at rearing pups as 3-year-olds, and thus it is not surprising that the benefit of breeding at age 3 did not outweigh the cost. This conclusion is consistent with field data from both colonies, where the modal age at first reproduction is 4 years. Also, if P3 is selected against, the frequency of 3-year-old breeders should be declining (if age of primiparity is heritable). Notably, on SEFI, fewer than 15% of the animals are primiparous at age 3 (Huber et al. 1991; Sydeman et al. 1991). We would expect a similar decline in P3 females at AN for recent years. Finally, Reiter and Le Boeuf's analysis points to the adaptive value of dispersal and immigration to small or newly formed colonies when high-density conditions prevail at a natal rookery.

SIGNIFICANCE OF INDIVIDUAL VARIATION FOR LIFE HISTORY PATTERNS

Understanding the evolution of reproductive strategies requires understanding *variance*, in addition to the *mean*, in reproductive success. Hence, an analysis of individual differences in weaning success is called for (Gillespie 1977; Seger and Brockmann 1988). Here we address the possibility of differences among females as well as differences within individuals between years. To this end, in addition to an analysis of mean weaning success, we provide an analysis of the distribution of successes and failures for each individual. Many investigators have looked at the total number of young weaned (or for birds, fledged), but no one, to our knowledge, has examined the pattern of successes and failures during an individual's lifetime.

First, as an introduction to this analysis, we consider what proportion of the variation in weaning success is attributable to female age and breeding experience and what proportion to differences among individuals. We restricted our data set to those females with a history of at least three years of reproductive success. Without controlling for age or experience, we calculated that the "individual effect" accounted for approximately 26% of the variation in weaning success (F = 1.47, df = 120, 500, p = .003, r^2 = .261). Controlling for age and experience, we found that the individual effect accounted for approximately 24% of the variation in weaning success (F = 1.49, df = 120, 495, p = .0019, partial r^2 = .235). Age and experience, combined, explained an additional 12% of the variation in weaning success (F = 17.3, df = 4, 493, p < .001, partial r^2 = .115). This analysis indicates that some females are consistently better and others consistently worse at raising their pups to weaning age. Additionally, the effect of differences among individual females on weaning success is independent of the effects of age and experience, and, notably, between-individual differences in weaning success had greater predictive powers than age or experience in determining reproductive success.

Next, we compared the distribution of successes and failures with a null model that assumes that all individuals of a given "longevity" class have the same probability of successful reproduction. Biologically, this implies that there is no variation (i.e., homogeneity) among individuals in the probability of success in a given year. Differences among individuals (i.e., heterogeneity) would be expressed as excess numbers of individuals that are either very successful or very unsuccessful in rearing pups to weaning age. Second, the null model assumes that there is independence among the reproductive attempts within an individual. Biologically, this implies that success or failure in one attempt does not affect success or failure in a following attempt. If there is within-individual dependence in pup-rearing ability, it could be of a *compensating* or *depensating* nature. For example, T. H. Clutton-Brock et al. reported that individual red deer, *Cervus elaphus*, hinds that were barren or unsuccessful at rearing a calf in one year were more likely to successfully rear a calf in the following year. Interannual dependence of this nature is compensating and would lead to less variation in the distribution of successful and unsuccessful weaning among females than expected under a null model. Alternatively, Le Boeuf, Condit, and Reiter (1989) found that female northern elephant seals that failed to wean a pup

TABLE 8.2 Matrix of the number of successful reproductive attempts by longevity class for female northern elephant seals on SEFI, California, 1978–1989. Longevity class is an index to longevity as it represents the number of records per individual in our SEFI data set for which we have information on weaning success.

Longevity class	Number of successful attempts										
	0	1	2	3	4	5	6	7	8	9	10
1	53	58									
2	5	19	19								
3	1	6	7	11							
4	0	4	4	13	9						
5	0	1	4	5	7	9					
6	0	0	0	0	4	7	3				
7	0	0	0	0	2	3	4	1			
8	0	0	0	0	0	0	2	0	6		
9	0	0	0	0	0	0	1	2	2	0	
10	0	0	0	0	0	0	1	0	1	1	0

in one year were *less* likely to rear a pup in the next year. Interannual dependence of this nature is depensating and would lead to excess variation in the distribution of successful and unsuccessful weaning among females than expected under the null model. Thus two factors, heterogeneity among individuals and depensation, lead to excess variance, while only one factor, compensation, leads to a deficit of variance relative to that expected in a history of successes and failure of a given animal. Depensation is essentially a positive feedback mechanism between successive breeding attempts, while compensation represents a negative feedback mechanism.

We first revisit the question, are there differences among individuals in the likelihood of success? We use the Farallon distributional data set to answer it. We summarized successes and failures in relation to the number of breeding records per female from 1978 to 1989 in table 8.2. Each individual female appears only once in the summary. The number of records per individual is an index of longevity because it represents the number of records per individual for which we have information on successful or unsuccessful weaning, not the true reproductive life span of females. Only for longevity classes 2, 3, 4, and 5 (table 8.3) do we have adequate data to test individual females by comparing observed and expected numbers of successes and failures. For longevity class 2, we found that the variation in weaning success per individual was equal to that predicted by the null model. For longevity classes 3, 4, and 5 considered separately, we found that the observed distribution of success and failure tended to be slightly excessive

TABLE 8.3   Stratified longitudinal analyses of weaning success for female northern elephant seals with 3, 4, and 5 years of reproductive history on SEFI, California.

	Number of successful reproductive attempts					
	0	1	2	3	4	5
Longevity class = 3 ($\chi^2 = 2.46$, df = 2, NS)						
Observed number	1	6	7	11	—	—
Expected number	0.63	4.36	10.99	8.82	—	—
Longevity class = 4 ($\chi^2 = 3.53$, df = 3, NS)						
Observed number	0	4	4	13	9	—
Expected number	0.172	1.809	7.156	12.576	8.288	—
Longevity class = 5 ($\chi^2 = 5.30$, df = 3, p = .151)						
Observed number	0	1	4	5	7	9
Expected number	0.19	0.307	1.991	6.452	10.455	6.777

TABLE 8.4   Longitudinal analysis of female northern elephant seals: Mean and variance in weaning success. Expected variance under the assumption of a binomial null model (see text).

Longevity class	N	Mean	Observed variance	Expected variance	% Difference
1	111	.52	0.250	0.250	—
2	43	.66	0.452	0.447	+1
3	25	.71	0.826	0.622	+25
4	30	.73	0.957	0.798	+16
5	26	.75	1.428	0.947	+45
6	14	.82	0.495	0.880	−44
7	10	.77	0.840	1.234	−32
8	8	.94	0.750	0.469	+38
9	5	.80	0.560	1.440	−61
10	3	.77	1.555	1.789	−13

relative to expected values $(0.20 > p > .10)$. By pooling the data for females with three, four, and five years of reproductive history, we found significant heterogeneity among individuals $(\chi^2 = 7.66$, df = 3, p = .054). There were more animals that succeeded completely or that failed totally than expected by the null model (fig. 8.2).

To examine the significance of heterogeneity among individuals, compensation, or depensation, we calculated the observed and expected *variance* in weaning success for each longevity class assuming homogeneity among

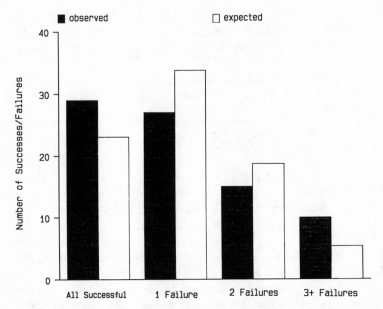

Fig. 8.2. The observed and expected number of successes and failures for female northern elephant seals on SEFI, California, for which we had 3 to 5 years of reproductive history. For example, "one failure" represents females who were successful in weaning a pup 2 out of 3, 3 out of 4, and 4 out of 5 times.

individuals and no compensation or depensation within individuals (table 8.4). The final column indicates the direction and magnitude of the deviation between the observed and expected variance. Up to longevity class 5, the difference between observed and expected variance is positive, whereas after class 5, the difference is negative. The change in the sign indicates excess variation, relative to the null model, for females with five or fewer years of reproductive history and minimal variation for animals with six or more years of information. The negative discrepancy between expected versus observed variance for animals appearing often in the data set can only be explained by dependence of successes and failures within an individual that is of a compensating nature; that is, failure in one year promotes success in the next and vice versa (Clutton-Brock, Albon, and Guinness 1989). However, the positive discrepancy in the observed versus the expected variance for animals appearing relatively infrequently in the data set could be attributed to either heterogeneity among individuals or depensation.

One hypothesis, that excess variation among "longevity" classes 3, 4, and 5 is attributable to heterogeneity among individuals, predicts that "poor quality" individuals, that is, those most likely to fail, and those that contribute much of the excess variation in longevity classes 3, 4, and 5, will

TABLE 8.5 Reproductive success of "short-lived" females (1–5 breeding seasons) versus "long-lived" females (6–10 breeding seasons), controlling for breeding experience, for 3rd, 4th, and 5th breeding seasons.

Weaning success	Mean[a]	SE	N
"Short-lived" females	0.894	0.027	169
"Long-lived" females	0.863	0.032	117

[a] Difference in means NS (p > .6, ANOVA).

drop out of the data set late in life. If so, longevity should be significantly correlated with weaning success. Indeed, weaning success increased significantly with increasing "longevity" (table 8.4; logistic regression analysis; $LRS_{linear} = 33.63$, $p < .001$; $LRS_{quadratic} = 3.92$, $p = .048$); however, after controlling for age, the effect of "longevity" was no longer significant ($LRS = 1.57$, $p = .210$). The same negative result was obtained after controlling for experience ($LRS_{longevity} = 0.74$, $p = .390$). Therefore, the longevity effect can be attributed to age and/or experience; longevity adds little predictive power beyond the effect of age or experience in determining weaning success. Supporting this result is a comparison of the reproductive success of "short-lived" females and "long-lived" females during second and third breeding attempts only. Females destined to be long-lived were no more successful as young breeders than females destined to be short-lived (table 8.5). Together, these results provide evidence against the prediction that poor quality individuals are less likely to survive to the oldest longevity classes in our data set. Instead, these results support the concept of depensation, rather than heterogeneity among individuals, as an explanation for the excess variation observed in weaning success for animals appearing five or fewer times in the Farallon data set. For northern elephant seal females, we conclude that (1) early in life there appears to be depensating dependence between reproductive attempts (as suggested by Le Boeuf, Condit, and Reiter 1989), but later in life there is compensating dependence (as in Clutton-Brock, Albon, and Guinness 1989), and (2) heterogeneity among individuals is not related to longevity.

CONCLUSION

In the literature, there is considerable evidence of trade-offs in the life history of female northern elephant seals. These trade-offs involve both future fecundity and survival. In terms of future fecundity, the cost of reproduction is related to continuous rather than intermittent reproduction and reproduction at a young age versus deferred breeding (Sydeman et al. 1991). Conceptually, the cost may be detailed as follows: late in life, individuals who have bred often for their age, that is, those that show high levels of breeding intensity, have poorer reproductive success than those who have bred fewer times for their age. We believe that age of primiparity plays a more significant role than continuous reproduction in determining future fecundity, but additional research is needed to address this issue. The exact mechanism associated with reduced weaning success for elderly females is unknown, but a decrease in aggressiveness or dominance may play a role in declining pup survival. In terms of survival, the cost of reproduction appears to be related to age at first breeding (Reiter and Le Boeuf 1991) and possibly natality (Huber 1987). Maternal survival was inversely related to age of primiparity (Reiter and Le Boeuf 1991), although we have remarked that additional data are needed to confirm statistical significance of differences in survival rates. Calculations of reproductive value and the intrinsic rate of population increase (Reiter and Le Boeuf 1991) and estimates of annual population growth rates (this chap.) indicate that deferred breeding to age 4 is optimal. Based on this model, P3 females should be decreasing in frequency in the California population, a suggestion consistent with field observations from the Farallones (Huber et al. 1991). However, due to the frequency of ENSO events in the 1980s and changes in population density, arguments concerning the match between optimal age of primiparity and field observations remain speculative.

We also investigated the potential "individual female" effect on weaning success directly in this chapter. Results indicate that some females are better and others are worse at raising their pups to weaning age. In fact, the individual female effect accounted for more of the variation in mean weaning success than maternal age and breeding experience combined. The individual female effect may be due to choice of parturition site or correlates of phenotypic variation, such as body size. Excess variation in the distribution of successes and failures among relatively short-lived females on the Farallones was probably related to depensation rather than heterogeneity among individuals, although direct between-year, within-individual comparisons are needed to address this hypothesis. The concept of depensation in northern elephant seals is supported by the findings of Le Boeuf, Condit, and Reiter that resightings of females that had successfully reared a pup were significantly greater than resightings of females that failed to rear a pup and that females that had experienced reproductive failure in one year were no more likely to be successful in the subsequent year. The deficit in variation observed for females appearing six or more times in the Farallon data set can only be explained by compensation.

The individual female effects described herein also have application to life history theory and estimates of cost of reproduction for this species. Models of alternative reproductive strategies need to be viewed in the context of differences among individuals because such variation may affect estimates of the cost of reproduction (Nur 1988; Partridge 1989). For example, if females of superior quality show higher reproductive rates than females of inferior quality, and if reproductive effort is optimized in relation to quality, then a positive correlation between maternal survival and weaning success could arise despite there being a survival cost associated with increased reproductive effort (Nur 1988). If individual factors influence weaning success, it is likely that these same factors will influence female survival. The result of such confounding is that a survival cost associated with reproductive effort may be masked by differences in female quality. Indeed, the lack of statistical significance in survival rates between P3 and P4 females, as determined by us using the CPHM, may be explained by this type of confounding by individual variation.

Individual female northern elephant seals may be following mixed reproductive strategies (Maynard Smith 1982) governed by individual quality. Breeding late in life or intermittently may be part of a life history characterized by low reproductive effort and longevity. Conversely, early breeding and continuous reproduction may characterize a life history of high reproductive effort at the expense of longevity. Direct modeling of fitness curves for these or other alternative strategies would offer insight. With continued investigations on marked populations of northern elephant seals, the causes and consequences of variation in life history patterns in a varying environment may be better understood.

ACKNOWLEDGMENTS

We thank the Farallon Patrol of the Point Reyes Bird Observatory for providing transportation to and from the Farallon Islands. We acknowledge and thank the staff of San Francisco Bay National Wildlife Refuge–U.S. Fish and Wildlife Service, managers of the Farallon Islands, for their help and encouragement in conducting these studies. Harriet Huber and Steve Emslie deserve special recognition for their supervision of fieldwork and data management and David Ainley, for his interest and foresight in establishing the Farallon project. Numerous field biologists participated in this study; to all, we offer our sincere appreciation. Results and interpretations of this study were sharpened by critical reviews and discussions with Burney Le Boeuf, Joanne Reiter, Jim Estes, Roger Gentry, and Gerry Kooyman. This is PRBO contribution no. 529.

REFERENCES

Clutton-Brock, T., ed. 1988. *Reproductive Success: Studies of Individual Variation in Contrasting Breeding Systems.* Chicago: University of Chicago Press.

Clutton-Brock, T. H., S. D. Albon, and F. E. Guinness. 1989. Fitness costs of gestation and lactation in wild mammals. *Nature* 337: 260–262.

Gillespie, J. H. 1977. Natural selection for variances in offspring numbers: A new evolutionary principle. *American Naturalist* 111: 1010–1014.

Glantz, M. H., and J. D. Thompson, eds. 1981. *Resource Management and Environmental Uncertainty: Lessons from Coastal Upwelling Fisheries.* New York: John Wiley and Sons.

Huber, H. R. 1987. Natality and weaning success in relation to age of first reproduction in northern elephant seals. *Canadian Journal of Zoology* 65: 1131–1316.

Huber, H. R., A. C. Rovetta, L. A. Fry, and S. Johnston. 1991. Age-specific natality of northern elephant seals at the South Farallon Islands, California. *Journal of Mammalogy* 72: 525–534.

Kalbfleisch, J., and R. L. Prentice. 1980. *The Statistical Analysis of Failure Time Data.* New York: John Wiley and Sons.

Le Boeuf, B. J., and K. T. Briggs. 1977. The cost of living in a seal harem. *Mammalia* 41: 167–195.

Le Boeuf, B. J., R. Condit, and J. Reiter. 1989. Parental investment and secondary sex ratio in northern elephant seals. *Behavioral Ecology and Sociobiology* 25: 109–117.

Le Boeuf, B. J., and J. Reiter. 1988. Lifetime reproductive success in northern elephant seals. In *Reproductive Success*, ed. T. Clutton-Brock, 334–362. Chicago: University of Chicago Press.

Maynard Smith, J. 1982. *Evolution and the Theory of Games.* Cambridge: Cambridge University Press.

Nur, N. 1988. The cost of reproduction in birds: An examination of the evidence. *Ardea* 76: 155–168.

———. 1990. The cost of reproduction in birds: Evidence from manipulative and non-manipulative studies. In *Population Biology of Passerine Birds: An Integrated Approach*, ed. J. Blondel, A. Gosler, J. D. Lebreton, and R. McCleery, 281–296. Heidelberg: Springer Verlag.

Partridge, L. 1989. Lifetime reproductive success and life-history evolution. In *Lifetime Reproductive Success in Birds*, ed. I. Newton, 21–440. London: Academic Press.

Reiter, J., and B. J. Le Boeuf. 1991. Life history consequences of variation in age of primiparity in northern elephant seals. *Behavioral Ecology and Sociobiology* 28: 153–160.

Reiter, J., K. J. Panken, and B. J. Le Boeuf. 1981. Female competition and reproductive success in northern elephant seals. *Animal Behavior* 29: 670–687.

Schaffer, W. M., and M. L. Rosenzweig. 1977. Selection for optimal life histories. II. Multiple equilibria and the evolution of alternative reproductive strategies. *Ecology* 58: 60–72.

Seger, J., and H. J. Brockmann. 1988. What is bet-hedging? In *Oxford Surveys in Evolutionary Biology*, vol. 4, ed. P. H. Harvey and L. Partridge, 182–211. Oxford: Oxford University Press.

Stearns, S. C. 1976. Life-history tactics: A review of the ideas. *Quarterly Review of Biology* 51: 3–47.

Sydeman, W. J., H. R. Huber, S. D. Emslie, C. A. Ribic, and N. Nur. 1991. Age-specific weaning success of northern elephant seals in relation to previous breeding experience. *Ecology* 72: 2204–2217.

Williams, G. C. 1966. Natural selection, costs of reproduction, and a refinement of Lack's principle. *American Naturalist* 100: 687–690.

NINE

Sexual Selection and Growth in Male Northern Elephant Seals

Walter L. Clinton

ABSTRACT. I studied the interactions between body size, growth, and life history in male northern elephant seals, *Mirounga angustirostris*, by constructing a growth curve and comparing the characteristics of growth to the pattern of male life history. The aim of this study was to determine the relationship between growth rate and age-specific male mortality rates. Four related exponential functions and a two-component logistic function were fitted to age-length data by nonlinear least squares regression. The two-component logistic curve fit the age and length data better than the best fitting exponential curve, the Richards function; however, both functions indicated a peak in growth rate around 3 to 5 years of age. Growth rates were high from 2 to 6 years of age, with relative growth rates of about 10% per year. Standard length of males increased each year of life until 9 years of age, and measurements of actual yearly growth indicated that after physical maturity, males stopped growing.

The peak in growth rate around 3 to 5 years of age and the end of growth by 9 years of age were related to important characteristics in the life history of males. The growth spurt may be associated with delayed maturity and a consequence of sexual selection for large body size. The timing of the growth spurt coincided with the lowest age-specific mortality rates over the life span of males and with the ages when increased growth rate was matched by longer periods of foraging at sea. Thus, the ages when males were exposed to the survival disadvantages of high growth rates were actually a period of low mortality. The high mortality among males occurred at 9 to 10 years of age after growth has ended and appeared to be associated with competition for mates.

Growth rates and body size are correlated with life history parameters such as age at maturity and age-specific fertility (Calder 1984). Where sexual selection has produced sexual dimorphism with males larger in size, male life history characteristics have also changed: males mature more slowly than females, delay breeding to older ages, and die at a higher age-specific rate (Trivers 1985). To study the interactions between body size, growth, and life history, I collected data on size and age of male northern elephant seals, *M. angustirostris*, and constructed a growth curve to compare the characteristics of growth with those of male life history (Clinton 1990; Clinton and Le Boeuf 1993).

Modifications of pinniped male growth patterns due to sexual selection have been analyzed mainly by comparison with the growth pattern of females (Laws 1953; Scheffer and Wilke 1953; Carrick, Csordas, and Ingham 1962; Bryden 1972; Innes, Stewart, and Lavigne 1981). These studies show that males undergo an adolescent growth spurt, a type of pattern that was first described in humans (Bogin 1988; Harrison et al. 1988). In southern elephant seals, *M. leonina*, growth curves differ early between the sexes (Carrick, Csordas, and Ingham 1962). Males are larger than females after 1 year of age, and from 2 to 4 years of age, males grow at a faster rate than females. Growth curves indicate the female growth rate steadily decreases after 1 year of age. After a period of decreasing growth rate, the males' growth rate increased at 6 to 7 years of age, which is one to two years after puberty. This growth pattern in pinniped males was also found in northern fur seals, *Callorhinus ursinus*, in which the growth rate accelerated one to two years after puberty (Scheffer and Wilke 1953).

In polygynous species in which males are larger than females, the high male growth rates seen near puberty may be associated with increased male mortality rates (Ralls, Brownell, and Ballou 1980). The aim of this study was to determine the relationship between growth rate and age-specific male mortality in northern elephant seals. First, I review male life history to establish the age-specific pattern of male mortality; second, I present the growth curve for males; and finally, to consider whether increased growth rate is associated with increased mortality, I examine the concurrence between increased growth rate and changes in age-specific male mortality.

REVIEW OF MALE LIFE HISTORY

The life history of male northern elephant seals has been strongly shaped by sexual selection (Le Boeuf and Reiter 1988; Clinton 1990; Clinton and Le Boeuf 1993). Compared to females, males delay the age of first breeding and live shorter lives. Most important, male reproductive success depends on the ability to compete for matings. Males that obtain the most matings presumably incur the greater risks and expend more energy during breeding competition (Deutsch, Haley, and Le Boeuf 1990); thus, increases in male reproductive success should be associated with decreased survival and decreased future mating success.

The life table indicates the key periods during the life of males and reveals the effects of sexual selection (table 9.1) (Clinton 1990; Clinton and

TABLE 9.1   Life table for male northern elephant seals at Año Nuevo.

Age	Survivorship (l_x)	Mortality (q_x)	Fecundity (m_x)	Reproductive value	Life expectancy
W	1.000	0.510	0	0.503	2.848
1	0.490	0.193	0	1.026	4.287
2	0.396	0.180	0	1.272	4.193
3	0.324	0.128	0	1.552	4.006
4	0.283	0.064	0	1.781	3.523
5	0.264	0.237	0.016	1.903	2.730
6	0.202	0.310	0.038	2.475	2.425
7	0.139	0.253	0.134	3.531	2.288
8	0.104	0.333	0.642	4.546	1.894
9	0.069	0.411	2.413	5.857	1.590
10	0.041	0.667	2.345	5.847	1.351
11	0.014	0.174	2.886	10.506	2.053
12	0.011	0.267	5.914	9.224	1.380
13	0.008	0.800	4.513	4.513	0.700
14	0.002	1.000	0	0	0.500

NOTE: From Clinton and Le Boeuf 1993.

Le Boeuf 1993). Males reached phenotypic maturity and began to mate at ages ranging from 5 to 10 years, with a mean age of 8 years. Male fecundity increased until 12 years of age, which is very late in the short fourteen-year male life span. After an initial decrease over the first four years of life, male mortality rates increased steadily from 5 to 10 years of age, a period that included the ages when males are growing in size and developing their secondary sexual characteristics. But at 11 to 12 years of age mortality rates dropped, even though fecundity continued to increase.

The second drop in mortality rates produced a characteristic "hump" in the middle of the age-specific mortality curve. Two factors possibly contributing to this increased mortality are relatively young and inexperienced males entering reproductive competition and nutritional stress due to high male growth rates (Ralls, Brownell, and Ballou 1980). Male elephant seals fast during the breeding season, and the length of the fast increases sharply at 6 years of age (fig. 9.1). Part of the increase in male mortality may be attributed to young males increasing the length of their breeding haul-out and changing its timing so that it overlaps with the period of female estrus.

In male northern elephant seals, reproduction early in life is costly in terms of survival and contributes to the increase in mortality in males from 6 to 8 years of age. A phenotypic cost of mating is shown by the negative relationship between current mating success and future mating success at these ages (table 9.2) Males that mated at 7 to 8 years of age had poor

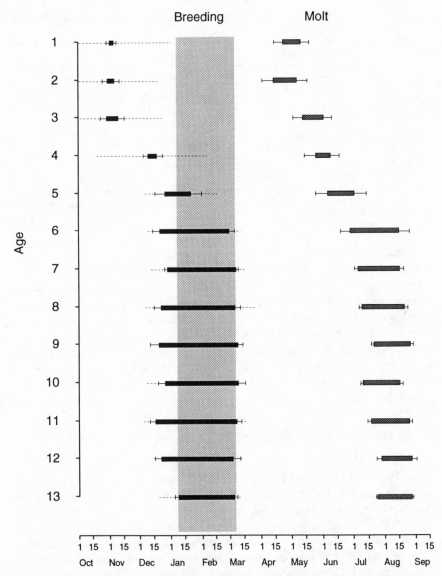

Fig. 9.1. The annual cycle of males from 1 to 13 years of age. The black and gray bars show the dates and duration of mean haul-out periods for the breeding season and molt season. The broad gray band indicates the period when females are in estrus. The error bars are the standard deviation of the haul-out duration, and the dashed lines indicate the standard deviation of the arrival and departure dates (when these exceeded the standard deviation of the haul-out duration).

TABLE 9.2   Correlations between current mating success and future mating success.

Age	Correlation coefficient	Probability level	Sample size
6	−0.300	p = .10	31
7	−0.343*	p = .03	41
8	−0.339*	p = .04	37
9	0.046	p = .80	30
10	−0.003	p = .99	19
11	0.249	p = .43	11
12	0.638	p = .07	9

NOTE: From Clinton 1990. Correlation coefficients are Spearman rank-order coefficients. Probabilities are for two-tailed tests for difference from zero. Significant correlations are marked with asterisks.

prospects because their mortality was higher during the year after breeding compared to males who did not mate (Clinton and Le Boeuf 1993). At 9 years of age, the relationship between current mating and future mating success diminished, then shifted to a positive but not significant relationship at the ages when mortality rates decreased. The cost of mating was evident during the ages when males were not yet phenotypically mature, and high male growth rates before maturity may have contributed further to the increased mortality rate. To determine the growth rate of males, I collected length data from a cross section of males from weaning to 14 years of age.

METHODS OF MEASURING LENGTH AND GROWTH

Measuring the Length of Male Elephant Seals in the Field

From 1983 to 1988, serial nose-to-tail length measurements of 117 known-age males and 103 tagged males of unknown age were obtained at Año Nuevo, California, during the fall haul-out or winter breeding season (Clinton 1990). Tagged males were measured each season they were present whenever this was possible. Individual seals were measured more than once in a season so that measurement error could be assessed. Attempts to obtain multiple measurements focused primarily on known-age males. Altogether, 437 length measurements were obtained in the field.

Male seals were measured with a tape measure when lying at rest with a straight body posture on flat sand or gravel. Straight line measurements from nose to tail were taken as males lay on their bellies or sides. Body position (lying on belly or side) and body posture (straight or tail curved) were recorded. According to the American Society of Mammalogist's (1967) definition of standard length for seals, the males ought to have been measured while lying on their backs. This was not possible with unrestrained seals, and males on their sides usually appeared to have the same general posture as an animal on its back, so measurements of individuals in both positions, lying on the belly and lying on the side, were obtained to measure the effect of body position on the length measurement.

Field measurements were adjusted to a set of standard conditions that were equivalent to the measurement of length that would have been obtained if a seal was on its side. Based on multiple measurements in different positions, males measured while lying on their bellies had 4.7 cm added to their length to compensate for the slight decrease in length obtained due to this body position. These adjusted standard lengths are simply referred to as standard lengths in the text.

Means and standard errors of standard length, growth rate, and relative growth rate were calculated for individual males grouped by age. Actual yearly growth of individual males was measured by subtraction of consecutive standard lengths for each seal measured in consecutive seasons, and relative growth was calculated by dividing annual growth by the initial standard length. Based on repeated measurements of individuals during the same season, I estimated the error of length measurements to be ±10.8 cm, which is 2 to 3% of a male's length (Clinton 1990); the magnitude of measurement error limited the accuracy of measured yearly growth for individual males, especially as growth rates neared zero.

To increase the sample size for males at older ages, 14 tagged adult males, 10 to 13 years old, were included in the cross-sectional data. These males were tagged as subadults and could be aged to within ±1 year based on the developmental categories that each seal was assigned as they returned to Año Nuevo each season (Le Boeuf 1974; Clinton 1990; Clinton and Le Boeuf 1993).

Exponential Growth Curves

A family of four exponential functions—the Richards, the logistic, the Gompertz, and the Brody curves (table 9.3)—were fitted to these data to estimate the parameters of the growth curve (Brody 1945; Richards 1959; Fitzhugh 1975). These functions are commonly used because their parameters have useful biological interpretations (Richards 1959). The equations are written so that length (L) is a function of age, but since these curves are based on the idea that growth rate depends on an organism's current size, growth and relative growth are functions of length. The important features of the male growth pattern, such as the changes in growth rate with age, the age of highest growth rate, and the age when growth ends, were obtained from the parameters of the function that best fitted the data. The important parameters are m, k, and A because my objectives are to describe the growth pattern of male seals to adult size (which is related to m

TABLE 9.3　Exponential growth functions.

	Richards	Logistic	Gompertz	Brody
Length	$A(1 \pm be^{-kt})^{\frac{1}{1-m}}$	$\dfrac{A}{1 + be^{-kt}}$	$A(e^{-be^{-kt}})$	$A(1 - be^{-kt})$
Growth	$\dfrac{kL[(\frac{A}{L})^{1-m} - 1]}{1 - m}$	$\dfrac{kL(A - L)}{A}$	$kL\ln(\frac{A}{L})$	$k(A - L)$
Relative growth	$\dfrac{k[(\frac{A}{L})^{1-m} - 1]}{1 - m}$	$k(1 - \frac{L}{A})$	$k\ln(\frac{A}{L})$	$k(\frac{A}{L} - 1)$
Length at maximum growth rate	$Am^{\frac{1}{1-m}}$	$\dfrac{A}{2}$	$\dfrac{A}{e}$	undefined

NOTE: This family of equations is written so that length is a function of time, but since these curves are based on the idea that growth rate depends on an organism's current size, growth and relative growth ($\frac{growth}{length}$) are functions of length.

and k) and to estimate when growth ends (which is related to A). The value of m determines the shape of the curve and the size (or age) when growth rate is maximal. If $m > 0$ the age-length curve will be sigmoid, and the growth rate will have a maximum at some age after the beginning of the curve. The value of k determines the inhibition of growth as age increases; basically, if m, b, and A are constant, larger k values mean larger size at given ages during the period of growth, faster growth at young ages, and a faster approach to asymptotic size. Because b is a time scaling factor, its value is not biologically significant.

A sigmoidal curve describes growth in most birds and mammals. This growth pattern is usually modeled with the Richards curve, or two of the special cases of this curve—the logistic and the Gompertz curves (Richards 1959; Case 1978; O'Connor 1984; Sibly and Calow 1986). The Richards equation is a four-parameter function that encompasses the three-parameter functions, since each of the other curves can be derived by setting m to specific values: $m = 2$ for the logistic, and $m \to 1$ for the Gompertz. Where growth decreases steadily after birth, the best fit will be obtained when $m = 0$, which is the Brody function. Each of these functions has a different shape mainly because each has a different point of inflection, which is the length (or age) at which growth rate is highest.

Multicomponent functions based on the logistic function have been used widely in recent longitudinal studies of growth curves that featured growth spurts (Bock et al. 1973; Koops 1986). The change in growth rate is modeled by an additional set of parameters that begin to contribute to the shape of the growth curve at the transition between stages in a multicomponent function. In this study, the cross-sectional data for male *M. angustirostris* were fitted to a two-component function consisting of the sum of two logistic functions:

$$L = \frac{A_1}{1 + e^{-k_1(t-d_1)}} + \frac{A_2}{1 + e^{-k_2(t-d_2)}}.$$

The first part of this equation governs growth early in life, and the second part begins to contribute to size later in life. The parameters d_1 and d_2 are the ages of the maximum growth rate for the first and second components; these parameters also eliminate the scaling parameter b from the function. For the cross-sectional data in this study, maximum growth in the first phase was assumed to occur near birth when pups are nursing from their mothers, so d_1 was set to zero initially before estimating the other parameters.

The exponential growth functions were fitted to the cross-sectional age-length data by nonlinear least squares regression using the SAS NLIN procedure (SAS Institute 1985). All standard length measurements obtained for aged males were included in the data set. Strictly speaking, the data were mixed cross-sectional–longitudinal since some individuals were measured at more than one age (Fitzhugh 1975). Although data for these seals violated the least squares assumption that the observations are independent, the estimates of age-specific size and growth rate are usually improved by collecting mixed data instead of obtaining purely cross-sectional data (Tanner 1951).

GROWTH OF MALE NORTHERN ELEPHANT SEALS

Standard length of males increased each year of life until 9 years of age. Table 9.4 shows the age distribution of males measured and the descriptive statistics for standard length, growth, and relative growth. Yearly growth rates were high from 2 to 6 years of age, and the relative growth rates indicate males increased their linear size by adding about 10% per year to their length until growth slowed at 7 years of age. But changes in growth rate in young males are not reflected in these data since 2- to 5-year-old males were difficult to measure in consecutive seasons, which limited the sample sizes for growth at each age to one to two males. The curves fitted to the length and age data had larger samples for these ages, and the growth rates of young males could be calculated from the parameters of the age-length curve.

To determine the exponential function that best described the rela-

TABLE 9.4   Standard length, measured growth rate, and relative growth rate.

Age	Standard length	N	Growth rate		N
			Annual	Relative $\left(\dfrac{\text{growth}}{\text{length}}\right)$	
1	182.9 ± 3.3	5	—	—	
2	207.4 ± 2.4	9	22.6 ± 5.6	0.11 ± 0.03	2
3	232.7 ± 3.3	10	29.8	0.12	1
4	272.6 ± 5.7	12	32.0	0.12	1
5	304.7 ± 4.5	10	39.6 ± 8.4	0.13 ± 0.03	2
6	334.2 ± 2.9	27	29.9 ± 5.0	0.09 ± 0.02	12
7	356.7 ± 3.0	35	17.4 ± 3.7	0.05 ± 0.01	15
8	370.5 ± 3.9	23	6.7 ± 4.1	0.02 ± 0.010	7
9	390.6 ± 4.1	20	5.3 ± 5.7	0.014 ± 0.014	6
10	394.8 ± 5.7	13	4.4 ± 14.0	0.012 ± 0.036	2
11	395.4 ± 10.7	7	−2.5 ± 5.8	−0.006 ± 0.015	3
12	397.2 ± 3.6	7	−9.9 ± 8.9	−0.024 ± 0.022	2
13	395.7 ± 3.0	2	—	—	

NOTE: Figures are mean ± SE in cm. At ages 9–13, figures include estimated age males.

tionship between age and standard length, the four related exponential curves were fitted to the cross-sectional age and length data by nonlinear least squares regression. Since the variances in standard length of males at each age were not equal (Bartlett's test: B = 23.951, df = 13, p < .05), the data for each age were weighted by the inverse of the standard deviation of length at that age. The Richards function with an m value close to 3 was the best fitting function of the exponential family of curves (table 9.5). The next best exponential function was the logistic curve, where m = 2, and 3 parameters are estimated by least squares rather than 4 parameters. A likelihood ratio test (Gallant 1987) of the Richards function versus the logistic function indicated the Richards function was a better fit of the data (L = 3.393, P(L) = P[$F_{1,171}$] = 0.933, P is the likelihood that the full model is a better fit).

The two-component logistic curve fitted the age and length relationship of the cross-sectional data better than the Richards function according to several criteria. The two-component logistic curve had a higher R-square value (table 9.5). Although a reliable F test evaluating the significance of the difference between the fit of these two curves is not available, a pseudo-F test (variance ratio) indicates an improved fit of the two-component logistic curve versus the Richards function ($F_{1,171}$ = 18.233). The plot of cross-sectional data with the two-component curve showed that this curve closely

TABLE 9.5 Parameters (estimate ± SE) for growth functions.

Function		A	b	k	R^2
Brody		440.9 ± 7.5	0.70 ± 0.007	0.172 ± 0.009	.969
Gompertz		418.6 ± 4.7	1.12 ± 0.017	0.264 ± 0.010	.973
Logistic		407.1 ± 3.6	1.88 ± 0.045	0.359 ± 0.011	.975
Richards	$m = 2.97 \pm 0.52$	400.6 ± 4.2	6.43 ± 3.68	0.453 ± 0.053	.975
Two-component logistic[a]					
First part		141.6 ± 14.8	—	7.83 ± 1.01	.977
Second part	$d_2 = 3.87 \pm 0.262$	256.0 ± 16.8	—	0.525 ± 0.043	

NOTE: All parameter values and the asymptotic standard errors were obtained by weighted least-squares estimation using the DUD (doesn't use derivatives) method (Gallant 1987; SAS Institute 1985).

[a] Five parameters, including d_2, were estimated for the two-part logistic function, and $d_1 = 0$ was assumed. The scale parameter, b, is not needed in this form of the logistic curve, since the inflection point occurs when $age = d$.

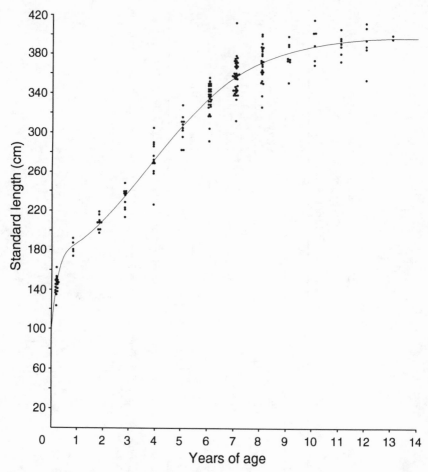

Fig. 9.2. The two-component logistic growth curve plotted with the cross-sectional data for male age and standard length.

followed the data (fig. 9.2). In comparison with the yearly growth data that were collected, the two-component logistic function was closer to measured growth rates among young males, and both functions were good predictors of growth after 5 years of age. Since the growth rate predictions of the functions differed most at these younger ages, the better fit of the two-component logistic curve indicates that it was a better model of male growth during the ages of high growth rates.

At 9 years of age, both exponential functions for the age and length curves were within 95% of asymptotic size, and the instantaneous yearly growth rate was about $8 \frac{cm}{year}$. For several years the fitted functions continue to approach the zero growth rate, but male seals appear to stop growing by 9 years of age (table 9.4). Measurements of yearly growth indicated that 7 years of age was the last year of significant measurable growth (H_0: $\mu = 0$, $\bar{x} = 17.4\frac{cm}{year}$, t = 4.703, df = 14, p < .05) and that growth of 8-year-old males was near zero ($\bar{x} = 6.7\frac{cm}{year}$, t = 1.634, df = 6, p = .153). Yearly growth was near zero among males from 9 to 12 years of age (table 9.4), but the sample sizes were small. A larger sample, including both aged males and males of unknown age that were initially tagged as adults, showed that yearly growth measured from adult males was zero ($\bar{x} = -0.8\frac{cm}{year}$, t = -0.0573, df = 39, p = .955). Thus, after physical maturity, males did not increase in standard length during a year. Like virtually all mammals, male elephant seals have a determinant growth pattern in which growth ends near the age of physical maturity. The mean standard length of all adult males measured, including those of unknown age, was 384 ±17.5 cm (±SD; N = 127, range = 345–419 cm).

LIFE HISTORY AND GROWTH

The Richards and the two-component logistic growth curves both showed two important features of the growth pattern for male northern elephant seals, which were the peak in growth rate around 3 to 5 years of age and the end of growth after 8 years of age. These two characteristics in the growth pattern relate to the life history of males in this species. First, increased growth rates of males near puberty have been found in other polygynous seals and in anthropoid primates in which males are larger than females in size, males mature and reproduce later in life than females, and the mating systems are polygynous. This suggests that particular modifications in the growth patterns, such as a growth spurt near puberty, may be associated with sexual selection for large body size in males and that the acceleration in growth rate around puberty may be connected to delayed maturity. Second, growth in length ends when maturity is reached at 8 to 10 years of age and males make the transition to the ages of successful social competition for mates (Clinton 1990; Clinton and Le Boeuf 1993).

High growth rates around puberty were first noted in domesticated animals and in humans by S. Brody (1945). He suggested that the postnatal growth rate increased until puberty and then decreased, producing a sigmoidal growth curve such as the Richards curve. Except for the primate growth models with an adolescent growth spurt, the commonly used growth models have been simple exponential functions with a sigmoidal shape; primate growth models have required multicomponent functions to account for the more complicated pattern of growth featured by some primates (van Wagenen and Catchpole 1956; Bock et al. 1973; Froehlich, Thorington, and Otis 1981; Glassman et al. 1984). In some pinniped species (Laws 1953;

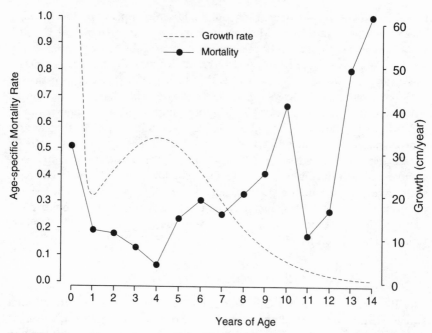

Fig. 9.3. The male growth curve versus the age-specific male mortality curve (modified from Clinton and Le Boeuf 1993), showing that there is no direct relationship between growth rate and mortality rate.

Scheffer and Wilke 1953; Carrick, Csordas, and Ingham 1962; Bryden 1972), the male growth pattern also features a growth spurt, and in this study of *M. angustirostris*, a two-component logistic growth function yielded the best model of male growth. In a recent reanalysis of growth studies in pikas, mice, and rabbits, W. J. Koops (1986) found that the multicomponent logistic functions provided a "nearly perfect" fit to longitudinal growth data that had originally been modeled with simple exponential functions. This raises the question, if growth patterns in mammals are better modeled with more complex functions, then why have simple exponential functions provided realistic and reliable models for so many studies? Koops suggests that part of the reason is the difficulty in collecting the data needed to distinguish the number of growth phases. Also, the simple exponential functions have fitted growth data well enough to model growth in species other than those with large, obvious growth spurts. In pinnipeds, where large size is strongly favored, selection for high growth rates in males appears to have accentuated the differences between growth phases by increasing the rates at the peak periods of growth during later stages of growth.

At 4 years of age, the timing of the growth spurt in northern elephant seals coincided with the age when age-specific mortality rates were at their lowest over the life span of males (fig. 9.3). The high growth rate during the growth spurt does not appear to affect survival of males. Instead, the relationship between the life history of males and growth rates indicates that high growth rates occur when the energetic costs of growth may be at their lowest. Growth rates are very high during nursing, when the pup is provided milk with a very high energy content by the mother (Ortiz, Costa, and Le Boeuf 1978). The high growth rates from 3 to 5 years of age coincide with the period in the annual cycle of males that has short periods of fasting onshore during the fall and, conversely, longer periods of foraging at sea. By the ages when males begin to fast for long periods during the breeding season, the growth rate has diminished. Thus, increased growth rates coincided with periods when the increased energetic costs of higher growth rates were more likely to be matched by increased feeding rates. This result is surprising, since higher growth rates have been reasonably assumed to be a costly disadvantage of large male size, which is thought to be favored by sexual selection and opposed by increased mortality. Such an effect is not likely in northern elephant seals. The period when males ought to be most exposed to the survival disadvantages of high growth rates is in fact a period of low mortality, and the level of natural selection is at a low point during this period. Instead, the highest mortality among males occurs at the ages after growth has ended. Selection mostly takes place on males that have attained full size. Thus, the size attained during growth would be the character potentially affected by selection rather than the growth rate.

REFERENCES

American Society of Mammalogists. 1967. Standard measurements of seals. *Journal of Mammalogy* 48: 459–462.

Bock, R. D., H. Wainer, A. Petersen, D. Thissen, J. Murray, and A. Roche. 1973. A parameterization for individual human growth curves. *Human Biology* 45: 63–80.

Bogin, B. 1988. *Patterns of Human Growth.* Cambridge: Cambridge University Press.

Brody, S. 1945. *Bioenergetics and Growth.* New York: Hafner Publishing Company. Reprinted 1968.

Bryden, M. M. 1972. Growth and development of marine mammals. In *Functional Anatomy of Marine Mammals*, ed. R. J. Harrison, 1–79. London: Academic Press.

Calder, W. A., III. 1984. *Size, Function, and Life-History.* Cambridge: Harvard University Press.

Carrick, R., S. E. Csordas, and S. E. Ingham. 1962. Studies on the southern elephant seal, *Mirounga leonina* (L.). IV. Breeding and development. *CSIRO Wildlife Research* 7: 161–197.

Case, T. J. 1978. On the evolution and adaptive significance of postnatal growth rates in the terrestrial vertebrates. *Quarterly Review of Biology* 53: 243–282.

Clinton, W. L. 1990. Sexual selection and life history of male northern elephant seals. Ph.D. dissertation, University of California, Santa Cruz.

Clinton, W. L., and B. J. Le Boeuf. 1993. Sexual selection's effects on male life history and the pattern of male mortality. *Ecology* 74: 1884–1892.

Deutsch, C. J., M. P. Haley, and B. J. Le Boeuf. 1990. Reproductive effort of male northern elephant seals: Estimates from mass loss. *Canadian Journal of Zoology* 68: 2580–2593.

Fitzhugh, Jr., H. A. 1975. Analysis of growth curves and strategies for altering their shape. *Journal of Animal Science* 42: 1036–1051.

Froehlich, J. W., R. W. Thorington, Jr., and J. S. Otis. 1981. The demography of howler monkeys (*Alouatta palliata*) on Barro Colorado Island, Panama. *International Journal of Primatology* 2: 207–236.

Gallant, A. R. 1987. *Nonlinear Statistical Models*. New York: John Wiley and Sons.

Glassman, D. M., A. M. Coelho, Jr., K. D. Carey, and C. A. Bramblett. 1984. Weight growth in savannah baboons: A longitudinal study from birth to adulthood. *Growth* 48: 425–433.

Harrison, G. A., J. M. Tanner, D. R. Pilbeam, and P. T. Baker. 1988. *Human Biology*. 3d ed. Oxford: Oxford University Press.

Innes, S., R. E. A. Stewart, and D. M. Lavigne. 1981. Growth in Northwest Atlantic harp seals, *Phoca groelandica*. *Journal of Zoology, London* 194: 11–24.

Koops, W. J. 1986. Multiphasic growth curve analysis. *Growth* 50: 169–177.

Laws, R. M. 1953. The elephant seal (*Mirounga leonina*, Linn.). I. Growth and age. *Falkland Islands Dependencies Survey, Scientific Reports* 8: 1–62.

Le Boeuf, B. J. 1974. Male-male competition and reproductive success in elephant seals. *American Zoologist* 14: 163–174.

Le Boeuf, B. J., and J. Reiter. 1988. Lifetime reproductive success in northern elephant seals. In *Reproductive Success*, ed. T. H. Clutton-Brock, 344–362. Chicago: University of Chicago Press.

O'Connor, R. J. 1984. *The Growth and Development of Birds*. Chichester: John Wiley and Sons.

Ortiz, C. L., D. Costa, and B. J. Le Boeuf. 1978. Water and energy flux in elephant seal pups fasting under natural conditions. *Physiological Zoology* 51: 166–178.

Ralls, K., R. L. Brownell, and J. Ballou. 1980. Differential mortality by sex and age in mammals, with specific reference to the sperm whale. *Report of the International Whaling Commission Special Issue* 2: 233–243.

Richards, F. J. 1959. A flexible growth function for empirical use. *Journal of Experimental Botany* 10: 290–300.

SAS Institute. 1985. *SAS User's Guide: Statistics*. Ver. 5. Cary, N.C.: SAS Institute.

Scheffer, V. B., and F. Wilke. 1953. Relative growth in the northern fur seal. *Growth* 17: 129–145.

Sibly, R. M., and P. Calow. 1986. *Physiological Ecology of Animals*. Oxford: Blackwell Scientific Publications.

Tanner, J. M. 1951. Some notes on the reporting of growth data. *Human Biology* 23: 93.

Trivers, R. L. 1985. *Social Evolution*. Menlo Park, Calif.: Benjamin/Cummings Publishing Company.

van Wagenen, G., and H. R. Catchpole. 1956. Physical growth of the rhesus monkey (*Macaca mulatta*). *American Journal of Physical Anthropology* 14: 245–273.

TEN

Sex- and Age-Related Variation in Reproductive Effort of Northern Elephant Seals

Charles J. Deutsch, Daniel E. Crocker, Daniel P. Costa, and Burney J. Le Boeuf

ABSTRACT. The aim of this study was to determine how reproductive effort (RE) varies with age and sex in the northern elephant seal, *Mirounga angustirostris*. RE is an important feature of life history that links the proximate costs with the fitness costs of reproduction. Percentage of mass lost over the breeding season provided an energetic index of RE because both sexes fast on the rookery. We summarize research conducted over an 11-year period at Año Nuevo, California. Seventy-three females ranging in age from 3 to 12 years were chemically immobilized and weighed during the lactation period, providing 22 measurements of maternal mass loss. Males ranging in age from 5 to 13 years were weighed (N = 56), or their mass was estimated using a photogrammetric technique (N = 94), yielding 87 measurements of breeding mass loss.

The principal findings were as follows. *1.* The magnitude of RE was similar between males and females. Adults of both sexes lost slightly more than one-third of their mass, on average, despite large sex differences in reproductive strategy and body size. Adult males were an average of 1.4 times longer and 3 to 4 times heavier (mean ± SD = 1,814 ± 233 kg, range = 1,430–2,550 kg) than adult females (488 ± 80 kg, range = 360–710 kg) at the start of the breeding season. Male mating effort was more variable than female parental effort, and some bulls lost up to half of their arrival mass. The timing and intensity of RE differed between the sexes: males fasted three times longer than females, but they incurred less than one-third the mass-specific mass loss rate of females. Males suffered vastly more external injuries from conspecifics than did females, yet females were probably at greater risk of receiving life-threatening internal injuries. *2.* Female effort during lactation was not correlated with maternal age or mass, but gestation effort (i.e., relative neonatal mass) declined with increasing maternal mass. Absolute measures of investment in offspring—including maternal mass loss, neonatal mass, weanling mass, and pup mass gain—were directly proportional to maternal mass, which increased from 3 to 6 years of age before reaching an asymptote. Maternal investment in sons was similar to that in daughters. *3.* Males delayed serious efforts at mating until age 6, two to three years later than females but still physically immature, and mean RE was subsequently constant with age. Dominant males devoted a greater proportion of body stores to breeding and obtained higher mating success than did subordinates.

The similarity in RE of growing "subadults" and physically mature breeders for both sexes was contrary to theoretical expectations, considering the poorer reproductive success and higher fitness costs of the former. Possible explanations include the existence of a threshold RE below which fitness costs are minimal, the higher fitness benefit of early-born offspring in an expanding population, and the benefit of experience to future reproductive performance.

One can, in effect, treat the sexes as if they were different species, the opposite sex being a resource relevant to producing maximum surviving offspring. —R. L. Trivers (1972)

Males and females reproduce in fundamentally different ways, and the resulting divergent selective pressures have led to numerous sex differences in morphology, physiology, life history, and behavior (Trivers 1972, 1985; Glucksman 1974; Clutton-Brock, Guinness, and Albon 1982). A crucial aspect of mammalian life history that has rarely been compared quantitatively across the sexes is reproductive effort (RE). Typically defined as the proportion of available time and energy that an organism allocates to reproduction over a specified period of time (Gadgil and Bossert 1970; Hirshfield and Tinkle 1975; Tuomi, Hakala, and Haukioja 1983), RE also includes a component of risk that is important but more difficult to quantify (Calow 1979; Warner 1980). Reproductive effort can be divided into two categories: mating effort, which involves expenditure of time, energy, or risk to obtain matings (e.g., searching for mates, fighting for dominance or territory necessary for access to mates); and parental effort, which involves such expenditure to produce, nurture, and raise offspring (Low 1978). Herein lies a major difference between the RE of males and of females. In the majority of mammals and in most polygynous species, males exhibit little or no parental care, so that virtually all male RE is mating effort (Alexander and Borgia 1979). In contrast, females invest considerably in their young, especially in mammals, which exhibit extended postnatal care and nurture offspring via the energetically expensive process of lactation (see Loudon and Racey 1987). Consequently, the selective pressures that have shaped the pattern of RE should differ between the sexes. Female RE is thought to have evolved primarily in response to ecological and demographic factors, whereas the evolved pattern of male RE has apparently been strongly influenced by the mating system and the intensity of male-male competition for mates (Trivers 1972; Warner 1980; Thornhill 1981).

The principal reason for the paucity of empirical studies contrasting male and female RE in polygynous vertebrates is that male mating effort has been neglected (Stearns 1976; Warner 1980; Gittleman and Thompson

1988). Furthermore, sex differences in energy intake during the breeding season and in risks associated with reproducing (e.g., Ryan 1985) often make it difficult to measure RE in a comparable way for each sex (Knapton 1984). Ideally, one would study a species in which breeding is temporally separated from foraging and other activities related to growth and survival. Then essentially all activity would serve a reproductive function, all energy expended above maintenance levels would be devoted to reproduction, and energy intake would be nil or absent. Such conditions are found in some phocids, such as elephant seals and gray seals, in which both sexes fast and remain on land during the breeding season, relying on stored energy reserves to fuel their metabolic and reproductive needs. Studies on phocid reproductive energetics were initially conducted on females and pups due to their smaller size, greater site predictability, and relevance of the work to parental investment theory (Fedak and Anderson 1982; Ortiz, Le Boeuf, and Costa 1984; Stewart and Lavigne 1984; Costa et al. 1986; Stewart 1986; Bowen, Boness, and Oftedal 1987; Hill 1987; Tedman and Green 1987; McCann, Fedak, and Harwood 1989; Kovacs, Lavigne, and Innes 1991; Hammill et al. 1991; Bowen, Oftedal, and Boness 1992; Fedak et al., this volume; see review by Oftedal, Boness, and Tedman 1987). Logistic difficulties involving the capture and weighing of large, aggressive, and mobile animals have hindered research on male reproductive energetics in pinnipeds until recently (Anderson and Fedak 1985; Reilly 1989; Deutsch, Haley, and Le Boeuf 1990; Boyd and Duck 1991; Reilly and Fedak 1991; Twiss 1991; Bartsh, Johnston, and Siniff 1992; Fedak et al., this volume; Walker and Bowen, in press).

We studied sex- and age-related variation in reproductive effort of the highly polygynous and sexually dimorphic northern elephant seal, *M. angustirostris*, in central California. Sex differences in life history and behavior are numerous in this species (see Le Boeuf 1981; Le Boeuf and Reiter 1988; Le Boeuf and Laws, this volume). Age at primiparity ranges from 2 to 6 years, with most females initially giving birth at age 3 (36%) or age 4 (54%) at Año Nuevo (Reiter and Le Boeuf 1991). Females produce one pup annually thereafter, with occasional skipping (Huber 1987), until death at a maximum age of 19 years. Male growth is extended by two to three years, and reproduction is delayed compared to females. Males attain sexual maturity by age 5 and rarely skip breeding seasons subsequently (Clinton 1990); yet due to intrasexual competition, most do not mate until physical maturity at age 8 or 9 (Clinton, this volume), five years later than the age of first mating for most females. The maximum life span recorded for a male is 14 years.

Considerable interest in optimal life history theory has focused on identifying the age-specific pattern of reproduction that maximizes fitness for a given set of demographic conditions and constraints (see Charlesworth 1980). Assuming that reproduction imposes a cost, measured as a reduction in future fitness (Bell 1980; Bell and Koufopanou 1985), there should be a level of RE at each age that optimizes trade-offs between current and future reproduction (Williams 1966a, 1966b; Gadgil and Bossert 1970; Pianka and Parker 1975). A frequently asserted prediction is that RE should increase with age as residual reproductive value declines; that is, since older individuals generally have less to lose in terms of future offspring than younger ones, investment should be high late in life (Williams 1966b; Gadgil and Bossert 1970; Pianka 1976; Caswell 1982). This trend is favored by low population growth rate, low extrinsic adult mortality rate, and continuous increase in body size or reproductive efficiency during adulthood (Charlesworth 1980). The opposite conditions theoretically favor a decline in RE with age. In rapidly expanding populations, for example, breeding should occur as soon as physiologically possible (Lewontin 1965; Cody 1971; Stearns 1976), and young adults should exhibit the greatest effort (Charlesworth and León 1976). Though studies on this point among long-lived birds and mammals are few, there is some empirical support for increasing (Clutton-Brock 1984; Pugesek 1981, 1983, 1984; Hamer and Furness 1991; Pärt, Gustafsson, and Moreno 1992; but see Nur 1984), decreasing (Stewart 1986), and constant (Reid 1988) RE with age; other studies have yielded mixed results (Green 1990).

The expected ontogenetic pattern of resource allocation to reproduction must be considered separately for each sex. Reproductive value of female northern elephant seals declines gradually from primiparity to age 11 and then drops steeply to the end of the life span (Reiter and Le Boeuf 1991), suggesting that the fitness cost of reproduction should also decrease with age, especially among the oldest animals. Weaning success increases with age and experience, particularly from sexual maturity to physical maturity at age 6 (Reiter, Panken, and Le Boeuf 1981; Reiter and Le Boeuf 1991; Sydeman et al. 1991). This combination of rising benefits and declining potential costs of breeding with age suggests that maternal effort should increase with age. However, the current high population growth rate (Cooper and Stewart 1983; Stewart et al., this volume) enhances the fitness value of early-born offspring relative to those born later in life, and this will favor high effort early in life. Given these opposing forces, it is difficult to predict how female RE should change with age, although we expect a high terminal investment when residual reproductive value is low (i.e., >10 years old). Since natality rate is high and relatively constant with age after the initiation of reproduction (Le Boeuf and Reiter 1988; Huber et al. 1991), our study focused on the RE of parous females and did not consider the frequency of barren years.

For male elephant seals, sexual selection is likely to have shaped the age-specific pattern of RE. Intense intrasexual competition among males in highly polygynous species should favor a lifetime mating strategy involving little investment when young and a high mating effort after attaining a reproductively competitive size (Warner 1980; Maher and Byers 1987). Male elephant seals have a relatively short effective breeding life span (age 8–13), with mating success rising precipitously on reaching physical maturity (Le Boeuf and Reiter 1988; Clinton and Le Boeuf 1993). Since subadult males are growing, any allocation of time and energy to reproduction might reduce their growth rate (e.g., Green and Rothstein 1991), final adult size (e.g., Boyce 1981), and, hence, future fecundity. Therefore, we predicted that male mating effort should be low during the subadult period (age 4–7) when the fitness benefit of reproductive investment is nil and potential costs are high and that effort would increase abruptly at adulthood when bulls are able to fight effectively for high dominance rank and access to mates. Contrary to expectation, C. J. Deutsch, M. P. Haley, and B. J. Le Boeuf (1990) found that older subadults (6–7 years old) expended the same RE (i.e., percent mass loss) as low-ranking adult males.

In this chapter, we compare reproductive effort between male and female northern elephant seals as indicated by mass loss while breeding (energy component), duration of stay on the rookery (time component), and incidence of external injuries (risk component). Our objectives are to: (1) summarize existing data on body mass as a function of age for both sexes, much of which is currently unpublished or scattered across a number of publications; (2) compare the magnitude of male mating effort with female parental effort, in terms of relative mass loss, time commitment, and apparent risk; and (3) describe age-specific patterns of RE for both sexes and relate them to changes in reproductive success and mortality with age.

METHODS

General Methods and Study Area

The studies summarized here were conducted from 1981 to 1991 at Año Nuevo State Reserve, 70 km south of San Francisco, California. Most work was done on the expanding mainland colony (see Le Boeuf and Kaza 1981 for a description of the area), which increased over the study period from 300 to 1,500 breeding females distributed among 8 to 16 harems. The annual maximum in number of sexually mature males present on the Año Nuevo mainland during the breeding season (December to March) fluctuated between 300 and 500, of which about one-third were adults. Individuals were identified over the course of the breeding season by applying cream bleach or black hair dye to the pelage. Numbered plastic tags placed in the webbing of the hind flippers provided information on age and permitted identification of individuals from year to year.

Exact ages were unknown for most males in the study. Males were assigned to one of five age classes based on their overall size and on the development of secondary sexual characters (neck shield and proboscis), both of which correspond with age to within about one year (Le Boeuf 1974; Cox 1983; Clinton 1990): subadult-one (SA1) males = 4 years old; subadult-two (SA2) = 5 years; subadult-three (SA3) = 6 years; subadult-four (SA4) = 7 years; and adult males (AD) = 8–14 years. Most observations concern the last three age classes, since there were few SA1 and SA2 males present during the breeding season. Known-age adult males and those with known histories were further classified as either young adult (first- or second-year adult, approximately 8–9 years of age) or old adult (at least third-year adult, 10 years and older).

Copulations and dominance interactions among males were recorded in the field using standard methods, and the estimated number of females inseminated provided a measure of male mating success (Le Boeuf 1974). Dominance ranks of adult males were determined in each year using the Bradley-Terry method (Boyd and Silk 1983; Haley, Deutsch, and Le Boeuf, in press), and these values were used to assign males to one of three rank classes.

Measurement of Body Mass and Mass Loss

Females. One or more weights were obtained on 73 females and their pups near the beginning and/or the end of the female's lactation period. Sixty-one mothers were of known age, ranging from 3 to 12 years old. The sample includes female-pup pairs weighed during the course of studies on diving behavior (Le Boeuf et al. 1986, 1988, 1989) and reproductive energetics (Costa et al. 1986; Crocker 1992; D. Costa and B. Le Boeuf, unpubl. data). An additional 18 pups were weighed within three days of birth and again within six days of weaning, 17 of whose mothers were of known age (Kretzmann 1990; Kretzmann, Costa, and Le Boeuf 1993). The 91 mother-pup pairs included 9 females that were experimental subjects in two or more years, giving a total of 79 different females. After chemically immobilizing a lactating female, she and her pup were weighed with a tripod scale (accurate to ± 2.5 kg) and standard length measurements were taken (Costa et al. 1986). For the energetic studies, 22 different female-pup pairs were weighed 0 to 5 days postpartum, and the procedure was repeated an average of 21.2 ± 2.0 days later (range = 18–25 days), within a few days of weaning. Methods are given in more detail in D. P. Costa et al. (1986) and D. E. Crocker (1992). Data on energy expenditure, body composition, and milk intake are presented elsewhere (Ortiz, Le Boeuf, and Costa 1984; Costa et al. 1986; Crocker 1992; Kretzmann, Costa, and Le Boeuf 1993).

Females that abandoned or lost their pup after experimental manipulation (N = 5), or who regularly nursed more than one pup (N = 2), were excluded from calculations of maternal mass loss, departure mass, lactation duration, pup mass gain, and pup mass and length at weaning.

Males. Body mass was obtained or estimated for 140 different males (n = 247 measurements of mass) ranging in age from 5 to 13 years and in dominance status from low-ranking to alpha male. Estimates of daily and seasonal mass loss are presented for 82 males (n = 87 measurements) and 71 males (n = 75 measurements), respectively. The data on male body mass come from two studies.

1. In 1988 and 1989, 56 males were weighed on a platform scale (accurate to + 0.25%), yielding 95 weights and 30 measures of breeding mass loss over a mean ± SD interval of 32.7 ± 18.7 days (range = 9–83 days). One male was anesthetized; the rest were either lured onto the scale using a model of a female elephant seal and playback of female vocalizations or moved onto the scale using tarpaulins and playback of male threat vocalizations. Details of weighing methods and results appear in Deutsch, Haley, and Le Boeuf (1990).

2. Body mass was estimated for an additional 94 males at the start and/or the end of the 1989 breeding season using a photogrammetric technique that predicts mass with a 95% confidence interval of ±14% (r^2 = .93; Haley, Deutsch, and Le Boeuf 1991). This provided 152 estimates of mass and 57 measures of mass loss over a mean interval of 58.0 ± 13.3 days (range = 42–96 days) (Deutsch 1990).

Since there were no significant differences in daily or seasonal mass loss (absolute and mass-specific) of males between the direct weighing and photogrammetric methods (t-tests for each age class, p > .05), the two data sets were pooled for presentation here. Photogrammetric estimates of adult male body mass at arrival and departure were significantly greater, however, than were the extrapolated figures based on direct measurements (arrival: 1,851 ± 236 vs. 1,710 ± 200 kg, t = 2.09, df = 61, p < .05; departure: 1,227 ± 144 vs. 1,098 ± 115 kg, t = 3.72, df = 75, p < .001). This probably reflects the fact that large, high-ranking males (which were difficult to weigh) comprised a greater proportion of the photographic data set than the direct weight data set. The two methods of measuring mass did not differ significantly for SA3 males (p > .05), but photogrammetric estimates were significantly greater than direct estimates for SA4 males (arrival: 1,539 ± 234 vs. 1,274 ± 140 kg, t = 2.26, df = 11, p < .05; departure: 1,106 ± 158 vs. 870 ± 107 kg, t = 3.81, df = 18, p < .01). This mass difference was apparently not due to a methodological discrepancy, because the photographic sample of SA4 males was indeed larger in standard length than the weighed sample ($p < .05$). Standard length was measured photogrammetrically in both studies (Haley, Deutsch, and Le Boeuf 1991).

Incidence of Injuries

Injuries provide one index of risk taken in reproduction because they represent the degree to which the animal was exposed to potentially damaging or lethal interactions while reproducing (Deutsch 1990; Le Boeuf and Mesnick 1991). Wounds can get infected and, if severe, can reduce fecundity (e.g., Clutton-Brock et al. 1979; Le Boeuf, Riedman, and Keyes 1982) or lead to death (e.g., Laws 1953; Wilkinson and Shank 1976). The incidence of external injuries was measured by recording the number, location, and severity of fresh wounds on males (1989) and females (1991) during the main mating period in February. Wounds were considered fresh if they were open, bleeding, or oozing and thus represented recent injuries. This analysis was limited to injuries inflicted by conspecifics and excluded shark-inflicted wounds.

Data Analysis

Values of body mass are presented at arrival (or parturition for females) and departure dates to standardize comparisons across ages and sexes. Only direct measurements of mass and breeding mass loss are used in the sex comparison. Descriptive statistics are presented as mean ± one standard deviation (SD). For nine females and nine males that were sampled in two or more years, mean values of size, mass change, and other variables were calculated for each individual; these values were then used in the calculation of overall means by sex or age class (e.g., table 10.3). For correlations between RE variables and size or age, however, each measurement was treated as an independent data point (e.g., table 10.4). In these analyses, P-values were determined after reducing the degrees of freedom to equal the number of individuals (rather than measurements) in the sample minus the number of estimated parameters (see Sydeman et al. 1991). The level of statistical significance was set at $\alpha = .05$ for all tests.

Measures of the effort devoted to reproduction (e.g., mass loss, pup mass gain) are presented here in absolute and mass-specific terms as a function of age and sex. RE should be expressed as a fraction of the organism's ability to invest (e.g., proportion of energy stores or body mass). However, since proportions can sometimes be misleading when the dependent variable varies allometrically with body mass (Packard and Boardman 1987), we adjusted for the potentially confounding effect of body mass on indexes of RE using one of three additional methods: (1) analysis of covariance (ANCOVA) for comparison of male mass loss by age class and for comparison of maternal investment by sex of pup; (2) partial correlations between

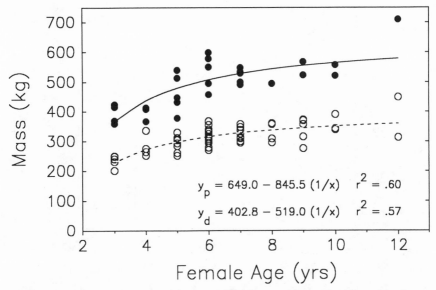

Fig. 10.1. Breeding female mass at parturition (solid circles, N = 28) and at departure from the rookery (i.e., weaning) (open circles, N = 54) as a function of maternal age. Both regressions are significant (p < .0001). The heavier departure mass at age 12 is a rough estimate (see footnote c in table 10.3).

maternal investment variables and age, controlling for female mass; and (3) correlation of residuals (from the regression of maternal investment indexes against parturition mass) with maternal age. Plots of residuals were inspected visually for homoscedasticity (i.e., constant variance of residuals around the regression line), and the normality of residuals was checked with the univariate procedure in SAS (SAS Institute 1985). An arcsine-square-root transformation was applied to percentages before performing parametric statistical tests (Sokal and Rohlf 1981). Descriptions of how mass, mass loss, and other variables were calculated are presented in appendix 10.1.

RESULTS

Sexual Size Dimorphism

The most striking difference between the sexes in elephant seals is their size. At arrival, adult males weighed an average of 3.5 (range = 2–7) times more than adult females (table 10.1; see figs. 10.1 and 10.2). Estimates of arrival mass for adult males varied from 1,430 kg to 2,265 kg for weighed animals and up to 2,550 kg for a photographed animal. Departure mass of

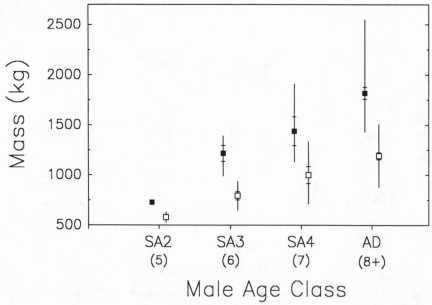

Fig. 10.2. Male mass at arrival (solid squares) and at departure (open squares) from the breeding rookery as a function of age class. Approximate age in years of each class is shown in parentheses. Symbols show the mean (squares), 95% confidence interval of the mean (hatch marks), and range; a 95% confidence interval is not shown for the SA2 class due to the small sample size. Sample sizes for arrival and departure mass, respectively, are as follows: SA2 (N = 1, 2); SA3 (N = 12, 18); SA4 (N = 11, 18); Adult (N = 62, 75).

adult males ranged from 895 to 1,500 kg (fig. 10.2). Female mass at parturition (i.e., immediately postpartum) ranged from 360 to 710 kg, and departure mass ranged from 200 kg to at least 400 kg (fig. 10.1; table 10.3). Given that males and females arrive at different times and subsequently lose mass until departure, the mass ratio between an adult male and a lactating female at any given time could range from about 1.5 to 10. However, a three- to sixfold dimorphism in mass would be most common for a typical copulating pair in early February.

Standard length (SL) gives a more straightforward measure of sexual size dimorphism because it does not fluctuate over the annual cycle as does mass. Mean SL of 89 physically mature males measured photogrammetrically in this study was 382 ± 15 cm (range = 350–420 cm), which is similar to the mean of 385 ± 18 cm found by W. L. Clinton (this volume). Physically mature females (i.e., 6+ years of age) measured in this study had a mean SL of 265 ± 8 cm (range = 248–282 cm, N = 32 individuals). Thus, mature males were an average of 1.44 times longer (range = 1.24–1.69) than mature females.

Both sexes became reproductively active while still growing. Females giving birth for the first time at age 3 had attained approximately 75% of mature mass (fig. 10.1, table 10.3), whereas 6-year-old males (SA3 class) were roughly two-thirds the mass of mature bulls (fig. 10.2, table 10.5). The cross-sectional data on size and a limited amount of longitudinal data indicate that females stopped growing in length and mass at age 6 (parturition mass = 541 ± 58 kg, N = 16, ages 6–12 years) and males reached a physically mature size at 8 to 9 years, growing little or not at all in length thereafter (Clinton, this volume).

Sex Comparison of Reproductive Effort
A comparison of the energetic component (breeding mass loss) and the time component (duration of the breeding fast) of RE for adult male and female elephant seals is presented in table 10.1. The most important finding was that the mean proportion of mass lost over the breeding season was very similar for both sexes (36–37%), suggesting that the energetic component of RE while ashore was also similar. A proximate reason for this outcome was that the greater intensity of the females' parental effort (i.e., percent body mass lost per day) compared to the males' mating effort was balanced by a longer duration of stay for males. Duration of the breeding fast for adult males was about three times longer than that of females (table 10.1). Relative mass loss over the season was significantly more variable for adult males (range for weighed animals = 26–46%) than for females (range = 31–41%) (variance-ratio test, F = 5.78, df = 12, 21, p < .01).

Males clearly incurred a much greater risk of external injury than females. Nearly 90% of males surveyed in February possessed fresh wounds, compared to less than 20% of females (table 10.2). The mean number of fresh wounds per animal was 20 times greater for males than females (table 10.2); those females with open wounds usually had only one to three small cuts. Although most injuries appeared superficial for both sexes, the potential for serious injury was reflected in the percentage of seals with fresh wounds on the head (including face, nose, and eye), which again was significantly higher for males than females.

Female Reproductive Effort in Relation to Maternal Age, Mass, and Pup Sex
Energetic measures of the two components of maternal investment—gestation and lactation—are plotted as a function of maternal age, mass, and pup sex in figure 10.3 and summarized as a function of maternal age in table 10.3.

TABLE 10.1 Body mass, mass loss rate, total mass loss, and duration of stay on the rookery for adult male and female northern elephant seals during the breeding season.

	N	Body mass (kg)[a]	Mass loss per day		Mass loss over season		Duration of stay (days)
			kg	% Body mass[a]	kg	% Body mass[a]	
Male	13[b]	1704 ±213	7.08 ±1.54	0.41 ±0.06	622.0 ±171.5	36.2 ±6.6	91.0 ±14.8
Female	22	489 ±64	7.22 ±0.99	1.48 ±0.11	180.4 ±29.6	36.8 ±2.8	25.0 ±2.1

NOTE: Numbers shown are mean ± 1 SD. Data for males include only direct weight measurements. Data for females do not include the 6.5-day terrestrial period prior to parturition and are based on females that successfully weaned a pup. Including this preparturition period, females lost an estimated 38.7% of their arrival mass (excluding mass of pup and placenta) over the 31.5-day breeding fast (see Deutsch, Haley and Le Boeuf 1990 for calculations).

[a] Estimated body mass at arrival (males) or at parturition (females); see appendix 10.1.

[b] Sample size was 17 males for absolute mass loss per day.

TABLE 10.2   Sex difference in the incidence of external injuries during the mating season (February).

	Male (N = 161)	Female (N = 121)	Significance
% with fresh wounds	86.3	16.5	p < .0001[a]
% with fresh head wounds	17.4	0.8	p < .0001[a]
# of fresh wounds/animal			
Mean ± 1 SD	10.2 ± 12.1	0.5 ± 1.3	
Median	6.0	0.0	p < .01[b]

[a] Chi-square test of independence.
[b] Median two-sample test.

Effect of Pup Sex on Maternal Investment.   Theoretical predictions (Maynard Smith 1980) and some empirical studies (e.g., Anderson and Fedak 1987; Oftedal, Iverson, and Boness 1987) indicate that females invest more in sons than in daughters in polygynous mammals. Therefore, we initially examined the data, using two-tailed t-tests, for differences in the mother's size, age, and investment or the pup's size and growth rate as a function of pup sex. Of all variables in table 10.3, only percent daily maternal mass loss was significantly different between mothers with sons ($1.53 \pm 0.10\%$, N = 10) and mothers with daughters ($1.43 \pm 0.10\%$, N = 12; t = 2.43, df = 20, p = .025). Absolute mass loss per day did not vary significantly with sex of the pup (t = 0.71, df = 20, p = .48); after adjusting for the effect of maternal parturition mass, however, mothers of male pups showed a higher mass loss rate than mothers of female pups (ANCOVA: F = 5.34, df = 1, 19, p = .032). This finding is not consistent with our other results because none of the other t-tests revealed a significant effect of pup sex on maternal investment (e.g., total mass loss, pup growth rate); analyses of covariance on the nine other absolute measures of maternal investment in table 10.3, with parturition mass as the covariate, confirmed these results (p > .10 for all tests). Furthermore, recent studies have shown a lack of differential investment in sons versus daughters in both northern and southern elephant seals (McCann, Fedak, and Harwood 1989; Le Boeuf, Condit, and Reiter 1989; Campagna et al. 1992; Kretzmann, Costa, and Le Boeuf 1993; Fedak et al., this volume). Our interpretation is that this one significant result was probably due to chance, since one comparison is expected to be significant at the p = .05 level when over 20 tests are performed simultaneously. None of the slopes defining the linear regressions of maternal investment variables on maternal mass varied significantly with pup sex (general linear models, p > .25), but the correlation coefficients were usually lower for sons than for daughters, as reflected in the greater scatter of points among small mothers of sons (see fig. 10.3e, g, i). T. S. McCann, M. A. Fedak, and J. Harwood (1989) made a similar observation for the southern elephant seal. We con-

TABLE 10.3 Measures of age-specific maternal investment and body size of female northern elephant seals.

	Female age (years)										All ages combined[a]		
	3	4	5	6	7	8	9	10	11	12	Mean±SD	Min.	Max.
Female													
Standard length[b] (cm)	228±11 (7)	244±11 (5)	253±9 (10)	261±8 (14)	268±5 (10)	266±12 (4)	267±8 (6)	270±9 (3)	— (0)	276±8 (2)	255±15 (62)	213	282
Parturition mass (kg)	392±32 (4)	397±25 (3)	462±65 (5)	536±58 (5)	520±24 (5)	494 (1)	545±31 (2)	539±26 (2)	— (0)	710 (1)	488±80 (34)	359	710
Departure mass (kg)	229±23 (6)	283±37 (4)	294±26 (8)	320±32 (14)	324±23 (8)	331±34 (4)	339±41 (5)	358±28 (3)	— (0)	313[c] (1)	303±46 (56)	201	>400
Initial mass[d] (kg)	360±40 (4)	381±35 (3)	429±44 (5)	518±55 (5)	493±25 (5)	479 (1)	530±42 (2)	516±13 (2)	— (0)	684 (1)	466±81 (34)	313	684
Interval from birth to initial treatment (days)	5.5±4.0 (4)	2.7±2.1 (3)	5.0±4.3 (5)	2.2±1.9 (5)	3.4±0.6 (5)	2.0 (1)	2.0±1.4 (2)	3.0±1.4 (2)	— (0)	3.0 (1)	3.2±2.6 (34)	0	10
Mass loss per day (kg)	6.15±0.46 (2)	6.75±0.02 (2)	6.11±1.12 (3)	7.78±1.10 (5)	8.08±0.50 (3)	7.71 (1)	8.08 (1)	7.73±0.46 (2)	— (0)	—[c] (0)	7.22±0.99 (22)	5.24	9.58
Mass loss per day (% FPM)[e]	1.55±0.04 (2)	1.64±0.02 (2)	1.34±0.06 (3)	1.45±0.10 (5)	1.59±0.03 (3)	1.56 (1)	1.42 (1)	1.43±0.02 (2)	— (0)	—[c] (0)	1.48±0.11 (22)	1.28	1.66
Mass loss over lactation (kg)	152±32 (2)	152±5 (2)	159±44 (3)	197±24 (5)	197±16 (3)	185 (1)	210 (1)	198±28 (2)	— (0)	— (0)	180±30 (22)	126	230
Mass loss over lactation (% FPM)[e]	38.1±4.4 (2)	37.0±1.6 (2)	34.6±3.6 (3)	36.7±1.9 (5)	38.7±2.3 (3)	37.4 (1)	37.1 (1)	36.6±3.5 (2)	— (0)	— (0)	36.8±2.8 (22)	31.2	41.2
Lactation duration (days)[f]	23.6±2.4 (7)	25.2±3.3 (5)	26.2±1.6 (22)	25.7±1.8 (13)	25.0±1.6 (8)	25.6±1.3 (5)	26.5±0.7 (2)	26.2±1.5 (4)	26.0 (1)	26.5±0.7 (2)	25.5±2.0 (66)	20	30
Pup													
Standard length at birth (cm)	115 (1)	121±5 (4)	126±7 (5)	121±6 (3)	124±6 (3)	— (0)	129 (1)	124±2 (2)	135 (1)	123 (1)	123±6 (25)	110	135
Standard length at weaning (cm)	142±8 (5)	142±6 (5)	142±7 (14)	144±5 (9)	150±4 (6)	148±7 (4)	149±3 (4)	144±5 (4)	155 (1)	148 (1)	144±6 (52)	129	156
Birth mass (kg)	38.0±10.0 (2)	33.8±7.1 (4)	37.5±3.4 (15)	38.2±3.3 (5)	40.2±2.3 (6)	42.1±1.7 (2)	40.4±4.4 (2)	42.0±3.5 (3)	48.8 (1)	45.9 (1)	38.3±4.6 (43)	25.2	48.8
Birth mass (% FPM)[e]	9.48±1.59 (2)	8.61±1.66 (2)	8.09±0.74 (3)	7.18±0.79 (5)	7.71±0.50 (5)	8.27 (1)	7.40±0.37 (2)	7.43±0.21 (2)	— (0)	6.47 (1)	7.70±1.06 (27)	6.28	10.61
Initial mass[d] (kg)	40.2±10.0 (2)	37.1±4.9 (4)	40.4±3.6 (15)	41.3±6.8 (5)	45.3±1.9 (6)	44.3±1.7 (2)	42.6±1.3 (2)	45.0±2.9 (3)	51.0 (1)	50.3 (1)	41.4±5.0 (44)	31.3	51.7

Initial mass (% FPM)[c]	10.04±1.54 (2)	9.58±0.98 (2)	8.82±1.36 (3)	7.75±1.28 (5)	8.73±0.51 (5)	8.72 (1)	7.82±0.22 (2)	8.24±0.33 (2)	— (0)	7.09 (1)	8.50±1.39 (28)	6.28	12.03
Interval from birth to initial treatment (days)	2.0±0.0 (2)	2.5±1.7 (4)	2.3±0.8 (15)	2.2±1.9 (5)	3.3±0.5 (6)	2.0±0.0 (2)	2.0±1.4 (2)	2.3±1.5 (3)	2.0 (1)	3.0 (1)	2.4±1.1 (44)	0	5
Weaning mass (kg)[b]	112±34 (5)	123±13 (5)	127±13 (21)	142±16 (9)	143±11 (7)	149±7 (4)	139±18 (5)	146±21 (4)	154 (1)	156±22 (2)	132±19 (61)	93	172
Weaning mass (% FPM)[c]	33.0±10.4 (2)	30.0±0.6 (2)	25.9±2.1 (5)	25.6±2.8 (5)	27.1±2.3 (3)	29.4 (1)	28.2 (1)	27.6±3.6 (2)	— (0)	24.2 (1)	27.2±3.8 (27)	20.0	40.4
Weaning mass (% FDM)[g]	49.6±12.0 (5)	45.1±3.3 (4)	40.4±2.1 (7)	43.0±5.9 (9)	44.3±3.9 (6)	43.9±3.1 (3)	41.5±6.5 (5)	39.9±8.7 (3)	45.1 (1)	45.1 (1)	43.5±6.6 (42)	29.0	68.8
Mass gain per day (kg)	4.03±1.32 (2)	3.72±0.62 (3)	3.84±0.51 (17)	4.20±0.72 (5)	4.46±0.47 (4)	4.47±0.24 (2)	4.66 (1)	4.55±0.62 (3)	4.3 (1)	5.05 (1)	4.02±0.63 (43)	2.70	5.34
Mass gain per day (% FPM)[c]	1.00±0.23 (2)	0.99±0.01 (2)	0.79±0.10 (5)	0.78±0.09 (5)	0.89±0.08 (3)	0.94 (1)	0.82 (1)	0.85±0.12 (2)	— (1)	0.71 (1)	0.84±0.12 (27)	0.62	1.17
Total mass gain (kg)	95.3±44.0 (2)	82.1±5.9 (3)	91.4±12.7 (15)	98.6±14.2 (5)	99.4±12.6 (4)	103.0±1.8 (2)	116.6 (1)	110.9±18.5 (3)	105.2 (1)	125.7 (1)	96.0±16.6 (38)	61.9	127.1
Total mass gain (% FPM)[c]	23.6±8.8 (2)	20.6±1.6 (2)	18.5±3.0 (3)	18.4±2.3 (5)	19.5±2.6 (3)	21.1 (1)	20.6 (1)	20.1±3.8 (2)	— (0)	17.7 (1)	19.4±3.4 (22)	13.3	29.8
Efficiency of mass transfer[h]	65.0±16.6 (2)	60.3±1.1 (2)	63.5±3.2 (3)	54.0±5.3 (5)	55.7±4.2 (3)	60.2 (1)	57.7 (1)	59.6±7.6 (2)	— (0)	— (0)	57.5±7.0 (22)	44.6	76.7

NOTE: Numbers shown are mean ± 1 SD. Sample sizes are in parentheses.

[a] Totals include an additional 3 to 10 observations on females of unknown age. Sample size refers to the number of different mothers; since up to 8 females are included in more than one age group for some variables (e.g., pup weaning mass), the total N is sometimes less than the sum of N's for each age group.

[b] See Reiter, Panken, and Le Boeuf (1981) for additional data on lactation length.

[c] The large 12-year-old female for which we have parturition and midlactation mass ("Bertha") is not included here because she departed before a final weight was obtained. Given that she is the heaviest female ever recorded, it seems useful to have an estimate of her departure mass, which can be done in several ways: (1) extrapolating her actual mass loss rate (8.6 kg/day from 3 to 15 days postpartum) to departure yields a departure mass of 478 kg and a mass transfer efficiency (MTE) of 74.1% during the latter half of lactation; (2) using the observed MTE (43.7% from 3 to 15 days postpartum) to estimate mass loss rate during the second half of lactation (14.6 kg/day) yields a departure mass of 406 kg; (3) assuming an MTE of 58% (the mean for all females) during the unobserved period of lactation gives an estimated mass loss rate of 11.0 kg/day and a departure mass of 449 kg (plotted in fig. 10.1); and (4) finally, her estimated rate of mass loss during the second half of lactation, based on the regression on initial mass (appendix 1), is 9.9 kg/day, giving a departure mass of 461 kg. We estimate that Bertha's departure mass was between 400 and 480 kg.

[d] Mass of female and pup at initial treatment is given to allow comparisons with other studies that did not extrapolate mass to parturition.

[e] Mass-specific values are given as percentages of female parturition mass (% FPM).

[f] Data on lactation duration and all other variables include only the mother-pup pairs in which either the female or the pup (or both) were weighed.

[g] Weaning mass expressed as a percentage of female mass at departure.

[h] Efficiency of mass transfer = (pup mass gain per day / female mass loss per day) × 100.

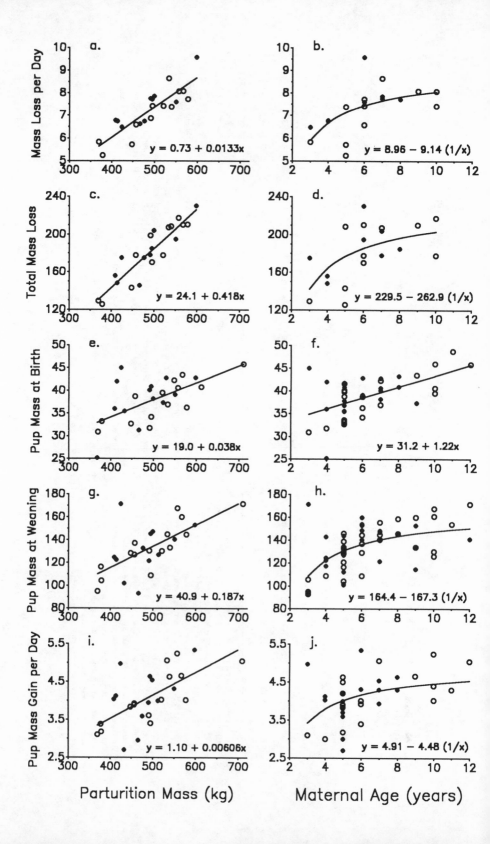

a. $y = 0.73 + 0.0133x$

b. $y = 8.96 - 9.14 (1/x)$

c. $y = 24.1 + 0.418x$

d. $y = 229.5 - 262.9 (1/x)$

e. $y = 19.0 + 0.038x$

f. $y = 31.2 + 1.22x$

g. $y = 40.9 + 0.187x$

h. $y = 164.4 - 167.3 (1/x)$

i. $y = 1.10 + 0.00606x$

j. $y = 4.91 - 4.48 (1/x)$

Mass Loss per Day

Total Mass Loss

Pup Mass at Birth

Pup Mass at Weaning

Pup Mass Gain per Day

Parturition Mass (kg)

Maternal Age (years)

clude that the results do not support the hypothesis that maternal effort varies as a function of pup sex. Therefore, further analyses use the pooled data set.

Gestation Effort: Pup Mass at Birth. Pup birth mass increased significantly with maternal mass and age (fig. 10.3e, f; table 10.4). Maternal investment during the period of gestation, as indicated by pup birth mass expressed as a fraction of mother's parturition mass, declined significantly with increasing maternal mass ($y = 11.44 - 0.0075x$) and maternal age ($y = 9.36 - 0.234x$) (tables 10.3 and 10.4). A similar negative relationship between relative neonatal mass and maternal mass is also found in interspecific comparisons among phocids (Kovacs and Lavigne 1986) and in other mammals (Millar 1977; Robbins and Robbins 1979; Gittleman 1986). Due to their smaller size, young mothers (3–5 years) put forth a greater gestation effort (pup birth mass was $8.64 \pm 1.28\%$ of maternal mass, N = 8) than physically mature, older mothers ($7.43 \pm 0.63\%$, N = 16; Wilcoxen two-sample test, $z = 2.358$, $p < .05$). Maternal age had no significant effect on pup birth mass, however, after accounting for the effect of maternal body mass (table 10.4). Analyses of pup mass at initial treatment (0–5 days postpartum) yielded similar results, although the relationship with age appeared more asymptotic.

Lactation Effort: Maternal Mass Loss. Daily mass loss of lactating females was strongly and positively correlated with parturition mass and, to a weaker degree, with female age (fig. 10.3a, b; table 10.4). This reflects both increased absolute maintenance costs and increased mass transfer to the pup with larger maternal size (maternal mass loss per day vs. pup mass gain per day: $r = .73$, N = 22, $p < .0001$). Independent of maternal mass, age was not significantly correlated with rate of mass loss during lactation (table 10.4).

Total mass loss over the lactation period increased with female mass and age in a similar fashion (fig. 10.3c, d; table 10.4). Total percentage of body mass lost over lactation, our principal index of female reproductive effort, varied between 31 and 41% and showed no significant correlation with maternal mass or age (table 10.4, fig. 10.4). Partial correlation and residual analyses yielded the same results: age had no discernible effect on breeding mass loss after controlling for the effects of maternal mass (table 10.4).

◁——

Fig. 10.3. Absolute measures of maternal investment as a function of female parturition mass and age for male (closed circles) and female (open circles) pups. (*A, B*) Average daily mass loss of mother. (*C, D*) Total mass loss of mother over lactation period. (*E, F*) Pup mass at birth. (*G, H*) Pup mass at weaning. (*I, J*) Average daily mass gain of pup. All units of mass are in kg. See table 10.4 for sample sizes, correlation coefficients, and levels of significance.

TABLE 10.4 Pearson product-moment correlation coefficients (r) between absolute and relative measures of maternal investment (MI) and maternal parturition mass and age.

| | Absolute measures of MI | | Mass-specific measures of MI[a] | | Adjusted MI[b] |
	Mass vs.	Age[c] vs.	Mass vs.	Age vs.	Age vs. Age
Mother					
Mass loss per day	0.86**** (22)	0.59** (19)	−0.21 (22)	−0.18 (19)	−0.10 (19)
Mass loss per season	0.91**** (22)	0.59** (19)	0.25 (22)	0.04 (19)	−0.13 (19)
Lactation duration	0.46* (29)	0.29* (69)	N/A	N/A	−0.10 (24)
Pup					
Birth mass	0.62*** (27)	0.56*** (41)	−0.58** (27)	−0.52** (24)	−0.03 (24)
Weaning mass	0.66*** (27)	0.56**** (63)	−0.33 (27)	−0.26 (22)	0.03 (22)
Mass gain per day	0.65*** (27)	0.41* (39)	−0.27 (27)	−0.29 (22)	−0.04 (22)
Total mass gain	0.66*** (22)	0.44** (37)	−0.10 (22)	−0.17 (20)	−0.03 (20)
Mass transfer efficiency[d]	−0.19 (22)	−0.37 (19)	N/A	N/A	−0.08 (19)

NOTE: Sample sizes are in parentheses.

[a] Mass-specific measures of MI are expressed per kg of maternal parturition mass.

[b] Maternal investment variables were adjusted for maternal age by regressing them against parturition mass for known-age females and then correlating the residuals with maternal age. Partial correlation coefficients of MI versus age, controlling for parturition mass, yielded similar results (i.e., r's differed by no more than ± .06).

[c] All absolute MI indices were regressed on the inverse of age, except for birth mass in which a linear regression on age was used (see fig. 10.3).

[d] Pup mass gain per day/female mass loss per day.

*p < .05.   **p < .01.   ***p < .001.   ****p < .0001.   N/A = Not applicable.

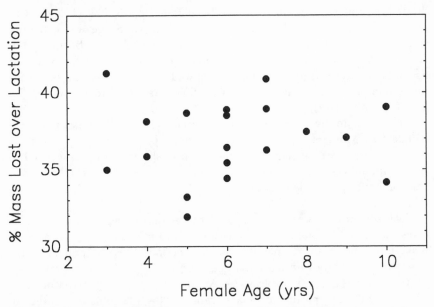

Fig. 10.4. Percentage of female parturition mass lost over the lactation period as a function of maternal age (r = .04, N = 19, p = .87).

Lactation Effort: Pup Mass Gain.   Mass gain of pups provided another energetic index of maternal investment (e.g., Ortiz, Le Boeuf, and Costa 1984). Pup mass gain per day and over the entire lactation period rose significantly with female mass and age (fig. 10.3i, j; table 10.4). Once again, maternal age had no apparent effect on mass gain of pups after adjusting for maternal mass (table 10.4).

On average, pups gained 57.5% of the mass lost by their mothers (table 10.3). We expected that the efficiency of mass transfer would increase with maternal size since a female's mass-specific energy expenditure for maintenance metabolism (i.e., her "metabolic overhead," Fedak and Anderson 1982) should decline with increasing mass. Thus, a larger mother should be able to allocate more energy to milk production and relatively less to her own maintenance costs. The data did not support this prediction; in fact, there was a nonsignificant tendency in the opposite direction, with younger mothers being somewhat more efficient (tables 10.3, 10.4).

Lactation Duration.   Duration of the lactation period averaged 25 to 26 days and was positively correlated with maternal mass (table 10.4). Mean lactation duration increased from age 3 to 5 and then remained approximately constant among older females (table 10.3). Chemical immobilization of the female near the end of lactation sometimes triggered a premature departure, and so these figures may be biased slightly low (1–2 days at most). The relationship between duration of stay and age was generally similar to that found by J. Reiter, K. J. Panken, and B. J. Le Boeuf (1981). A comparison of experimental and control groups matched for maternal age, year of study, and breeding site showed no significant effects of manipulation on lactation duration or weanling mass (D. Crocker and B. Le Boeuf, unpubl. data).

There was an inverse relationship between the duration and intensity of female RE. Mothers that lost mass at a high daily rate, relative to their body size, nursed their pups for a significantly shorter period compared to those with a less intense maternal effort (partial r = −0.50, N = 22, p < .025, controlling for parturition mass).

Total Maternal Investment: Weanling Mass.   A pup's mass at weaning indicates the mother's total absolute investment in the young, the sum of birth mass plus mass gain during the nursing period. Weaned pup mass increased significantly with maternal mass (fig. 10.3g) and with maternal age up to 6 years (fig. 10.3h, table 10.4). At weaning, pups weighed a mean of 27.2% of mother's parturition mass (range = 20–31%, plus one outlier of 40%), and this was not significantly associated with maternal age (table 10.4). Likewise, controlling for the effect of maternal body mass on weanling mass with partial correlation and residual analyses demonstrated no effect of age independent of maternal mass (table 10.4). The outlier mentioned above may reflect milk stealing by the pup, a behavior that confounds measurements of maternal investment that rely on pup energy gain (see appendix 10.2).

Summary: Effects of Maternal Mass, Age, and Pup Sex on Female RE.   Larger mothers invested absolutely but not relatively more in offspring during lactation than smaller mothers. Female RE during gestation, as reflected in relative neonatal mass, was negatively correlated with maternal age and mass. Changes in absolute measures of maternal investment with age generally tracked age-specific changes in body mass, increasing from age 3 to 6 and then reaching a plateau. Age had no discernible effect on any index of female RE (including gestation and lactation components) when the effect of maternal mass was controlled statistically. With the exception of female mass loss rate, all other indexes of maternal investment did not vary significantly with pup sex.

Male Reproductive Effort in Relation to Age, Mass, and Dominance Rank
For all ages combined (N = 71 individuals), male elephant seals lost a mean of 0.37 ± 0.08% of their arrival mass per day and 33.1 ± 7.0% of their

mass over the three-month breeding fast. Daily and seasonal mass loss increased significantly with body mass at arrival ($r_s = .78$, $N = 75$ measurements, $p < .0001$; $r_s = .82$, $N = 75$, $p < .0001$, respectively), with standard length ($r_s = .65$, $N = 87$, $p < .0001$; $r_s = .61$, $N = 75$, $p < .0001$, respectively), and with age class (table 10.5), as expected. On a mass-specific basis, however, breeding mass loss was weakly correlated with body mass (daily: $r_s = .21$, $N = 75$, $p = .08$; seasonal: $r_s = .31$, $N = 75$, $p < .01$) and was not correlated wlth length (daily: $r_s = .04$, $N = 75$, $p = .71$; seasonal: $r_s = .05$, $N = 75$, $p = .71$) or age (table 10.5). The lack of an age effect for SA3 and older classes was confirmed with an ANCOVA that showed no significant age-related variation in daily mass loss ($F = 1.335$, $df = 2, 67$, $p = .27$) or seasonal mass loss ($F = 1.694$, $df = 2, 67$, $p = .19$) when controlling for the effect of arrival mass; similar results were obtained using length as the covariate. Note that the above positive correlations between measures of mass loss and arrival mass are probably inflated due to the autocorrelation of measurement errors (i.e., an overestimate of arrival mass would cause an overestimate of both absolute and relative mass loss). Length was measured with greater accuracy than mass and showed no correlation with daily or seasonal RE (i.e., percent mass loss).

The mean duration of stay at the Año Nuevo rookery was 91.0 ± 10.9 days (range = 51–109, $N = 71$, excluding the SA2 male) and did not vary significantly with age among the SA3 to adult classes (table 10.5). This value may be somewhat high because our sample was probably biased against seals that departed early and were therefore less likely to be weighed or photographed a second time. There is no reason to believe that such a bias would vary with age class; Clinton (this volume) showed that average breeding haul-out duration remains fairly constant from age 6 onward.

There is a dramatic increase in male mating effort from 4 to 6 years of age, as demonstrated by the following behavioral measures: increasing duration of stay on the rookery; shifting of the haul-out period so that it overlaps the estrous period; and increasing time spent in proximity to harems (Cox 1983; Clinton 1990; Deutsch 1990; see fig. 9.1, chap. 9). Only one pubescent male (SA2 class) was weighed in this study, and he lost 27% of his mass over a 60-day period (table 10.5). Multiplying the average haul-out duration for the SA2 class (Clinton 1990: a minimum mean of 27 ± 20 days, $N = 53$; Deutsch 1990: mean of 56 ± 12 days, $N = 6$) by the mass-specific rate of mass loss for the weighed SA2 male (table 10.5) yields an estimate of 12 to 25% of arrival mass lost over the breeding fast for this age group. This suggests that young subadults devote a smaller proportion of their body stores to reproductive activity, compared to older males, by fasting for a shorter period on the rookery.

TABLE 10.5 Body size, absolute and mass-specific breeding mass loss, duration of stay on the rookery, and mating success of male northern elephant seals as a function of age class.

	Male age class				Significance[a]
	SA2	SA3	SA4	Adult	
Approximate age (years)	5	6	7	8–13	
Arrival mass (kg)	727 (1)	1217 ± 125 (12)	1446 ± 251 (11)	1814 ± 233 (62)	p < .0001
Departure mass (kg)	579 ± 71 (2)	793 ± 83 (18)	1002 ± 179 (18)	1192 ± 149 (75)	p < .0001
Standard length (cm)[b]	295 ± 5 (2)	340 ± 13 (24)	361 ± 16 (21)	382 ± 15 (89)	p < .0001
Mass loss per day (kg)	3.30 (1)	4.2 ± 0.87 (11)	4.59 ± 1.30 (14)	6.99 ± 2.02 (58)	p < .0001
% Mass loss per day	0.45 (1)	0.36 ± 0.07 (10)	0.34 ± 0.08 (10)	0.38 ± 0.08 (52)	NS
Total mass loss (kg)	198 (1)	392 ± 79 (10)	409 ± 113 (10)	631 ± 186 (52)	p < .0001
% Mass loss over season	27.2 (1)	32.0 ± 4.5 (10)	29.5 ± 7.4 (10)	34.1 ± 7.1 (52)	NS
Duration of stay (days)[c]	60 (1)	89.9 ± 10.7 (10)	87.8 ± 6.7 (10)	92.2 ± 11.3 (52)	NS
Mating success (ENFI)[c]	0.0 (1)	0.1 ± 0.3 (10)	0.0 ± 0.0 (10)	13.0 ± 18.5 (51)	p < .0001

NOTE: Numbers shown are mean ± 1 SD. Sample sizes are in parentheses. Mass-specific values are given as percentages of estimated arrival mass. Direct measurements and photogrammetric estimates of mass are combined. NS = Not significant (p > .05).

[a] A one-way analysis of variance test that does not require equal group variances (Rice and Gaines 1989, 1993) was used to determine whether variation among the three oldest age classes was significant for each variable.

[b] See Clinton (1990, this volume) for age-specific standard length and growth.

[c] Sample sizes correspond to data on total mass loss over the breeding season.

The highest levels of RE that we measured were exhibited by some adult males that lost nearly half of their body mass over the breeding season (maximums of 46.3% for a weighed animal and 50.5% for a photographed animal). This was considerably greater than the highest RE shown by younger males (SA3: 38.5%, weight; 35.1%, photograph; SA4: 38.8%, weight; 35.0%, photograph), perhaps because adults had the opportunity to achieve much higher mating success than subadults. The difference in sample size between the age groups provides an alternative explanation that cannot be completely ruled out, but it is unlikely that subadults ever lose much over 40% of their mass.

What can account for the twofold variation in RE among adult males? Total percent mass loss over the breeding season was slightly but not significantly higher for old adult males than for young adults (table 10.6). Age is positively correlated with dominance rank and mating success in males (Le Boeuf 1974; Le Boeuf and Reiter 1988; Clinton and Le Boeuf 1993), and the tendency for older bulls to invest more in reproduction, can be attributed to their higher rank. Although sample sizes were small, table 10.6 shows that the RE of young and old adults was quite similar for a given rank or mating success class. Males in the top third of the dominance hierarchy lost a significantly greater proportion of their arrival mass over the season ($36.0 \pm 7.8\%$, $N = 23$) than did bulls in the middle third ($32.2 \pm 5.8\%$, $N = 18$) and the bottom third of the hierarchy ($32.9 \pm 6.9\%$, $N = 9$) ($p < .05$, table 10.6). Likewise, males that achieved high mating success (ENFI > 15) devoted a greater effort to reproduction ($36.3 \pm 6.1\%$ of mass lost, $N = 14$) compared to adults with moderate sexual success (ENFI of 5–15; $32.6 \pm 9.1\%$, $N = 13$) or to those who copulated relatively few times (ENFI < 5; $33.8 \pm 6.5\%$, $N = 24$). The difference in percent mass loss between males with high mating success (median ENFI = 28.8) and those with low to moderate success (median ENFI = 2.0) was not significant, however ($p = .098$, table 10.6).

Summary: Effects of Age Class and Dominance Status on Male RE.   Error inherent in the photogrammetric measurement of body mass created considerable variation in mass loss variables among individuals, which made it more difficult to detect the real effects of potential factors such as age or dominance status on male mating effort. Nevertheless, the following patterns stood out. 1. The energetic component of male RE increased from age 5 (SA2) to age 6 (SA3), due to increased time spent on the rookery, and then remained constant to adulthood; preliminary data indicate no effect of age on RE of adults. 2. Sexually successful, high-ranking males lost a greater proportion of body mass than did subordinate adults and subadults, but the difference in RE was small compared to the discrepancy in mating success.

TABLE 10.6   Adult male reproductive effort (% mass loss over breeding season) as a function of age, dominance rank, and mating success (ENFI).

	Young adult (8–9 years)	Old adult (10–14 years)	All adults (8–14 years)
Dominance class			
High-rank (top 1/3)	38.2 ± 8.8	37.5 ± 5.4	36.0 ± 7.8
	(2)	(8)	(23)
Low- to mid-rank (bottom 2/3)	31.4 ± 4.8	31.3 ± 5.3	32.4 ± 6.1
	(8)	(3)	(27)
Mating success class			
High (ENFI > 15)	32.0	38.5 ± 5.8	36.3 ± 6.1
	(1)	(7)	(14)
Low to moderate	32.8 ± 6.2	33.3 ± 5.1	33.4 ± 7.4
(ENFI = 0–15)	(9)	(5)	(37)
All dominance and ENFI classes	32.7 ± 5.9	36.4 ± 5.9	
	(10)	(12)	

NOTE: Numbers shown are mean ± 1 SD. Sample sizes are in parentheses. Direct measurements and photogrammetric estimates of mass are combined. Results of Student's t tests: high vs. low dominance class (t = 1.81, df = 48, p = .038, one-tailed test); high vs. low ENFI class (t = 1.31, df = 49, p = .098, one-tailed test); and young vs. old adult (t = 1.43, df = 19, p = .17, two-tailed test). Excluding the young adult-high ENFI class (N = 1), t tests comparing adult age groups for a given dominance or ENFI class were not significant (p. > .70), whereas those comparing dominance or ENFI class for a given adult age group approached significance (.05 < p < .10, one-tailed tests).

DISCUSSION

Comparison of Reproductive Effort between the Sexes

This study reveals important similarities and differences in the magnitude and timing of reproductive effort between male and female northern elephant seals. The most striking similarity is in the energetic index of RE, adults of both sexes losing slightly more than one-third of their body mass, on average, while on the breeding rookery. This finding suggests that a common physiological mechanism, perhaps related to changing body composition or increasing protein catabolism (Groscolas 1986; Cherel, Robin, and Le Maho 1988), may underlie the termination of the breeding fast in both sexes.

There is no a priori reason why male and female RE should be equivalent, since the factors affecting reproductive success differ so widely between the sexes in elephant seals. In males, mating success achieved for a given effort depends on the number, competitive ability, and RE of other males on the rookery (Trivers 1972; Warner 1980). Intense male intrasexual competition for mates has been a strong selective force favoring continued growth and delayed maturity, thereby molding the level and developmental timing of mating effort. Consequently, young male elephant seals invest little in reproduction and much in increased growth during the first two or three years that same-age females are pupping, a common phenomenon in polygynous species (Selander 1972; Clutton-Brock, Guinness, and Albon 1982). Compared to males, competitive social interactions exert less influence on female RS (see Reiter, Panken, and Le Boeuf 1981) and, therefore, on the magnitude and ontogeny of maternal effort. Natural selection favors an investment pattern in females that optimizes the opposing effects of energy transfer to the pup on maternal and offspring fitness, relationships that are affected by fasting physiology, foraging ecology, thermoregulatory requirements at sea, and other ecological factors.

Data on mass change for four other phocid species allow comparison of RE between the sexes. The mean percentage of body mass lost over lactation is in the range of 30 to 40% for females of all five species (table 10.7). Breeding mass loss was similar for the two sexes in both species of elephant seals but higher for females than males in gray, Weddell, and harbor seals (table 10.7). The energetic cost of reproduction in the sexually monomorphic harp seal, *Phoca groenlandica*, appears to be much greater for females than males based on changes in blubber thickness over the breeding season (Sergeant 1973). These data suggest that only in the most polygynous seal species does male mating effort approach the level of female parental effort in terms of energetic costs.

Energy allocation to reproduction (i.e., breeding activity, gonads, gametes, and young) is far greater in female golden-mantled ground squirrels, *Spermophilus saturatus*, than in males (Kenagy 1987; Kenagy, Sharbaugh, and Nagy 1989), and this is probably the case for most mammals. But is this the best measure of RE? Our use of relative mass loss as an index of RE in fasting pinnipeds included energy devoted to maintenance metabolism while breeding, and so it was not a measure of reproductive energy expenditure per se. Nevertheless, since elephant seals and some other pinniped species must fast in order to breed (see Riedman 1990), it is reasonable to include this maintenance metabolic expenditure as part of the total energy cost of reproduction. Because the mating strategy of males in polygynous species often entails reduced food intake during a period of high activity, males may deplete their energy reserves to a greater extent during the mating period than females do during pregnancy and lactation (e.g., Leader-Williams and Ricketts 1981; Michener and Locklear 1990), even though actual reproductive energy expenditure may be much higher in females. Thus, we consider percent mass loss (an index of percent energy loss) while breeding to be a more appropriate measure of RE in fasting seals than energy allocation to strictly reproductive functions.

TABLE 10.7   Comparison of reproductive effort (mean percent mass loss over the breeding season) of adult male and female phocids.

Species	Male	Female	Sources
Northern elephant seal	36.2 (13)	36.8 (22)	1,2,3,4
(*M. angustirostris*)	[34.1 (52)][a]		
Southern elephant seal	33.5 (4)	34.4 (96)	5,6
(*M. leonina*)		36.3 (15)	6,7
Gray seal	16.8 (33)	38.9 (43)	8,9,10,11
(*Halichoerus grypus*)	29.6 (59)	30.4–39.5 (54)	12,13,14,20
Weddell seal[b]	30.0 (6)	37 (21)	15,16
(*Leptonychotes weddellii*)	28.0 (17)[c]		16
Harbor seal[b]	11–14 (94)[c]	37 (67)[c]	17
(*Phoca vitulina*)	25 (21)[d]	> 33 (13)[c,e]	18,19

NOTE: Sample sizes are in parentheses.
SOURCES OF DATA: 1 = This study; 2 = Costa et al. 1986; 3 = Deutsch, Haley, and Le Boeuf 1990; 4 = Crocker 1992; 5 = Fedak et al., this volume; 6 = T. Arnbom, pers. comm.; 7 = McCann, Fedak, and Harwood 1989; 8 = Fedak and Anderson 1982; 9 = Anderson and Fedak 1985; 10 = Anderson and Fedak 1987; 11 = Fedak and Anderson 1987; 12 = Bowen and Stobo 1991; 13 = Iverson et al. 1993; 14 = W. D. Bowen, pers. comm.; 15 = Hill 1987; 16 = Bartsh, Johnston, and Siniff 1992; 17 = Härkönen and Heide-Jørgensen 1990; 18 = Walker and Bowen, in press; 19 = Bowen, Oftedal, and Boness 1992; 20 = Twiss 1991.
[a] The bracketed figure includes photogrammetric estimates and direct measurements of male mass, whereas all other values in the table are based on animals weighed directly with a scale.
[b] Female Weddell and harbor seals may forage during lactation (Testa, Hill, and Siniff 1989; Bowen, Oftedal, and Boness 1992); Boness, Bowen, and Oftedal, in press), thus violating the assumption of fasting necessary for a valid comparison.
[c] These figures were based on cross-sectional samples, whereas all other values were calculated from longitudinal measurements of mass change of marked individuals.
[d] Plasma lipid data on adult male harbor seals show no indication of feeding during the first half of the mating period (Walker and Bowen, in press). The 25% figure would overestimate mass loss, however, if males initiate or increase feeding activity during the latter half of this period.
[e] Female harbor seals lost an estimated 33% of parturition mass over the initial 80% of the 24-day lactation period. Extrapolation of mass loss to weaning (42%) would probably overestimate the actual value because most females appear to forage during late lactation (Boness, Bowen, and Oftedal, in press).

Differences in the pattern of reproductive effort between male and female elephant seals are related to their divergent reproductive strategies, as follows:

1. Rate of investment in reproduction, measured as mass-specific daily mass loss, was over three times higher for females than males. This has the effect of reducing the metabolic overhead costs of females relative to males. About 40% of the energy lost by females during lactation is utilized for maintenance and activity metabolism, the rest being transferred to the pup as milk (Costa et al. 1986). On the con-

trary, most of the energy males expend during the breeding season is probably allocated to maintenance metabolism since 85 to 95% of their time is spent resting (Sandegren 1976; Deutsch 1990).

2. Duration of RE was three times longer for males than females, reflecting the importance of early male arrival to obtaining high rank and of a lengthy stay to maximize chances for copulation. A consequence of this longer effort is that males had two fewer months than females to recover lost energy stores and to gain additional mass before the next breeding season. Of course, females invest energy in gestation for eight months of the year, which is not reflected in these comparisons (see Deutsch, Haley, and Le Boeuf 1990).

3. Variation in RE was greater among males than females. Male mating effort was quite flexible, as evidenced by the more than twofold variation in duration of the breeding fast and by great variability in measures of the intensity of effort, such as the proportion of time spent competing near harems (Deutsch 1990) and mass-specific rate of mass loss (Deutsch, Haley, and Le Boeuf 1990). Consequently, some young subordinate males invested considerably less than any female did, and the RE of some high-ranking males soared well above the maximum for females. In contrast, three observations indicate that flexibility in adjusting maternal effort during lactation is limited by energetic considerations. First, mothers that invested at a higher relative rate weaned their pups sooner, suggesting the existence of a cost ceiling. Second, females that lose their pups and therefore invest relatively little energy in lactation remain on the rookery longer than those who rear pups successfully (Reiter, Panken, and Le Boeuf 1981). Third, a 3-year-old mother ("Bat") who nursed two pups simultaneously lost mass at a rate similar to other females of her size (observed = 6.35 kg/day; expected = 6.25 kg/day), and the pups gained mass at a combined rate similar to single pups (observed = 3.46 kg/day; expected = 3.62 kg/day). The minimum RE of parous females would have been lower than 30% mass loss if we had included those that lost their pups and did not nurse. But since 90 to 95% of females pupping on the Año Nuevo mainland are successful in raising their pups to normal weaning age (B. Le Boeuf, unpubl. data), including unsuccessful mothers in our sample would not have altered the above finding.

This sex difference in the variance in RE mirrors the great difference in the variance in reproductive success between the sexes (Le Boeuf and Reiter 1988). Females must invest some minimum RE to produce a healthy offspring, whereas males that invest even less effort have some chance of siring a pup. Extremely high levels of maternal effort (over 40% of body mass lost) probably benefit a pup relatively little compared to the future costs incurred by the mother, but alpha males can raise their current RS substantially through continued high mating effort. Thus, the sex difference in RE variation reflects the fundamentally different means by which male and female elephant seals maximize their RS.

Reproduction can be a risky activity, sometimes even leading to death (e.g., Wilkinson and Shank 1976; Brunton 1986), so a complete sex comparison must also examine this facet of RE. Frequency of external wounds, the measure of risk used in this study, was many times greater in males than females, reflecting the violent competition among males for high rank and access to females. Similar sex differences in frequency of injury have been found in other polygynous mammals (e.g., Smith 1966; Michener and Locklear 1990). We note, however, that the observed injury rate for females may be an underestimate because they were sampled in harems, but they are most likely to get injured during departure from the rookery. The risk of receiving a lethal injury during agonistic or sexual interactions is relatively small for both sexes at Año Nuevo; female mortality due to male-inflicted injuries (0.1% per year; Le Boeuf and Mesnick 1991) appears to be about twice as high as that for males (Deutsch 1990; C. Deutsch and B. Le Boeuf, unpubl. data). Female elephant seals probably incur a higher incidence of fatal injuries than males because of the large sex difference in mass, the large canines of males, the bites to the female's neck during typical sexual behavior, and the emaciated condition of the female on departure, when she is exposed to mating attempts by aggressive and sexually motivated peripheral bulls (Mesnick and Le Boeuf 1991).

Differences in life history parameters between the sexes may be due to differences in the timing and magnitude of RE. In Richardson's ground squirrels, *S. richardsonii*, for example, males deplete their fat stores to a greater degree during the mating period than females do during lactation, and this is associated with higher male mortality (Michener and Locklear 1990). The authors conclude that the costs of mating effort for males are greater than the costs of parental effort for females in this species. This pattern is taken to the extreme by the marsupial shrew, *Antechinus stuartii*, in which all males die soon after the mating season, whereas females are iteroparous (Lee, Wooley, and Braithwaite 1982). In northern elephant seals, the survivorship curves are roughly similar for both sexes (see fig. 22.2 in Le Boeuf and Reiter 1988), although longevity is somewhat greater for females. This is consistent with the similar magnitude of RE for the sexes found in this study.

Age-specific Reproductive Effort and Fitness Consequences

Variation in the energetic component of RE was not associated with age for males 6 years and older or for parous females. This finding, based on mass loss, is corroborated by complementary data showing no correlation between mass-specific energy expenditure and milk energy output of lactating females and age (Crocker 1992; D. Crocker, unpubl. data) and no age-related variation in male time-activity budgets (Deutsch 1990). Unfortunately, very old females, with the lowest reproductive value and the highest predicted RE, were not included in our sample, but a follow-up study (D. Crocker, unpubl. data) found that three 11- to 12-year-old mothers lost a similar proportion of their body mass over lactation (range = 31–34%) as younger mothers. W. J. Sydeman et al. (1991) suggested that age-specific increases in female aggressiveness, success in dominance interactions, and weaning success could be attributed to increasing maternal effort with age. The energetic data do not support this interpretation; instead we propose that the increase in size and experience with age accounts for the higher dominance status and weaning success of older mothers. Even though the proportion of resources allocated to reproduction did not vary with age, it is still possible that younger, smaller seals incurred a greater risk by depleting their fat reserves to a lower, perhaps marginal, level than larger animals, as R. E. A. Stewart (1986) found for female harp seals. This was apparently not the case in elephant seals, however, since females weaned their pups on reaching 21 to 25% body fat, regardless of maternal age, parturition mass, or initial body composition (Crocker 1992; D. Costa and B. Le Boeuf, unpubl. data). Female investment in offspring was in direct proportion to maternal mass at parturition. This is similar to the pattern of maternal investment in southern elephant seals (McCann, Fedak, and Harwood 1989; Fedak et al., this volume) but unlike harp seals, in which small, young mothers produce weanlings of the same size and lose mass at the same rate as larger, older mothers (Stewart 1986; Kovacs, Lavigne, and Innes 1991).

The fitness costs of reproduction (e.g., reduced survival) should be positively related to the proximate energetic costs (i.e., RE), though probably in a nonlinear fashion, and this relationship should be manifested in the age-specific patterns of RE and mortality. This expectation is partially supported by the fact that age-specific mortality rates rise sharply upon the initiation of reproduction—at primiparity for females (age 3–4; Reiter and Le Boeuf 1991) and at puberty for males (age 5; Clinton, this volume). For females, this survival cost is associated with lactation, since mortality remains low during the year of their first pregnancy. Similarly, Clutton-Brock, Albon, and Guinness (1989) concluded that the fitness costs of reproduction in red deer, *Cervus elaphus*, are attributable to the energetic demands of lactation and not to gestation.

Two lines of evidence led us to expect the RE of reproductively active but physically immature "subadults" (i.e., 3- to 5-year-old females and 5- to 7-year-old males) to be less than that of physically mature animals. First, RS increases with age in both sexes, especially from puberty to physical maturity, due largely to the effects of intrasexual competition (see references given on p. 173). Young small females obtain a lower benefit (in terms of weaning success and weanling mass) for the same proximate cost (relative mass loss) as older, larger mothers. Even more striking is the fact that sub-adult males obtain little immediate benefit for the same mass-specific cost as low-ranking adult males; only 6 to 15% of 6- to 7-year-olds copulate (Clinton 1990). Second, a given RE seems to impose a greater strain on the younger breeders of both sexes, manifested as higher fitness costs. Females giving birth at age 3 suffer reduced survivorship and future fecundity relative to those delaying reproduction to age 4 or 5 (Huber 1987; Reiter and Le Boeuf 1991; but see Sydeman and Nur, this volume). Current mating success is negatively correlated with subsequent survival and fecundity for young males (6–8 years old) but not for older males (Clinton and Le Boeuf 1993; Clinton, this volume). Also, C. R. Cox (1983) found that 6-year-old males that spend a greater amount of time on the breeding beach suffer higher mortality than those exhibiting lower effort. The higher fitness costs incurred by young breeders may result from the simultaneous allocation of energy to growth and reproduction, leaving less available for storage or critical maintenance functions.

Given that benefits of reproductive investment increase and fitness costs decrease with age during the period between sexual and physical maturity, selection should favor a relatively low level of RE for "subadult" animals. This probably explains why most females delay primiparity beyond age 3 (Huber 1987; Reiter and Le Boeuf 1991) and why pubescent males (4–5 years old) invest little time or energy in mating attempts (Clinton 1990; Deutsch 1990), despite the theoretical benefits of early breeding (Sibly and Calow 1986). Still, the pattern of constant mean RE with age for parous females and for males older than 5 years is perplexing. Theoretically, the optimal level of investment at each age is determined by the age-specific functions relating RE to fitness benefits (i.e., current RS) and to fitness costs (i.e., reduction in residual reproductive value). These functions are extremely difficult to determine empirically, and few data are available (Charlesworth 1980). It is noteworthy, however, that a constant age-specific RE is theoretically plausible, at least among adults, as stated by B. Charlesworth (ibid. 246): "It is not difficult, in principle, to find cases in which an optimal life-history exists which has constant reproductive effort among the adult age-classes, and hence constant adult survival and fecundity."

One could reasonably argue that a small, young female must expend a relatively large proportion of her energy stores during lactation to produce a weanling that is large enough to survive well (see Fedak et al., this volume). The point at which survival from weaning to age 2 appears to drop off is at a weaning mass of only 80 to 90 kg (Le Boeuf, Morris, and

Reiter, this volume). A hypothetical young mother of 400 kg would produce a 34-kg newborn (from fig. 10.3) and nurse it for 23 to 26 days with a mass transfer efficiency of 63% (from table 10.3). This female would lose 105 kg, or only 26% of her parturition mass, to produce a 100-kg weanling. Clearly, young mothers invest substantially more (36% of body mass) than is required to produce a viable pup. So this argument can probably be rejected; but the future fecundity of offspring may be enhanced by maternal investment above what is necessary for offspring survival. We offer three other nonexclusive explanations that may account for the lack of age-related variation in RE.

One explanation is that the seals are able to tolerate moderate energy deficiencies up to some threshold level, and only when that threshold is exceeded does a cost of reproduction become apparent (Tuomi, Hakala, and Haukioja 1983). Large fluctuations in body mass (e.g., loss of one-third mass) are a normal part of the elephant seal's annual cycle (Costa et al. 1986; McCann, Fedak, and Harwood 1989; Deutsch, Haley, and Le Boeuf 1990; Kretzmann, Costa, and Le Boeuf 1993; Worthy et al. 1992) and may entail little or no fitness costs in healthy animals. While costs associated with additional increments of investment above a critical level are likely to escalate (Burley 1988), the benefits should approach an asymptote, particularly for females. In fact, the survival benefit of large weanling mass is equivocal (Le Boeuf, Morris, and Reiter, this volume). Crocker (1992) postulated a physiological mechanism, based on changing body composition and rate of protein catabolism, that may account for a threshold relationship between maternal investment and fitness costs. Circumstantial evidence for such a threshold phenomenon includes the fact that female body fat at the end of lactation lies in a narrow range, despite the large variation in initial fat stores, in lactation duration, in absolute rates of mass loss, and in rates of lean and adipose tissue loss (Crocker 1992; see also Gales and Burton 1987). An animal exceeding the hypothesized threshold level of RE might be faced with the choice of either depleting its blubber layer beyond what is necessary for thermoregulation at sea or oxidizing muscle and other lean tissue. While this cost-based explanation is plausible, the trade-off between reproduction and growth (e.g., Robinson and Doyle 1985; Green and Rothstein 1991) would seem difficult to avoid, and so attempting to breed as a growing subadult could reduce size and, hence, fecundity as an adult (Clinton 1990; Haley, Deutsch, and Le Boeuf, in press). Furthermore, the evidence of a trade-off between reproduction and survival for young animals also argues strongly for the existence of substantial benefits, either immediate or delayed, to early reproductive attempts.

The second possibility draws on the idea that in expanding populations the fitness benefit of early-born offspring is greater than that of late-born young, thus favoring early maturity and declining RE with age (Lewontin

1965; Charlesworth and León 1976). This may partially explain why females mature at a relatively young age in *M. angustirostris* and in exploited populations of *M. leonina* (Carrick et al. 1962), although the costs of early reproduction would appear to outweigh the benefits under the prevailing demographic conditions (Huber 1987; Reiter and Le Boeuf 1991). The hypothesized effect seems too small, however, to compensate for the extremely low mating success of subadult males, unless we have grossly underestimated their success. Who mates with virgin females and where this occurs is poorly known (Huber et al. 1991) and could potentially affect our interpretation. Overall, we think that this can be a partial explanation at best.

The final explanation for the unexpectedly high RE among physically immature seals is that experience gained when young improves future reproductive performance (Caro 1988; Robinson 1988). The benefits of early maternal experience to subsequent weaning success have been documented in elephant seals (Reiter, Panken, and Le Boeuf 1981; Sydeman et al. 1991). C. J. Deutsch (1990) presents circumstantial evidence for the importance of fighting and sexual experience to males; this includes the observation that subadult males engage in frequent mock-fighting activity in contexts in which no immediate reproductive benefit is possible (e.g., during the molt period). Even male weanlings exhibit behaviors during social interactions that resemble the fighting behavior of adult males (Rasa 1971; Reiter, Stinson, and Le Boeuf 1978), suggesting that early social experience plays an important role in reproducing successfully later in life (Fagen 1981). This raises the interesting question of whether such social play among juvenile males should be considered a form of early RE with delayed benefits that has evolved via sexual selection on adult males.

In conclusion, we found that the magnitude of reproductive effort in northern elephant seals, as estimated by relative mass loss, was similar for males and females but that the temporal patterning of effort differed between the sexes both within a breeding season and over development. The similarity in RE between the sexes was unexpected, given the large size dimorphism, divergent reproductive strategies, and the generally held view that female mammals invest more in reproduction than males. RE did not vary with age in males (older than 5 years) or parous females, despite an increase in reproductive success and a decline in the fitness costs of breeding with age for both sexes. Future studies should investigate whether the substantial interindividual variation in RE is related to the amount of energy stores available at the start of breeding, thus potentially linking foraging success at sea with reproductive success on land.

ACKNOWLEDGMENTS

Numerous students and colleagues assisted in fieldwork over the decade covered by this study, and to them we are very grateful. We also thank

M. Haley and J. Williams for their collaboration in the mass loss studies; K. Richer for collecting data on male injuries; M. Kretzmann for contributing data on pup mass and mass change; and the rangers at Año Nuevo State Park for their cooperation in the field. We are grateful to S. Blackwell, M. Bryden, R. Gentry, G. Kooyman, and J. Reiter for making helpful comments on earlier versions of this manuscript. The research was funded in part by the National Science Foundation and by the Biology Board of Studies and the Institute of Marine Sciences at the University of California, Santa Cruz.

APPENDIX 10.1

Calculation of Body Mass and Mass Loss

Body Mass at Arrival (Males) or Parturition (Females). Calculated by using measured rates of mass loss for each animal to extrapolate from date of first weighing (initial mass) back to arrival date (males) or parturition date (females). For 14 males weighed (directly or photogrammetrically) only at the start of the breeding fast, initial mass was extrapolated to arrival date using the regression between mass loss per day (y) and length (where $x = 0.9*$ standard length): $y = -86.87 + 91.42x - 31.85x^2 + 3.84x^3$ ($r^2 = .68$, $N = 28$, $p < .0001$; Deutsch, Haley, and Le Boeuf 1990). For 12 females weighed only at the start of lactation, initial mass was extrapolated to parturition date using the regression between mass loss per day (y) and initial mass (x): $y = 1.058 + 0.01306x$ ($r^2 = .62$, $N = 22$, $p < .0001$). All units of mass are given in kg; length is given in meters.

Body Mass at Departure from the Rookery (Both Sexes). Calculated by using measured rates of mass loss for each animal to extrapolate from date of last weighing (final mass) forward to departure date. For 31 males weighed only at the end of the breeding fast, final mass was extrapolated to departure date using the regression above. For 40 females weighed only near the end of lactation, the corresponding calculation was based on the regression between mass loss per day (y) and final mass (x): $y = 1.486 + .01797x$ ($r^2 = .52$, $N = 22$, $p = .0002$). Similar mass estimates could have been obtained for these 40 females (and the 12 above) using standard length to estimate mass loss rate ($r^2 = .57$, $N = 22$, $p < .0001$).

Total Mass Loss over the Breeding Season (Males) or the Lactation Period (Females). Equals estimated arrival/parturition mass minus estimated departure mass for animals weighed two or more times over the season. Females weighed twice over an interval of less than 18 days were excluded from mass loss analyses. Total mass loss expressed as a percentage of arrival/parturition mass was used as an index of the energetic component of RE.

Pup Mass at Birth. Since different pups were initially weighed from 0 to 5 days postpartum, we standardized pup mass by estimating mass at birth. Initial pup mass (kg) was regressed against pup age (days) for 41 pups 0 to 5 days of age, yielding the following linear regression: Initial mass = 36.9 + 2.2 (age); $r^2 = .29$. For pups 0 to 1 day old, birth mass was taken to equal initial mass. For pups 2 to 5 days old, a mass gain of 2.2 kg/day was assumed, starting at day 1; this amount was subtracted from initial mass to obtain estimated birth mass. Since other studies typically report initial mass and since mass gain of pups can be quite variable during the first week (McCann, Fedak, and Harwood 1989), we also report initial mass.

Pup Mass at Weaning. Pups were usually weighed within a few days of weaning. If weighed before weaning, measured mass gain per day for each pup was used to extrapolate from date of final weighing to weaning date. If weighed after weaning, the following equation was used to estimate mass at weaning since weanlings fast and lose mass:

$$\textit{Mass at weaning} = \textit{Measured mass} \cdot (e^{k \cdot d}),$$

where k = .00596 and d = number of days between weaning and weighing (modified from Ortiz, Costa, and Le Boeuf 1978; P. M. Morris, pers. comm.). Pups weighed more than 10 days after weaning were excluded from analysis.

Total Pup Mass Gain. Equals estimated mass at weaning minus estimated mass at birth.

APPENDIX 10.2

Comments on an Outlier for Pup Mass Gain

The outlier was a male pup of a 424 kg, 3-year-old mother ("G22") and deserves special mention because it illustrates some of the problems inherent in interpreting this type of data. Values for this individual lie well outside the expected range for pup mass gain and weaning mass (fig. 10.3g–j). There are at least three possible explanations for this finding: (1) high maternal investment; (2) high efficiency of mass transfer; and/or (3) milk stealing by the pup from other females. G22 gave birth to an exceptionally large pup (10 kg heavier than expected for her size; fig. 10.3e, f), and she lost 41.2% of her parturition mass, the highest value measured for a female. This large mass loss was due primarily to an unusually long lactation period (27 days) for a 3-year-old. Even taking G22's high RE into consideration, however, the pup was weaned about 30 kg heavier than expected. This fact, plus the suspiciously high mass transfer efficiency of 77% (range was 45–67% for the rest), suggests that the pup stole milk from neighboring females during the nursing period, although this was not observed. For this reason, maternal mass or energy loss is probably a more reliable index of investment than pup mass or energy gain.

REFERENCES

Alexander, R. D., and G. Borgia. 1979. On the origin and basis of the male-female phenomenon. In *Sexual Selection and Reproductive Competition in Insects*, ed. M. S. Blum and N. A. Blum, 417–440. New York: Academic Press.

Anderson, S. S., and M. A. Fedak. 1985. Grey seal males: Energetic and behavioural links between size and sexual success. *Animal Behaviour* 33: 829–838.

————. 1987. Grey seal, *Halichoerus grypus*, energetics: Females invest more in male offspring. *Journal of Zoology, London* 211: 667–679.

Bartsh, S. S., S. D. Johnston, and D. B. Siniff. 1992. Territorial behavior and breeding frequency of male Weddell seals (*Leptonychotes weddellii*) in relation to age, size, and concentrations of serum testosterone and cortisol. *Canadian Journal of Zoology* 70: 680–692.

Bell, G. 1980. The costs of reproduction and their consequences. *American Naturalist* 116: 45–76.

Bell, G., and V. Koufopanou. 1985. The costs of reproduction. *Oxford Surveys in Evolutionary Biology* 3: 83–131.

Boness, D. J., W. D. Bowen, and O. T. Oftedal. In press. Evidence of a maternal foraging cycle resembling that of otariid seals in a small phocid, the harbor seal. *Behavioral Ecology and Sociobiology*.

Bowen, W. D., D. J. Boness, and O. T. Oftedal. 1987. Mass transfer from mother to pup and subsequent mass loss by the weaned pup in the hooded seal, *Cystophora cristata*. *Canadian Journal of Zoology* 65: 1–8.

Bowen, W. D., O. T. Oftedal, and D. J. Boness. 1992. Mass and energy transfer during lactation in a small phocid, the harbor seal (*Phoca vitulina*). *Physiological Zoology* 65: 844–866.

Bowen, W. D., and W. T. Stobo. 1991. Reproduction in primiparous and young multiparous grey seal females: Is it successful? In *Proceedings of the Ninth Biennial Conference on the Biology of Marine Mammals*, December 5–9, 1991, Chicago, 9 (Abstract).

Boyce, M. S. 1981. Beaver life-history responses to exploitation. *Journal of Applied Ecology* 18: 749–753.

Boyd, I. L., and C. D. Duck. 1991. Mass changes and metabolism in territorial male Antarctic fur seals (*Arctocephalus gazella*). *Physiological Zoology* 64: 375–392.

Boyd, R., and J. B. Silk. 1983. A method for assigning cardinal dominance ranks. *Animal Behaviour* 31: 45–58.

Brunton, D. H. 1986. Fatal antipredator behavior of a killdeer. *Wilson Bulletin* 98: 605–607.

Burley, N. 1988. The differential-allocation hypothesis: An experimental test. *American Naturalist* 132: 611–628.

Calow, P. 1979. The cost of reproduction—A physiological approach. *Biological Review* 54: 23–40.

Campagna, C., B. J. Le Boeuf, M. Lewis, and C. Bisioli. 1992. The Fisherian seal: Equal investment in male and female pups in southern elephant seals. *Journal of Zoology, London* 226: 551–561.

Caro, T. M. 1988. Adaptive significance of play: Are we getting closer? *Trends in Ecology and Evolution* 3: 50–54.

Carrick, R., S. E. Csordas, S. E. Ingham, and K. Keith. 1962. Studies on the southern elephant seal, *Mirounga leonina* (L.). IV. Breeding and development. *CSIRO Wildlife Research* 7: 161–197.

Caswell, H. 1982. Optimal life history and the age-specific costs of reproduction. *Journal of Theoretical Biology* 98: 519–529.

Charlesworth, B. 1980. *Evolution in Age-Structured Populations*. Cambridge: Cambridge University Press.

Charlesworth, B., and J. A. León. 1976. The relation of reproductive effort to age. *American Naturalist* 110: 449–459.

Cherel, Y., J.-P. Robin, and Y. Le Maho. 1988. Physiology and biochemistry of long-term fasting in birds. *Canadian Journal of Zoology* 66: 159–166.

Clinton, W. L. 1990. Sexual selection and the life-history of male northern elephant seals. Ph.D. dissertation, University of California, Santa Cruz.

Clinton, W. L., and B. J. Le Boeuf. 1993. Sexual selection's effects on male life history and the pattern of male mortality. *Ecology* 74: 1884–1892.

Clutton-Brock, T. H. 1984. Reproductive effort and terminal investment in iteroparous animals. *American Naturalist* 123: 212–229.

Clutton-Brock, T. H., S. D. Albon, R. M. Gibson, and F. E. Guinness. 1979. The logical stag: Adaptive aspects of fighting in red deer (*Cervus elaphus* L.). *Animal Behaviour* 27: 211–225.

Clutton-Brock, T. H., S. D. Albon, and F. E. Guinness. 1989. Fitness costs of gestation and lactation in wild mammals. *Nature* 337: 260–262.

Clutton-Brock, T. H., F. E. Guinness, and S. D. Albon. 1982. *Red Deer: Behavior and Ecology of Two Sexes*. Chicago: University of Chicago Press.

Cody, M. L. 1971. Ecological aspects of reproduction. In *Avian Biology*, vol. 1, ed. D. S. Farner and J. R. King, 461–512. New York: Academic Press.

Cooper, C. F., and B. S. Stewart. 1983. Demography of northern elephant seals, 1911–1982. *Science* 219: 969–971.

Costa, D. P., B. J. Le Boeuf, C. L. Ortiz, and A. C. Huntley. 1986. The energetics of lactation in the northern elephant seal. *Journal of Zoology, London* 209: 21–33.

Cox, C. R. 1983. Reproductive behaviour of sub-adult elephant seals: The cost of breeding. In *Behavioral Energetics: The Cost of Survival in Vertebrates*, ed. W. P. Aspey and S. I. Lustick, 89–115. Columbus: Ohio State University Press.

Crocker, D. E. 1992. Reproductive effort and age in female northern elephant seals. Master's thesis, University of California, Santa Cruz.

Deutsch, C. J. 1990. Behavioral and energetic aspects of reproductive effort in male northern elephant seals (*Mirounga angustirostris*). Ph.D. dissertation, University of California, Santa Cruz.

Deutsch, C. J., M. P. Haley, and B. J. Le Boeuf. 1990. Reproductive effort of male northern elephant seals: Estimates from mass loss. *Canadian Journal of Zoology* 68: 2580–2593.

Fagen, R. 1981. *Animal Play Behavior*. London: Oxford University Press.

Fedak, M. A., and S. S. Anderson. 1982. The energetics of lactation: Accurate measurements from a large wild mammal, the grey seal (*Halichoerus grypus*). *Journal of Zoology, London* 198: 473–479.

———. 1987. Estimating the energy requirements of seals from weight changes. In *Approaches to Marine Mammal Energetics*, ed. A. C. Huntley, D. P. Costa, G. A. J. Worthy, and M. A. Castellini, 205–226. Lawrence, Kan.: Allen Press.

Gadgil, M., and W. H. Bossert. 1970. Life historical consequences of natural selection. *American Naturalist* 104: 1–24.

Gales, N. J., and H. R. Burton. 1987. Ultrasonic measurement of blubber thickness of the southern elephant seal, *Mirounga leonina* (Linn.). *Australian Journal of Zoology* 35: 207–217.

Gittleman, J. L. 1986. Carnivore life-history patterns: Allometric, phylogenetic, and ecological associations. *American Naturalist* 127: 744–771.

Gittleman, J. L., and S. D. Thompson. 1988. Energy allocation in mammalian reproduction. *American Zoologist* 28: 863–875.

Glucksman, A. 1974. Sexual dimorphism in mammals. *Biological Review of Cambridge Philosophical Society* 49: 423–475.

Green, W. C. H. 1990. Reproductive effort and associated costs in bison (*Bison bison*): Do older mothers try harder? *Behavioral Ecology* 1: 148–160.

Green, W. C. H., and A. Rothstein. 1991. Trade-offs between growth and reproduction in female bison. *Oecologia* 86: 521–527.

Groscolas, R. 1986. Changes in body mass, body temperature and plasma fuel levels during the natural breeding fast in male and female emperor penguins, *Aptenodytes forsteri*. *Journal of Comparative Physiology* 156B: 521–527.

Haley, M. P., C. J. Deutsch, and B. J. Le Boeuf. 1991. A method for estimating mass of large pinnipeds. *Marine Mammal Science* 7: 157–164.

———. In press. Size, dominance and copulatory success in male northern elephant seals, *Mirounga angustirostris*. *Animal Behaviour*.

Hamer, K. C., and R. W. Furness. 1991. Age-specific breeding performance and reproductive effort in great skuas, *Catharacta skua*. *Journal of Animal Ecology* 60: 693–704.

Hammill, M. O., C. Lydersen, M. Ryg, and T. G. Smith. 1991. Lactation in the ringed seal (*Phoca hispida*). *Canadian Journal of Fisheries and Aquatic Science* 48: 2471–2476.

Härkönen, T., and M.-P. Heide-Jørgensen. 1990. Comparative life histories of East Atlantic and other harbour seal populations. *Ophelia* 32: 211–235.

Hill, S. E. B. 1987. Reproductive ecology of Weddell seals (*Leptonychotes weddelli*) in McMurdo Sound, Antarctica. Ph.D. dissertation, University of Minnesota, Minneapolis.

Hirshfield, M. F., and D. W. Tinkle. 1975. Natural selection and the evolution of reproductive effort. *Proceedings of the National Academy of Science, USA* 72: 2227–2231.

Huber, H. R. 1987. Natality and weaning success in relation to age of first reproduction in northern elephant seals. *Canadian Journal of Zoology* 65: 1311–1316.

Huber, H. R., A. C. Rovetta, L. A. Fry, and S. Johnston. 1991. Age-specific natality of northern elephant seals at the South Farallon Islands, California. *Journal of Mammalogy* 72: 525–534.

Iverson, S. J., W. D. Bowen, D. J. Boness, and O. T. Oftedal. 1993. The effect of maternal size and milk energy output on pup growth in grey seals (*Halichoerus grypus*). *Physiological Zoology* 66: 61–88.

Kenagy, G. J. 1987. Energy allocation for reproduction in the golden-mantled ground squirrel. *Symposium of Zoological Society of London* 57: 259–273.

Kenagy, G. J., S. M. Sharbaugh, and K. S. Nagy. 1989. Annual cycle of energy and time expenditure in a golden-mantled ground squirrel population. *Oecologia* 78: 269–282.

Knapton, R. W. 1984. Parental investment: The problem of currency. *Canadian Journal of Zoology* 62: 2673–2674.

Kovacs, K. M., and D. M. Lavigne. 1986. Maternal investment and neonatal growth in phocid seals. *Journal of Animal Ecology* 55: 1035–1051.

Kovacs, K. M., D. M. Lavigne, and S. Innes. 1991. Mass transfer efficiency between harp seal (*Phoca groenlandica*) mothers and their pups during lactation. *Journal of Zoology, London* 223: 213–221.

Kretzmann, M. B. 1990. Maternal investment and the post-weaning fast in northern elephant seals: Evidence for sexual equality. Master's thesis, University of California, Santa Cruz.

Kretzmann, M. B., D. P. Costa, and B. J. Le Boeuf. 1993. Maternal energy investment in elephant seal pups: Evidence for sexual equality? *American Naturalist* 141: 466–480.

Laws, R. M. 1953. The elephant seal (*Mirounga leonina* Linn.). I. Growth and age. *Falkland Islands Dependencies Survey, Scientific Reports* 8: 1–62.

Leader-Williams, N., and C. Ricketts. 1981. Seasonal and sexual patterns of growth and condition of reindeer introduced into South Georgia. *Oikos* 38: 27–39.

Le Boeuf, B. J. 1974. Male-male competition and reproductive success in elephant seals. *American Zoologist* 14: 163–176.

———. 1981. Elephant seals. In *The Natural History of Año Nuevo*, ed. B. J. Le Boeuf and S. Kaza, 326–374. Pacific Grove, Calif.: Boxwood Press.

Le Boeuf, B. J., R. Condit, and J. Reiter. 1989. Parental investment and the secondary sex ratio in northern elephant seals. *Behavioral Ecology and Sociobiology* 25: 109–117.

Le Boeuf, B. J., D. P. Costa, A. C. Huntley, and S. D. Feldkamp. 1988. Continuous, deep diving in female northern elephant seals, *Mirounga angustirostris*. *Canadian Journal of Zoology* 66: 446–458.

Le Boeuf, B. J., D. P. Costa, A. C. Huntley, G. L. Kooyman, and R. W. Davis. 1986. Pattern and depth of dives in northern elephant seals, *Mirounga angustirostris*. *Journal of Zoology, London* 208: 1–7.

Le Boeuf, B. J., and S. Kaza, ed. 1981. *The Natural History of Año Nuevo*. Pacific Grove, Calif.: Boxwood Press.

Le Boeuf, B. J., and S. Mesnick. 1991. Sexual behavior of male northern elephant seals. I. Lethal injuries to adult females. *Behaviour* 116: 143–162.

Le Boeuf, B. J., Y. Naito, A. C. Huntley, and T. Asaga. 1989. Prolonged, continuous, deep diving by northern elephant seals. *Canadian Journal of Zoology* 67: 2514–2519.

Le Boeuf, B. J., and J. Reiter. 1988. Lifetime reproductive success in northern elephant seals. In *Reproductive Success: Studies of Individual Variation in Contrasting Breeding Systems*, ed. T. H. Clutton-Brock, 344–362. Chicago: University of Chicago Press.

Le Boeuf, B. J., M. Riedman, and R. S. Keyes. 1982. White shark predation on pinnipeds in California coastal waters. *Fishery Bulletin* 80: 891–895.

Lee, A. K., P. Wooley, and R. W. Braithewaite. 1982. Life history strategies of dasyurid marsupials. In *Carnivorous Marsupials*, vol. 1, ed. M. Archer, 1–11. Chipping Norton, NSW: Surry Beatty and Sons.

Lewontin, R. C. 1965. Selection for colonizing ability. In *The Genetics of Colonizing Species*, ed. H. G. Baker and G. L. Stebbins, 77–91. New York: Academic Press.

Loudon, A. S. I., and P. A. Racey, eds. 1987. *Reproductive Energetics of Mammals.* Symposia of the Zoological Society of London no. 57. Oxford: Oxford University Press.

Low, B. S. 1978. Environmental uncertainties and the parental strategies of marsupials and placentals. *American Naturalist* 112: 197–213.

McCann, T. S., M. A. Fedak, and J. Harwood. 1989. Parental investment in southern elephant seals, *Mirounga leonina. Behavioral Ecology and Sociobiology* 25: 81–87.

Maher, C. R., and J. A. Byers. 1987. Age-related changes in reproductive effort of male bison. *Behavioral Ecology and Sociobiology* 21: 91–96.

Maynard Smith, J. 1980. A new theory of sexual investment. *Behavioral Ecology and Sociobiology* 7: 247–251.

Mesnick, S., and B. J. Le Boeuf. 1991. Sexual behavior of male northern elephant seals. II. Female responses to potentially fatal mating encounters. *Behaviour* 117: 262–280.

Michener, G. R., and L. Locklear. 1990. Differential costs of reproductive effort for male and female Richardson's ground squirrels. *Ecology* 71: 855–868.

Millar, J. S. 1977. Adaptive features of mammalian reproduction. *Evolution* 31: 370–386.

Nur, N. 1984. Increased reproductive success with age in the California gull: Due to increased effort or improvement of skill? *Oikos* 43: 407–408.

Oftedal, O. T., D. J. Boness, and R. A. Tedman. 1987. The behavior, physiology, and anatomy of lactation in the Pinnipedia. *Current Mammalogy* 1: 175–245.

Oftedal, O. T., S. J. Iverson, and D. J. Boness. 1987. Milk and energy intakes of suckling California sea lion, *Zalophus californianus*, pups in relation to sex, growth, and predicted maintenance requirements. *Physiological Zoology* 60: 560–575.

Ortiz, C. L., D. P. Costa, and B. J. Le Boeuf. 1978. Water and energy flux in elephant seal pups fasting under natural conditions. *Physiological Zoology* 51: 166–178.

Ortiz, C. L., B. J. Le Boeuf, and D. P. Costa. 1984. Milk intake: An index of maternal investment in elephant seal pups. *American Naturalist* 124: 416–422.

Packard, G. C., and T. J. Boardman. 1987. The misuse of ratios to scale physiological data that vary allometrically with body size. In *New Directions in Ecological Physiology*, ed. M. E. Feder, A. F. Bennett, W. W. Burggren, and R. B. Huey, 216–239. Cambridge: Cambridge University Press.

Pärt, T., L. Gustafsson, and J. Moreno. 1992. "Terminal investment" and a sexual conflict in the collared flycatcher (*Ficedula albicollis*). *American Naturalist* 140: 868–882.

Pianka, E. R. 1976. Natural selection of optimal reproductive tactics. *American Zoologist* 16: 775–784.

Pianka, E. R., and W. S. Parker. 1975. Age-specific reproductive tactics. *American Naturalist* 109: 453–464.

Pugesek, B. H. 1981. Increased reproductive effort with age in the California gull (*Larus californicus*). *Science* 212: 822–823.

———. 1983. The relationship between parental age and reproductive effort in the California gull. *Behavioral Ecology and Sociobiology* 13: 161–171.

———. 1984. Age-specific reproductive tactics in the California gull. *Oikos* 43: 409–410.

Rasa, O. A. E. 1971. Social interaction and object manipulation in weaned pups of the northern elephant seal *Mirounga angustirostris*. *Zeitschift Tierpsychologie* 29: 82–102.

Reid, W. V. 1988. Age-specific patterns of reproduction in the glaucous-winged gull: Increased effort with age? *Ecology* 69: 1454–1465.

Reilly, J. J. 1989. The water and energy metabolism of breeding male grey seals (*Halichoerus grypus*). In *Proceedings of the Eighth Biennial Conference on the Biology of Marine Mammals*, December 7–11, 1989, Pacific Grove, Calif., 53 (Abstract).

Reilly, J. J., and M. A. Fedak. 1991. Rates of water turnover and energy expenditure of free-living male common seals (*Phoca vitulina*). *Journal of Zoology, London* 223: 461–468.

Reiter, J., and B. J. Le Boeuf. 1991. Life history consequences of variation in age at primiparity in northern elephant seals. *Behavioral Ecology and Sociobiology* 28: 153–160.

Reiter, J., K. J. Panken, and B. J. Le Boeuf. 1981. Female competition and reproductive success in northern elephant seals. *Animal Behaviour* 29: 670–687.

Reiter, J., N. L. Stinson, and B. J. Le Boeuf. 1978. Northern elephant seal development: The transition from weaning to nutritional independence. *Behavioral Ecology and Sociobiology* 3: 337–367.

Rice, W. R., and S. D. Gaines. 1989. One-way analysis of variance with unequal variances. *Proceedings of the National Academy of Sciences, USA* 86: 8183–8184.

———. 1993. Calculating p-values for ANOVA with unequal variances. *Journal of Statistical Computation and Simulation* 46: 19–22.

Riedman, M. 1990. *The Pinnipeds: Seals, Sea Lions, and Walruses.* Berkeley, Los Angeles, and Oxford: University of California Press.

Robbins, C. T., and B. L. Robbins. 1979. Fetal and neonatal growth patterns and maternal reproductive effort in ungulates and subungulates. *American Naturalist* 114: 101–116.

Robinson, B. W., and R. W. Doyle. 1985. Trade-off between male reproduction (amplexus) and growth in the amphipod *Gammarus lawrencianus*. *Biological Bulletin* 168: 482–488.

Robinson, S. K. 1988. Anti-social and social behaviour of adolescent yellow-rumped caciques (Icterinae: *Cacicus cela*). *Animal Behaviour* 36: 1482–1495.

Ryan, M. J. 1985. *The Túngara Frog: A Study in Sexual Selection and Communication.* Chicago: University of Chicago Press.

Sandegren, F. E. 1976. Agonistic behavior in the male northern elephant seal. *Behaviour* 57: 136–158.

SAS Institute. 1985. *SAS User's Guide: Statistics*. Ver. 5. Cary, N.C.: SAS Institute.

Selander, R. K. 1972. Sexual selection and dimorphism in birds. In *Sexual Selection and the Descent of Man, 1871–1971*, ed. B. G. Campbell, 180–230. Chicago: Aldine.

Sergeant, D. E. 1973. Feeding, growth, and productivity of Northwest Atlantic harp seals (*Pagophilus groenlandicus*). *Journal of Fisheries Research Board of Canada* 30: 17–29.

Sibly, R., and P. Calow. 1986. Why breeding earlier is always worthwhile. *Journal of Theoretical Biology* 123: 311–319.

Smith, M. S. R. 1966. Injuries as an indication of social behaviour in the Weddell seal (*Leptonychotes weddelli*). *Mammalia* 30: 241–246.

Sokal, R. R., and F. S. Rohlf. 1981. *Biometry*. San Francisco: W. H. Freeman and Co.

Stearns, S. C. 1976. Life history tactics: A review of the ideas. *Quarterly Review of Biology* 51: 3–47.

Stewart, R. E. A. 1986. Energetics of age-specific reproductive effort in female harp seals, *Phoca groenlandica*. *Journal of Zoology, London* (A) 208: 503–517.

Stewart, R. E. A., and D. M. Lavigne. 1984. Energy transfer and female condition in nursing harp seals, *Phoca groenlandica*. *Holarctic Ecology* 7: 183–194.

Sydeman, W. J., H. R. Huber, S. D. Emslie, C. A. Ribic, and N. Nur. 1991. Age-specific weaning success of northern elephant seals in relation to previous breeding experience. *Ecology* 72: 2204–2217.

Tedman, R., and B. Green. 1987. Water and sodium fluxes and lactational energetics in suckling pups of Weddell seals (*Leptonychotes weddellii*). *Journal of Zoology, London* 212: 29–42.

Testa, J. W., S. E. B. Hill, and D. B. Siniff. 1989. Diving behavior and maternal investment in Weddell seals (*Leptonychotes weddellii*). *Marine Mammal Science* 5: 399–405.

Thornhill, R. 1981. *Panorpa* (Mecoptera: Panorpidae) scorpionflies: Systems for understanding resource-defense polygyny and alternative male reproductive efforts. *Annual Review of Ecology and Systematics* 12: 355–386.

Trivers, R. L. 1972. Parental investment and sexual selection. In *Sexual Selection and the Descent of Man, 1871–1971*, ed. B. Campbell, 136–179. Chicago: Aldine.

———. 1985. *Social Evolution*. Menlo Park, Calif.: Benjamin/Cummings Publishing Co.

Tuomi, J., T. Hakala, and E. Haukioja. 1983. Alternative concepts of reproductive effort, costs of reproduction, and selection in life-history evolution. *American Zoologist* 23: 25–34.

Twiss, S. D. 1991. Behavioural and energetic determinants of individual mating success in male grey seals (*Halichoerus grypus*, Fabricius 1791). Ph.D. thesis, University of Glasgow, Scotland.

Walker, B. G., and W. D. Bowen. In press. Changes in body mass and feeding behaviour in male harbour seals, *Phoca vitulina*, in relation to female reproductive status. *Journal of Zoology, London*.

Warner, R. R. 1980. The coevolution of life-history and behavioral characteristics. In *Sociobiology: Beyond Nature/Nurture?* ed. G. W. Barlow and J. S. Silverberg, 151–188. Boulder, Colo.: Westview Press.

Wilkinson, P. F., and C. C. Shank. 1976. Rutting-fight mortality among musk oxen on Banks Island, Northwest Territories, Canada. *Animal Behaviour* 24: 756–758.

Williams, G. C. 1966*a*. Natural selection, the costs of reproduction, and a refinement of Lack's principle. *American Naturalist* 100: 687–690.

———. 1966*b*. *Adaptation and Natural Selection.* Princeton: Princeton University Press.

Worthy, G. A. J., P. A. Morris, D. P. Costa, and B. J. Le Boeuf. 1992. Moult energetics of the northern elephant seal. *Journal of Zoology, London* 227: 257–265.

ELEVEN

Diet of the Northern Elephant Seal

*George A. Antonelis, Mark S. Lowry, Clifford H. Fiscus, Brent S. Stewart,
and Robert L. DeLong*

ABSTRACT. The diet of northern elephant seals, *Mirounga angustirostris*, at San Miguel Island, California, was examined by stomach lavage during the spring and summer from 1984 through 1990. Identifiable hard parts of prey species were recovered from the stomachs of 193 of 195 seals sampled. Cephalopods occurred in all stomachs containing identifiable remains of prey. Most prey species (70%) identified inhabit epi-, meso-, or bathypelagic oceanic zones, and relatively few (30%) occur in neritic or benthic regions. The five most frequently occurring prey species were the cephalopods *Octopoteuthis deletron* (58.0%), *Histioteuthis heteropsis* (43.0%), *Gonatopsis borealis* (40.9%), *Histioteuthis dofleini* (39.4%) and the teleost *Merluccius productus* (38.9%). The diets of males and females were similar, with the exception of a significantly higher occurrence of cyclostomes in the stomachs of subadult males. Variability in the diet of elephant seals was probably influenced by annual changes in the availability and abundance of prey.

The population of northern elephant seals, *Mirounga angustirostris*, has increased greatly during the past eight decades and may now number about 127,000 rangewide (Bartholomew and Hubbs 1960; Cooper and Stewart 1983; Stewart et al., this volume). Elephant seal behavior during brief periods ashore has been well documented (e.g., Bartholomew 1952; Le Boeuf 1974; Reiter, Panken, and Le Boeuf 1981), but their foraging ecology has only recently come under study (Hacker 1986; Antonelis et al. 1987; Le Boeuf et al. 1988, 1989; DeLong and Stewart 1991; Sakamoto et al. 1989; DeLong, Stewart, and Hill 1992; Le Boeuf, this volume; Stewart and De-Long, this volume). Here we describe the diet of northern elephant seals based on the examination of prey remains recovered from stomach contents and discuss the ecological relationships between elephant seals and their prey.

METHODS

We lavaged the stomachs of 11 adult male, 69 subadult male, and 115 female northern elephant seals when they hauled out on land to molt at San Miguel Island, California, in spring (late April/early May) and summer (July) from 1984 through 1990. Stomach lavage was accomplished by chemically immobilizing each seal with ketamine hydrochloride, intubating its stomach, and flushing out the remaining parts of prey with 3 to 5 liters of seawater (Antonelis et al. 1987; DeLong and Stewart 1991). Prolonged apnea associated with the use of ketamine hydrochloride and poor health condition resulted in the death of 6 of 195 elephant seals chemically immobilized. Portions of the dietary information from the 1984 field season were previously reported by G. A. Antonelis et al. (1987).

The percent occurrence of prey species found in the stomachs of elephant seals was used as an index of prey consumption. Those species occurring in over 30% of the stomachs were considered major prey items. Dietary information was determined by identifying remains of sagittal otoliths and bones (fish), mouthparts (cephalopods and cyclostomes), elasmobranch and cyclostome egg cases (Cox 1963), crustacean exoskeletons (Schmitt 1921), and tunicates (Thompson 1948).

We estimated age groups (<1 yr., 1 yr., 2 yrs., 3 yrs., 4–7 yrs.) of Pacific whiting, *Merluccius productus*, eaten by seals in 1984–1986 on the basis of size relationships between otoliths from stomach contents and known-age whiting reference specimens (Antonelis et al. 1987) and then estimated fish size using relationships among age, weight, and length (Dark 1975). Digestive erosion did not compromise our ability to categorize otoliths into age groups because Pacific Whiting size differences are conspicuous during the first three years of rapid growth (Dark 1975; Antonelis et al. 1987).

We estimated body sizes (mantle length) of individuals of one cephalopod prey, *Gonatopsis borealis*, from measurements of lower rostral beak length (Clarke 1986). These beak measurements were obtained from samples collected from 1984 to 1986. Sufficient data on relationships between body size and beak measurements were not available for the other cephalopods identified in this study. The annual occurrence of single and multiple prey taxa found in the stomachs of elephant seals (1984–1986) was calculated as percentages of the total number of stomachs containing identifiable food remains.

The null hypotheses (1) male and female elephant seals consumed the same prey and (2) diet did not differ among years were tested with a logistic regression analysis model (Aitkin et al. 1989). We compared the deviance explained by the variables (sex and year) to the deviance of the model without entering the variables. Changes in deviance were approximated by a chi-square distribution with degrees of freedom equal to the change in the degrees of freedom between the two models. Data from stomach lavages conducted from 1988 to 1990 were not tested statistically because of insufficient sample sizes.

RESULTS

We identified 53 prey taxa from stomach contents of 193 elephant seals (tables 11.1 and 11.2). No identifiable remains of prey were recovered from the stomachs of two subadult males. Thirty-seven (70%) of those taxa inhabit one or more of the pelagic habitats (epi-, meso-, or bathypelagic), whereas 16 taxa (30%) are restricted to neritic or benthic zones.

Cephalopods were the most commonly identified prey of elephant seals (table 11.1), consisting of 28 species, 6 of which have not been previously reported as prey of northern elephant seals (*Gonatus onyx, Japetella heathi, Architeuthis japonica, Megalocranchia sp., Octopus bimaculatus,* and *Ommastrephes bartrami*). We recovered cephalopod remains from all of the stomachs containing identifiable parts of prey species (11 adult males, 67 subadult males, and 115 adult females). *Octopoteuthis deletron, Histioteuthis heteropsis, Gonatopsis borealis,* and *Histioteuthis dofleini* were the most frequently occurring cephalopods found in the stomachs. We estimated mantle length of *G. borealis* eaten as 13.6–28.0 cm ($\bar{x} = 22.0$ cm, SD = 5.4 cm, n = 22).

Other prey groups were recovered from stomachs less often than cephalopods. Pacific whiting and pelagic red crabs, *Pleuroncodes planipes,* were the predominant fish and crustacean species (table 11.2). Two of the identified fish species, *Sebastolobus alascanus* and *Icichythys lockingtoni,* had not been previously reported as elephant seal prey.

Diets of male and female seals were not significantly different ($\chi^2 > 3.84$, df = 1, p > .05) with one exception; cyclostomes occurred more frequently in the stomachs of subadult males ($\chi^2 = 3.84$, df = 1, p < .05) than in adult females (fig. 11.1). Seals ate teleost fish more frequently ($\chi^2 = 5.99$, df = 2, p < .05) in 1985 (males = 42%, females = 54%) than they did in 1986 (males = 30%, females = 27%). Crustaceans were consumed more often ($\chi^2 = 5.99$, df = 2, p < .05) in 1984 (males = 80%, females = 76%) than in 1986 (males = 17%, females = 14%); cephalopods were eaten more frequently ($\chi^2 = 5.99$, df = 2, p < .05) in 1985 (males = 95%, females = 97%) than in 1984 (males = 76%, females = 79%). The diversity of prey eaten varied among years, with the greatest number of taxa consumed in 1984 and the fewest in 1986 (fig. 11.2).

The age composition of Pacific whiting that were eaten by elephant seals varied among years (fig. 11.3), but most fish (79% of 252) were less than 4 years old. Overall, 2-year-old Pacific whiting were the most often eaten age group (33%), followed by 1-year-old (21%), 4- to 7-year-old (21%), 3-year-

TABLE 11.1   Percent of occurrence and habitat of cephalopod prey lavaged from the stomachs of 193 northern elephant seals, 1984–1990.

	Frequency of occurrence		
	Number of stomachs	Percent occurrence in stomachs	Marine environmental zone[a]
Cephalopods			
Octopoteuthis deletron	112	58.0	Ba-M-E
Histiotheuthis heteropsis	83	43.0	Ba-M-E
Gonatopsis borealis	79	40.9	M-E
Histioteuthis dofleini	76	39.4	Ba-M-E
Onychoteuthis borealijaponica	41	21.2	E
Galiteuthis sp.	40	20.7	M-E
Gonatus sp.	36	18.7	M-E
Gonatus berryi	36	18.7	M-E
Gonatidae	25	13.0	M-E
Gonatus pyros[b]	23	11.9	M-E
Histioteuthis spp.	23	11.9	Be-M-E
Morotheuthis robusta	23	11.9	Be
Vampyroteuthis infernalis	22	11.4	Be-M
Chiroteuthis calyx	21	10.9	M-E
Taonius pavo	21	10.9	Ba-M
Abraliopsis felis	18	9.3	M-E
Chirotheuthidae	13	6.7	M-E
Loligo opalescens	9	4.7	M-E
Ocythoe tuberculata	9	4.7	M-E
Gonatus onyx[b]	7	3.6	M-E
Japetella heathi[b]	5	2.6	M-E
Octopus rubescens	5	2.6	N-Be
Taningia danae	5	2.6	M-E
Gonatus (type F)	4	2.1	M-E
Architeuthis japonica[b]	3	1.6	M-E
Cranchia scabra	3	1.6	E
Dosidicus gigas	3	1.6	M-E
Megalocranchia sp.[b]	1	0.5	M-E
Octopus bimaculatus[b]	1	0.5	N-Be
Octopus dofleini	1	0.5	Be
Ommastrephes bartrami[b]	1	0.5	M-E
Ommastrephidae	1	0.5	M-E

[a] E = epipelagic, M = mesopelagic, Ba = bathypelagic, Be = benthic, and N = neritic (Schmitt 1921; Thompson 1948; Cox 1963; Miller and Lea 1972; Roper and Young 1975; Butler 1980).

[b] Not previously reported as prey of the northern elephant seal.

TABLE 11.2   Percent of occurrence and habitat of noncephalopod prey lavaged from the stomachs of 193 northern elephant seals, 1984–1990.

	Frequency of occurrence		
	Number of stomachs	Percent occurrence in stomachs	Marine environmental zone[a]
Teleosts			
Merluccius productus	75	38.9	M-E
Sebastes sp.	14	7.3	N-Be
Coryphaenoides acrolepis	10	5.2	M-Ba
Macouridae	8	4.1	M-Ba
Sebastolobus alascanus[b]	2	1.0	M-E
Pleuronectidae	2	1.0	N-Ba
Icichthys lockingtoni[b]	1	0.5	E
Crustaceans			
Pleuroncodes planipes	67	34.7	E
Pasiphaea pacifica	18	9.3	E
Euphausia sp.	5	2.6	E
Unidentified	8	4.1	E
Elasmobranchs			
Cephaloscyllium ventriosum (egg case)	12	6.2	N-Be
Elasmobranch (vertebrae)	10	5.2	Be
Squalus acanthias (vertebrae)	2	1.0	N-Be
Apristurus brunneus (egg case)	1	0.5	N-Be
Hydrolagus colliei (egg case)	1	0.5	N-Be
Elasmobranch (egg case)	11	5.7	N-Be
Cyclostomes			
Eptaptretus stoutii (teeth)	13	6.7	Be
Eptaptretus stoutii (egg case)	2	1.0	Be
Lampetra tridentata (teeth)	5	2.6	A-Be
Tunicate			
Pyrosoma atlanticum	19	9.8	M-E

[a] E = epipelagic, M = mesopelagic, Ba = bathypelagic, Be = benthic, N = neritic, and A = anadromous (Schmitt 1921; Thompson 1948; Cox 1963; Miller and Lea 1972; Roper and Young 1975; Butler 1980).
[b] Not previously reported as prey of the northern elephant seal.

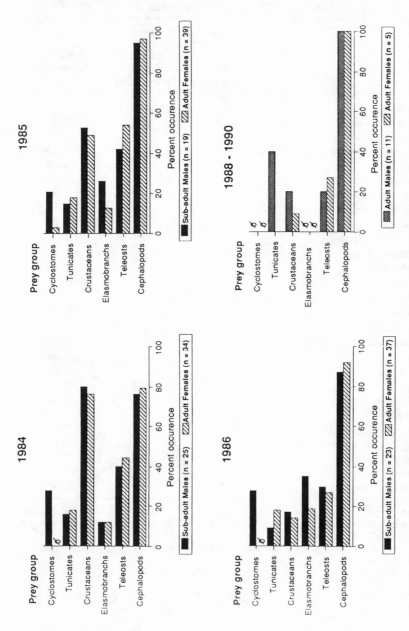

Fig. 11.1. Percent occurrence of the six major prey categories lavaged from the stomachs of adult male, subadult male, and adult female northern elephant seals (1984, 1985, 1986, and 1988–1990).

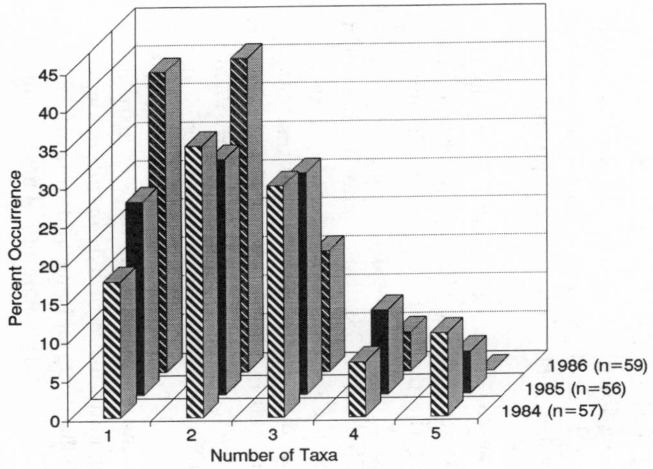

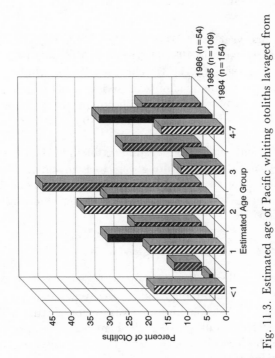

Fig. 11.3. Estimated age of Pacific whiting otoliths lavaged from the stomachs of northern elephant seals, 1984–1986.

Fig. 11.2. Occurrence of single and multiple prey taxa lavaged from the stomachs of northern elephant seals, 1984–1986.

old (15%), and first-year fish (10%). Pacific whiting that were less than 1 year old were rarely eaten in 1985 (1%) and in 1986 (7%) but were common prey (20%) in 1984 (fig. 11.3). The 1984 Pacific whiting cohort was also well represented in the diets of seals as 1-year-old fish in 1985 (27%) and as 2-year-old fish in 1986 (41%). The frequency of occurrence of 3- and 4-year-old Pacific whiting was greatest in 1986 and 1985, respectively (fig. 11.3).

The average length and weight estimates of individual 1-year-old, 2-year-old, and 3-year-old Pacific whiting were 15.4 cm and 0.038 kg, 28.0 cm and 0.163 kg, and 42.3 cm and 0.496 kg, respectively. The average size of individual 4- to 7-year-old Pacific whiting ranged from 44.6 cm (~ 0.570 kg) to 51.8 cm (~ 0.854 kg).

DISCUSSION

Our lavage studies indicate that elephant seals feed primarily on epi- and mesopelagic cephalopods, although other prey types that inhabit pelagic habitats are occasionally eaten. Recent data on the spatial and temporal distribution of foraging adult northern elephant seals (Stewart and DeLong 1990; Le Boeuf, this volume; Stewart and DeLong, this volume) reinforce these interpretations of the importance of resources in meso- and epipelagic habitats to northern elephant seals. Benthic organisms in the neritic zone may be more important prey for adult elephant seals when they forage over the continental shelf during migrations to and from breeding or haul-out locations. The foraging locations of juvenile elephant seals are unknown, but studies are currently under way (Le Boeuf, this volume). Limited dietary information (Hacker 1986) indicates that juveniles may feed frequently in the neritic zone.

All five major prey species make some type of diel vertical migration. *O. deletron* has been described as a vertical spreader that is found at 200 to 400 m depths during the day but disperses both upward (usually in the upper 100 m at night) and downward at night; the maximum reported depth is 1,600 m (Jefferts 1983; Roper and Young 1975). *H. heteropsis*, *H. dofleini*, and *G. borealis* are all second-order vertical migrators that move toward the surface at night (usually no shallower than 200 m) from daytime depths of 300 to 400 m (*H. heteropsis* and *H. dofleini*) to 400 to 700 m (*G. borealis*; Jefferts 1983, Roper and Young 1975). Pacific whiting migrate between daytime depths of 150 to 200 m and surface waters at night (Ermakov 1974); maximum daytime depths may reach 1,000 m (Best 1963). These depth ranges and diel variations of prey distribution are consistent with data on diving patterns recorded for adult elephant seals (e.g., Le Boeuf et al. 1986, 1988, 1989; DeLong and Stewart 1991; Stewart and De-Long, this volume; Le Boeuf, this volume). These results are also consistent

with the hypothesis that adult female and adult male seals dive to and forage on vertically migrating prey in the offshore mesopelagic zone (Le Boeuf
et al. 1988; DeLong and Stewart 1991). Additional studies of the diel movement patterns of these prey and their association with the deep scattering
layer may provide valuable insight into the foraging behavior of the northern elephant seal.

The luminescent characteristics of cephalopods may be an important
factor that facilitates their being visually detected and preyed on by elephant seals, especially under low-light conditions. Three of the four major
cephalopod prey (*O. deletron, H. heteropsis*, and *H. dofleini*) are highly luminescent relative to the other less frequently occurring cephalopod prey
(Nesis 1982). Unlike most of the other prey, these three cephalopods occur
in the darkness of the bathypelagic zone, where a high degree of lumines-
cence might make them more vulnerable to predation by elephant seals.

Only two other prey occurred in over 30% of the stomachs of elephant
seals: pelagic red crabs (*P. planipes*) and Pacific whiting. Pelagic red crabs
were transient prey, occurring in the diet of the seals only in 1984 (44%)
and in 1985 (38%). Pelagic red crabs have rarely been recovered during
lavage studies since this time. Their consumption was evidently linked to
the movement of large numbers of these crabs into the offshore waters
of California during the 1982–83 El Niño Southern Oscillation (Stewart,
Yochem, and Schreiber 1984; Hacker 1986). The influx of this prey into the
region and, perhaps, the change in availability of other prey (e.g., Bailey
and Incze 1985; Fiedler, Methot, and Hewitt 1986; Trillmich and Ono
1991) likely contributed to the high frequency of occurrence of *P. planipes* in
those years.

The variable patterns of occurrence of Pacific whiting in the diet of
northern elephant seals illustrate the relationship between the abundance
and availability of a prey resource and the frequency of its consumption
(fig. 11.3). Juvenile Pacific whiting (<1–3 years of age) are most commonly
found in the epi- and mesopelagic waters, near the central and southern
California coasts (Bailey, Francis, and Stevens 1982). Biomass estimates of
this resource vary annually and are greatly influenced by the strength of
each year class (Dorn et al. 1990). The great strength of the 1984 whiting
cohort (ibid.) was clearly reflected in the seals' diets. Relative to other
cohorts, the 1984 cohort occurred most often in elephant seals' diets each
year as first-year (1984), 1-year-old (1985), and 2-year-old fish (1986).
Similar relationships have been demonstrated between the abundance and
availability of prey resources and their consumption by California sea lions,
Zalophus californianus (Bailey and Ainley 1982; Antonelis, Fiscus, and De-
Long 1984; Lowry et al. 1990).

Most other noncephalopod prey were only occasionally eaten during the
spring and summer months. The inclusion of these prey in the diet of the

northern elephant seal does indicate, however, that these seals are capable of foraging on other prey species. For example, when resource availability was altered by the 1982–1983 El Niño Southern Oscillation (Trillmich and Ono 1991), the effects continued through 1984 when elephant seals foraged on a greater variety of prey taxa. G. A. Antonelis and C. H. Fiscus (1980) suggested that such dietary diversification might be expected during times of resource depletion.

The only detectable difference between the diets of subadult males and adult females was the more frequent occurrence of cyclostomes in the diet of the former. E. S. Hacker (1986) reported that juvenile males and females ate cyclostomes as well as other neritic and nearshore benthic animals. Further studies are needed to determine whether or not consistent differences exist between adult and immature elephant seals.

Northern elephant seals and other pinnipeds in the California Current eat some of the same prey (Antonelis and Fiscus 1980). The degree to which their foraging habitats overlap is minimized by the relatively short time elephant seals remain in nearshore waters. Most adult males and females move far offshore of their rookeries to feed in the Gulf of Alaska and the central/eastern North Pacific, respectively (Stewart and DeLong 1990; Le Boeuf, this volume; Stewart and DeLong, this volume). The degree to which juveniles forage in nearshore waters is poorly understood.

The size of prey consumed by northern elephant seals is not well documented. Our limited results on the size of Pacific whiting and *G. borealis* indicate that elephant seals tend to forage on prey that range from about 13 to 52 cm. *O. deletron, H. heteropsis,* and *H. dofleini* also occur in this size range (Nesis 1982), but additional information is needed to accurately describe the size of these and other prey consumed by northern elephant seals.

Assessing the importance of prey species on the basis of trace remains obtained by lavaging the stomachs of elephant seals must always be made with caution. Some of these prey, such as euphausiids, may be eaten incidental to other prey or may be secondary prey items (prey of the primary prey species) as suggested by W. F. Perrin et al. (1973). Biases may result also from individual differences in the time between feeding and lavage. In studies such as ours, most of the prey species of northern elephant seals are probably consumed relatively close to San Miguel Island and do not adequately represent those prey eaten while on the foraging grounds in the Gulf of Alaska or the central/eastern North Pacific (Stewart and DeLong 1990; Le Boeuf, this volume; Stewart and DeLong, this volume). Differential passage rates of cephalopod beaks versus otoliths in the stomachs of marine predators is another possible source of bias (Fiscus 1990; Miller 1978; Pitcher 1980). Such biases make it extremely difficult to accurately evaluate the energetic contribution of various prey species to the diet of northern elephant seals. Despite these limitations, we have demonstrated the ability to detect changes in the diet of elephant seals that are associated with documented changes in the availability of prey (e.g., Pacific whiting and pelagic red crabs). Similar predator-prey relationships cannot be made for cephalopods and elephant seals because so little information is available on the ecology and status of cephalopod stocks in the North Pacific. This lack of information emphasizes the need for future studies on the ecological relationships between northern elephant seals and their prey.

ACKNOWLEDGMENTS

We thank the following for their aid in this study: E. Berry, J. Barlough, D. DeMaster, S. Diamond, D. Hanan, S. Hawes, T. Loughlin, L. Hansen, J. Scholl, and D. Skilling for field assistance; J. Allen, S. Crockford, J. Dunn, D. Dwyer, B. Goetz, S. Hawes, F. G. Hochberg, D. Siefert, R. Wigen, and M-S. Yang for identifying prey remains; S. Melin for data base organization; B. Sinclair for assisting during the analysis of prey remains and for reviewing literature; A. York and L. Fore for statistical advice; and the staffs of the National Marine Mammal Laboratory, Alaska Fisheries Science Center, and Coastal Zone and Estuarine Studies Division Development Shop, Northwest Fisheries Science Center, Seattle, for help in constructing the lavage apparatus. J. Baker, B. Le Boeuf, R. Gentry, G. Kooyman, T. Loughlin, M. Perez, and B. Sinclair provided helpful critical reviews of the manuscript.

REFERENCES

Aitkin, M., D. Anderson, B. Francis, and J. Hinde. 1989. *Statistical Modeling in GLIM.* Oxford: Clarendon Press.

Antonelis, G. A., and C. H. Fiscus. 1980. The pinnipeds of the California current. *California Cooperative Oceanic Fisheries Investigations Report* 21: 68–78.

Antonelis, G. A., C. H. Fiscus, and R. L. DeLong. 1984. Spring and summer prey of California sea lions, *Zalophus californianus*, at San Miguel Island, California, 1978–1979. *Fishery Bulletin* 82: 67–76.

Antonelis, G. A., M. S. Lowry, D. P. DeMaster, and C. H. Fiscus. 1987. Assessing northern elephant seal feeding habits by stomach lavage. *Marine Mammal Science* 3: 308–322.

Bailey, K. M., and D. G. Ainley. 1982. The dynamics of California sea lion predation of Pacific whiting. *Fishery Research* 1: 163–176.

Bailey, K. M., R. C. Francis, and P. R. Stevens. 1982. The life history and fishery of Pacific whiting, *Merluccius productus. California Cooperative Oceanic Fisheries Investigations Report* 23: 81–98.

Bailey, K. M., and L. S. Incze. 1985. El Niño and the early life history of recruitment of fishes in temperate marine waters. In *El Niño North*, ed. W. S. Wooster and D. L. Fluharty, 143–165. Washington Sea Grant, University of Washington, Seattle.

Bartholomew, G. A. 1952. Reproductive and social behavior of the northern elephant seal. *University of California Publications in Zoology* 47: 369–472.

Bartholomew, G. A., and C. L. Hubbs. 1960. Population growth and seasonal movements of the northern elephant seal, *Mirounga angustirostris*. *Mammalia* 24: 313–324.

Best, E. A. 1963. Contribution to the biology of Pacific hake, *Merluccius productus*. *California Cooperative Oceanic Fisheries Investigations Report* 9: 51–56.

Butler, T. H. 1980. Shrimps of the Pacific coast of Canada. *Canadian Bulletin of Fisheries and Aquatic Sciences* 202.

Clarke, M. R., ed. 1986. *A Handbook for the Identification of Cephalopod Beaks*. Oxford: Clarendon Press.

Cooper, C. F., and B. S. Stewart. 1983. Demography of northern elephant seals, 1911–1982. *Science* 219: 969–971.

Cox, K. W. 1963. Egg-cases of some elasmobranchs and a cyclostome from California waters. *California Fish and Game* 49: 271–289.

Dark, T. A. 1975. Age and growth of Pacific hake, *Merluccius productus*. *Fishery Bulletin* 73: 336–355.

DeLong, R. L., and B. S. Stewart. 1991. Diving patterns of northern elephant seal bulls. *Marine Mammal Science* 7: 369–384.

DeLong, R. L., B. S. Stewart, and R. D. Hill. 1992. Documenting migrations of northern elephant seals using day length. *Marine Mammal Science* 8: 155–159.

Dorn, M. W., R. D. Methot, E. P. Nunnallee, and M. E. Wilkins. 1990. Status of the coastal Pacific whiting resource in 1990. Status of the Pacific coast groundfish fishery through 1990 and recommended acceptable biological catches for 1991. Stock assessment and fishery evaluation. Appendix Vol. I. Pacific Fishery Management Council, Portland, Ore.

Ermakov, Y. K. 1974. The biology and fishery of Pacific hake, *Merluccius productus*. Ph.D. dissertation, Pacific Scientific Institute of Marine Fisheries and Oceanography (TINRO), Vladivostok, USSR.

Fiedler, P. C., R. D. Methot, and R. P. Hewitt. 1986. Effects of California El Niño 1982–1984 on the northern anchovy. *Journal of Marine Research* 44: 317–338.

Fiscus, C. H. 1990. Notes on North Pacific gonatids: Identification of body fragments and beaks from marine mammals. Abstracts and Proceedings of the Annual Meeting held in Seattle, Wash., June 18–22, 1990, Western Society of Malacologists, Annual Report 23.

Hacker, E. S. 1986. Stomach content analysis of short-finned pilot whales (*Globicephala macrorhynchus*) and northern elephant seals (*Mirounga angustirostris*) from the southern California Bight. National Marine Fisheries Service, Southwest Fisheries Center Administrative Report LJ-86-08C.

Jefferts, K. 1983. Zoogeography and systematics of cephalopods of the northeastern Pacific Ocean. Ph.D. dissertation, Oregon State University, Corvallis.

Le Boeuf, B. J. 1974. Male-male competition and reproductive success in elephant seals. *American Zoologist* 14: 163–176.

Le Boeuf, B. J., D. P. Costa, A. C. Huntley, and S. D. Feldkamp. 1988. Continuous, deep diving in female northern elephant seals, *Mirounga angustirostris. Canadian Journal of Zoology* 66: 446–458.

Le Boeuf, B. J., D. P. Costa, A. C. Huntley, G. L. Kooyman, and R. W. Davis. 1986. Pattern and depth of dives in northern elephant seals, *Mirounga angustirostris. Journal of Zoology, London* 208: 1–7.

Le Boeuf, B. J., Y. Naito, A. C. Huntley, and T. Asaga. 1989. Prolonged, continuous, deep diving by northern elephant seals. *Canadian Journal of Zoology* 67: 2514–2519.

Lowry, M. S., C. W. Oliver, C. Macky, and J. B. Wexler. 1990. Food habits of California sea lions, *Zalophus californianus*, at San Clemente Island, California. *Fishery Bulletin* 88: 509–521.

Miller, D. S., and R. N. Lea. 1972. Guide to the coastal marine fishes of California. *California Department of Fish and Game, Fish Bulletin* 157: 1–235.

Miller, L. K. 1978. *Energetics of the Northern Fur Seal in Relation to Climatic and Food Resources of the Bering Sea.* Marine Mammal Commission, Washington, D.C. (Available U.S. Department of Commerce, National Technical Information Service, as PB-275296.)

Nesis, K. N. 1982. *Cephalopods of the World.* Translated from Russian by B. S. Levitov and edited by L. A. Burgess. Department of Nekton, P. P. Shirshov Institute of Oceanography, USSR Academy of Sciences, Moscow.

Perrin, W. F., R. R. Warner, C. H. Fiscus, and D. B. Holts. 1973. Stomach contents of porpoise, *Stenella* spp. and yellowfin tuna, *Thunnus albacares*, in mixed-species aggregations. *Fishery Bulletin* 71: 1077–1092.

Pitcher, K. W. 1980. Stomach contents and feces as indicators of harbor seal, *Phoca vitulina*, foods in the Gulf of Alaska. *Fishery Bulletin* 78: 797–798.

Reiter, J., K. J. Panken, and B. J. Le Boeuf. 1981. Female competition and reproductive success in northern elephant seals. *Animal Behaviour* 29: 670–687.

Roper, C. F. E., and R. E. Young. 1975. Vertical distribution of pelagic cephalopods. *Smithsonian Contributions to Zoology* 209: 1–51.

Sakamoto, W., Y. Naito, A. C. Huntley, and B. J. Le Boeuf. 1989. Daily gross energy requirements of a female northern elephant seal, *Mirounga angustirostris. Nippon Suisan Gakkaishi* 55: 2057–2063.

Schmitt, W. L. 1921. The marine decapod crustacea of California. *University of California Publications in Zoology* 23: 1–470.

Stewart, B. S., and R. L. DeLong, 1990. Sexual differences in migrations and foraging behavior of northern elephant seals. *American Zoologist* 30: 44A.

Stewart, B. S., P. K. Yochem, and R. W. Schreiber. 1984. Pelagic red crabs as a food source for gulls: A possible benefit of El Niño. *Condor* 86: 341–342.

Thompson, H. 1948. *Pelagic tunicates of Australia.* Commonwealth Council for Scientific and Industrial Research, Melbourne, Australia.

Trillmich, F., and K. Ono, eds. 1991. *Pinnipeds and El Niño.* Heidelberg and Berlin: Springer Verlag.

PART III

Diving and Foraging

TWELVE

Theory of Geolocation by Light Levels

Roger D. Hill

ABSTRACT. A technique for determining the location of elephant seals is described. This technique requires an accurate determination of time of dawn and dusk on a daily basis. The time midway between dawn and dusk, the local apparent noon, determines the seal's longitude, and the day length is used to determine latitude. The longitude determination is equally accurate throughout the year and at all latitudes except those with no dawn and dusk events; the latitude determination is most accurate at the solstices and useless at the equinoxes. Other sources of error are the accuracy of the light-level measurement, atmospheric aberration, and the seal's behavior.

Elephant seals present a particular challenge to the researcher who wants to know where they go. Elephant seals generally surface for inadequate periods of time for reliable tracking by the Argos satellite system (Stewart et al. 1989); however, some researchers have had success with this system by mounting the transmitter on the seal's head. Elephant seals also dive so deep that any instrumentation must be solid or have a substantial pressure housing. These two problems combine to make it difficult to track an elephant seal reliably for more than a few months without the transmitter becoming detached. The price of satellite-transmitters is also generally too high for large-scale studies. However, elephant seals are reasonably faithful to their molting and breeding beaches, so that deployments and recovery of memory-based instruments such as time-depth recorders (TDRs) have been quite successful. In early 1989, it was suggested that I attempt to incorporate a geolocation feature into our TDRs. By recording and storing light levels, times of dawn and dusk could retroactively be determined and used to calculate position using standard solar navigational equations.

A

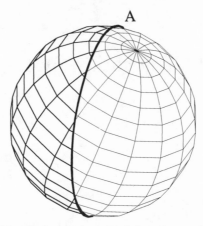

Fig. 12.1. Circle A is dawn/dusk interface at Time A.

THE THEORY

The principle of determination of location is reasonably straightforward: fig. 12.1 shows the earth, with the sun to the right. The bold circle that encompasses the earth is the line between day and night. Note that this circle does not pass through the north or south pole. Figure 12.2 shows the earth approximately 12 hours later. The sun is still to the right, but the earth has rotated by nearly 180° and Circle A from figure 12.1 has moved with it. Circle B is the line that currently divides day from night. If we consider a point on the earth for which the sun was rising in figure 12.1 and is now experiencing sunset, then the point must be on both Circle B and Circle A. The position is, therefore, an intersection of these two circles (one intersection is for A = dawn and B = dusk, the other is for A = dusk and B = dawn). If we know the times of dawn and dusk and the day of year (which affects the tilt of the earth and thus the position of the dawn-dusk circles), then we can theoretically calculate the location of their intersection and, hence, our position.

The standard equations used for solar navigation (Yallop and Hohenkerk 1985; Nautical Almanac Office 1991) predict the time at which a solar event occurs for a given day of year and location on the earth. The solar event of interest here is when the sun is at an azimuth of 96° (the center of the sun is 6° below the horizon). This is known as civil twilight and is when the sun's first light appears (dawn) or last light disappears (dusk). At these times light level is changing fastest, so that the times of these events can be determined from light-level measurements most accurately. Unfortunately, the standard equations yield the inverse of the required information, so an

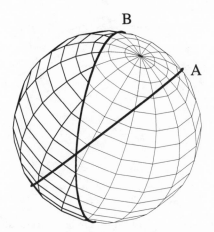

Fig. 12.2. Circle B is new dawn/dusk interface at Time B. Circle A has rotated with the earth from figure 12.1. Location that had dawn at Time A and dusk at Time B is at the intersection of the two circles.

iterative process is used to find the location for which predicted times of dawn and dusk coincide with measured values.

The iterative process starts with an estimate of longitude—longitude (°E) = (time of midnight) × 15, where time of midnight is halfway between dusk time and dawn time and all times are measured in Universal Time (UT) or Greenwich Mean Time (GMT)—and an estimate of latitude of −45° or +45°, depending on the previous position. This first guess of position is used to generate dawn and dusk times, which are compared with the observed dawn and dusk times to produce the next location estimate. This process continues until the observed and predicted dawn/dusk times match to the desired degree of accuracy.

SOURCES OF ERROR

Equinoxes

Our ability to determine latitude fails near the vernal or autumnal equinoxes, as shown in figure 12.3. The dawn/dusk circle now passes through or very close to the north and south poles. This means that for all places on the earth, the dawn and dusk circles (A and B in fig. 12.2) will now be very close or overlap, and the ability to determine latitude is lost. This does not affect the calculation of longitude. (The location is somewhere on the dawn/dusk circles, which are now also circles of longitude because they pass through the poles.) The usefulness of the solar equations for

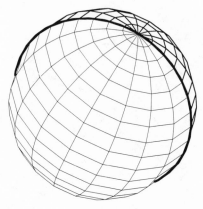

Fig. 12.3. At the equinoxes, dawn and dusk circles overlap, so that latitude determination is not possible.

calculating latitude is effectively determined by the variation of day length (the time between dawn and dusk) with latitude for each day of the year. If a small difference in day length generates a large difference in latitude, then position is difficult to estimate because any small error in assessing day length will cause a large error in the latitude determination. Conversely, if a large change in day length generates a small difference in latitude, then the position calculated will be accurate.

The relationship between time of year, latitude, and day length is summarized in figure 12.4. The accuracy of a latitude determination for a given day and latitude is determined by the slope of day length with latitude. For example, Day 1 shows a good slope between $-45°$ and $+45°$ latitude, indicating the potential for a good estimate of latitude, and a steeper slope between $-60°$ and $-45°$ and between $+45°$ and $+60°$, indicating the potential for even better latitude estimates. However, at about Day 265, there is almost no slope except very near the poles, indicating a complete inability to determine latitude from day length variations. Figure 12.5 shows an enlargement of figure 12.4 near the autumnal equinox. Note that between Day 252 and Day 281, the day length at both poles is 24 hours. This is because day length is measured from the time that the sun rises above $96°$ azimuth to when it sets below $96°$ azimuth (actually about 7 months at each pole). Had we chosen an azimuth of $90°$, day length would have changed from 0 to 24 hours (or vice versa) at both poles on the same day. A side effect of using an azimuth of $96°$ is that for Days 252 through 281 (and for the equivalent days near the vernal equinox), a measured dawn and dusk time will generate two locations. The correct location must be chosen by comparison to previous locations. Although the quality of latitude determinations near the equinoxes is generally poor, studying the

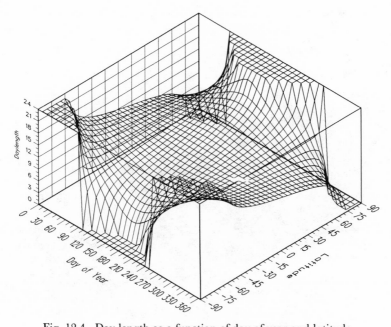

Fig. 12.4. Day length as a function of day of year and latitude.

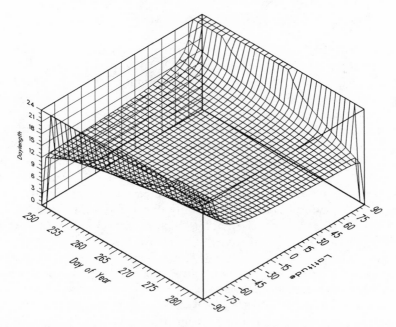

Fig. 12.5. Day length as a function of day of year and latitude near the autumnal equinox.

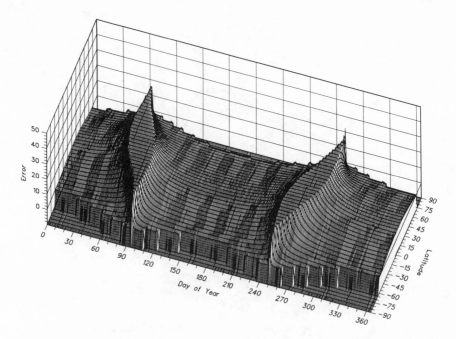

Fig. 12.6. Error in latitude determination caused by a 4-minute error in day length measurement as a function of day of year and latitude.

gradient of day length with latitude (fig. 12.5) shows that a latitude determination will be possible in some ranges of latitude and day of year (e.g., Day 249, latitude +45° to +75°).

An alternate way of displaying this relationship between day of year, latitude, and day length is shown in figure 12.6. A theoretical uncertainty in the accuracy of a location (error) is plotted against latitude and day of year. The error is the range of latitude that is generated by moving dawn and dusk times by ± 4 minutes. An accuracy of ±2 minutes is the limit of accuracy in observing dawn or dusk due to atmospheric phenomenon, and this has been doubled to reflect other likely errors. The error in latitude has been truncated at 20° for clarity. The slightly skewed nature of the two equinox "ridges" is again caused by using a dawn/dusk azimuth of 96°; had we used 90°, the ridges would not be twisted. The "plateaus" between the ridges show the days and locations where good locations can be expected from a measurement of light levels.

The simplest solution for reducing the ambiguity in latitude near the equinoxes is to use some other measurement to fix the latitude. The obvious choice is surface seawater temperature, which varies considerably with latitude and is available on a week-to-week basis from a combination of satellite imagery and oceanographic buoys. In the waters surrounding the United States, these data are compiled by the National Marine Fisheries Service and the National Weather Service. Latitude is found by determining which locations on the known line of longitude have the measured surface seawater temperature.

Accuracy of the Light-Level Measurements

There are several sources of potential error in recording the light-level measurements, only some of which can be controlled. The time at which the light-level measurements were taken must be known accurately; one minute of inaccuracy in the estimate of dawn and dusk times will generate an error of 0.25° of longitude and an error in latitude of about 25% of the error shown in figure 12.5. To minimize the timing errors, users must carefully set the recorder's clock before deployment and note any time error on retrieval. The analysis program must then use this error data and adjust all time measurements accordingly. The magnitude of the light level will change over many orders of magnitude between night and day, so it should be recorded on a logarithmic scale and carefully calibrated so that it will not "peg-out" in bright sunshine or complete dark. Since one will have no control over the orientation of the light sensor when it is collecting data, it should be responsive to light over wide angles and be positioned so that it will generally point up. Obviously, none of these suggestions will help if the study animal is on shore, on its back, and the light sensor is buried in sand. Such data points must be excluded at analysis time.

Atmospheric Aberration

Light does not generally pass through the earth's atmosphere in straight lines; it bends when it encounters thermal or pressure gradients. For this reason, it is generally considered impossible to measure the time of dawn or dusk to an accuracy of greater than 2 minutes, even if one is observing the sun directly rather than measuring ambient light levels (C. Acton pers. comm.). This is a major source of error in this type of navigation; 2 minutes of inaccuracy in the estimate of dawn and dusk times will generate an error of 0.5° of longitude and an error in latitude of about 50% of the error shown in figure 12.5. Some compensation for hot and cold and high and low pressure days can be applied to the navigational equations if one has these data.

Animal Diving at Dawn or Dusk

Elephant seals are known to perform 20-minute dives alternating with 3-minute surface times for many hours at a time (Le Boeuf et al. 1986, 1988,

1989; Stewart and DeLong 1990; DeLong and Stewart 1991). With behavior such as this, a light sensor will probably be at depth during the
actual dawn and dusk times. The analysis of light levels must cope with this
problem, and, generally, an interpolation technique around the times of
dawn or dusk will work well to determine the actual dawn or dusk times. If
there is an extended dive at dawn or dusk (greater than 30 minutes), then
determination of the dawn or dusk times will not be possible.

Animal Moving between Dawn and Dusk

The above analysis assumes that the animal does not change its location
between dawn and dusk. If it does move, the error induced depends on the
direction it moves. If the animal moves along Circle A in figures 12.1 and
12.2 between dawn and dusk, then the movement will have no effect on the
accuracy of the location, but the location given will be that of the animal at
dusk, not some median position. Other directions of movement will have
other results. Generally, large errors will only occur when the animal is
covering large distances per day, and under these circumstances, a larger
locational error is more acceptable. The analysis program could also minimize this error by performing the locational analysis from dusk to dawn
(rather than dawn to dusk) when day length is greater than 12 hours.

The Equations

The equations used to predict dawn and dusk times also contain some inherent inaccuracies, as much as ± 2 minutes under certain circumstances.
I have been unable to determine for which combinations of day and latitude
these inaccuracies are worst, but it seems reasonable that the inaccuracies
are going to be most severe where day length changes very rapidly with
latitude. These are the same circumstances that give us inherently better
locational accuracy. If this is true, then error from the equations will cancel
some (or all) of the improved accuracy generated by rapidly changing day
lengths. Until better equations can be provided to predict dawn and dusk
times, some allowance for errors in the equations must be made.

Presentation of Errors

Since both latitude and day of year will have a profound effect on the
accuracy of the latitude determination, it is important that positions generated from observed dawn/dusk times be provided with error estimates.
Ideally, all locations should be plotted on a map using rectangles that indicate the limits of the animal's position to a given level of certainty. It
should be noted that the center of such rectangles will not necessarily represent the likeliest location of the animal.

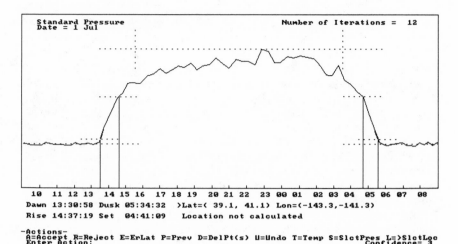

Fig. 12.7. Example of analysis program's graphical output. The light-level curve is shown with markers for derived dawn/dusk times and the calculated position range.

PRACTICAL CONSIDERATIONS

Wildlife Computers makes TDRs with temperature- and light-sensing options for use on diving animals. The TDR stores surface-seawater-temperature (SST) and light-level (LL) data when the instrument is at (or very near) the surface. These data are decoded and used by an analysis program to provide locations. The SST and LL data are stored whenever the study animal surfaces but generally no more frequently than every 15 minutes. These light-level data are extracted, plotted, and used to determine the times of dawn and dusk (see fig. 12.7). We have determined that dawn and dusk (sun is at an azimuth of 96°) correspond to a light level equal to the nighttime light level plus 5% of the difference between the night and day levels, when LL data are shown on a logarithmic scale.

It should come as no surprise that some of the errors inherent in attempting to geolocate a swimming or diving animal using light levels are unavoidable. Others, such as temperature effects and interpolation requirements, can be minimized in the analysis software. Our best efforts generally yield locations for elephant seals that are about ±1 degree in latitude and longitude. Although not as precise as one might like, these location data have greatly expanded our knowledge of the foraging migrations of both southern and northern elephant seals (DeLong, Stewart, and Hill 1992).

REFERENCES

DeLong, R. L., and B. S. Stewart. 1991. Diving patterns of northern elephant seal bulls. *Marine Mammal Science* 7: 369–384.

DeLong, R. L., B. S. Stewart, and R. D. Hill. 1992. Documenting migrations of northern elephant seals using day length. *Marine Mammal Science* 8: 155–159.

Le Boeuf, B. J., D. P. Costa, A. C. Huntley, and S. D. Feldkamp. 1988. Continuous, deep diving in female northern elephant seals, *Mirounga angustirostris*. *Canadian Journal of Zoology* 66: 446–458.

Le Boeuf, B. J., D. P. Costa, A. C. Huntley, G. L. Kooyman, and R. W. Davis. 1986. Pattern and depth of dives in northern elephant seals, *Mirounga angustirostris*. *Journal of Zoology, London* 208: 1–7.

Le Boeuf, B. J., Y. Naito, A. C. Huntley, and T. Asaga. 1989. Prolonged, continuous, deep diving by northern elephant seals. *Canadian Journal of Zoology* 67: 2514–2519.

Nautical Almanac Office. 1991. *Almanac for Computers*. Washington, D.C.: United States Naval Observatory.

Stewart, B. S., and R. L. DeLong. 1990. Sexual differences in migrations and foraging behavior of northern elephant seals. *American Zoologist* 30: 44A.

Stewart, B. S., S. Leatherwood, P. K. Yochem, and M.-P. Heide-Jorgensen. 1989. Prospects for tracking pinnipeds at sea using the Argos DCLS: Insights from studies of free-ranging harbor and ringed seals. In *Proceedings of the 1989 North American Argos Users Conference and Exhibit*, Landover, Maryland, 193–203.

Yallop, B. D., and C. Y. Hohenkerk. 1985. *Compact Data for Navigation and Astronomy for the Years 1986–1990*. London: Her Majesty's Stationery Office.

THIRTEEN

Variation in the Diving Pattern of Northern Elephant Seals with Age, Mass, Sex, and Reproductive Condition

Burney J. Le Boeuf

ABSTRACT. A principal aim of studies of northern elephant seals from the Año Nuevo rookery in central California has been to obtain complete descriptions of diving behavior for the various stages of life from weaning to adulthood. Studies were conducted during the period 1983–1991 using a variety of methods on known-age animals of both sexes at various stages of development: measurement of the free-ranging diving pattern using attached time-depth recorders, determination of mass before and after trips to sea, and measurement of physiological variables during homing experiments with juveniles.

Some of the principal findings and conclusions are (1) all elephant seals of both sexes and all ages dive deep, long, and continuously for the entire periods that they are at sea; (2) by age 2, the dive pattern is similar to that of adults; (3) dive duration increases slightly with mass in nonpregnant females, but when pregnant females and adult males are included, the relationship weakens, indicating that other variables influence dive duration; (4) dive depth is independent of age and mass in animals older than 2 years of age; (5) adult males migrate farther north and west than females to specific foraging areas along the continental margin, while females disperse more widely in the open ocean and forage en route; (6) pregnant females dive longer and migrate farther away from the rookery than postbreeding females; and (7) yearlings home reliably during the spring molt and fall rest period, revealing a dive pattern like that of free-ranging animals and thus offer the opportunity for short-term studies of diving and the measurement of physiological variables.

In-depth studies of the diving behavior of elephant seals throughout development are filling in gaps in our knowledge of the marine aspect of their natural history and enhancing our understanding of the biology of diving and marine foraging.

Complete dive descriptions for the various stages of life from initial water entry to adulthood do not exist for any diving mammal. This is a serious

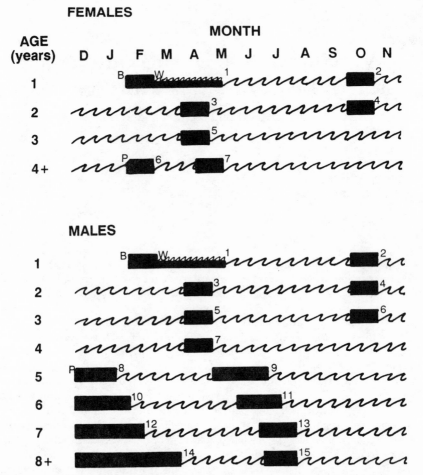

Fig. 13.1. Annual distribution of time spent on land and at sea for male and female northern elephant seals from birth to adulthood. Solid bars represent time on land, waves represent time at sea, bars and waves represent initial water experience prior to departure from the rookery. B = birth, W = weaning, P (for females) = parturition, P (for males) = puberty. Primiparity in females occurs at 3, 4, or 5 years of age. The numbers in the figure denote the consecutive trips to sea by each sex.

omission. A thorough description of this kind yields vital information on life history, facilitates the understanding of body functions during diving, sheds light on foraging economics, and puts other dive-related processes in perspective.

One aim of our diving studies at Santa Cruz has been to obtain a description of the diving pattern of northern elephant seals of both sexes dur-

TABLE 13.1   Free-ranging dive records obtained from northern elephant
seals at Año Nuevo, 1983–1991.

	Instrument type			
	Billups	Naito	Hill	Totals
Weanlings		1	1	2
Juveniles (1–2 years old)			8	8
Adult females				
Nonpregnant	12	5	10	27
Pregnant	2		5	7
Adult males			6	6
	14	6	30	50

ing all periods at sea throughout development. Except for a single trip to
sea during the year preceding puberty, individuals of both sexes spend two
periods at sea each year of their lives. The approximate time and duration
of these aquatic sojourns for each sex are shown in figure 13.1. This figure
makes clear that elephant seals are among the most pelagic of all seals.

I summarize data from 50 dive records obtained from elephant seals at
Año Nuevo from 1983 to 1991 (table 13.1), paying special attention to
variation in dive pattern with age, mass, reproductive condition, and sex. I
also present the main results of a translocation study of yearling elephant
seals that shows that it is feasible to study diving in elephant seals in the
short run. My aim is to give an overview of certain aspects of the diving research from this laboratory. The treatment will be topical because of space
limitations and because each topic will be treated in depth elsewhere.

The general method used to collect all free-ranging dive data consisted of
attaching a diving instrument and radio transmitter to a seal shortly before
it went to sea and then recovering the instrument when the animal returned
to the rookery months later (Le Boeuf et al. 1986, 1988, 1989). Instruments
were attached to the pelage above the shoulders with marine epoxy. Nearly
all seals were weighed before going to sea and on their return.

GENERALIZATIONS

The data support several generalizations about diving in elephant seals. All
northern elephant seals of both sexes and all age groups, from weanlings on
their first pelagic trip to sea to adults, exhibit a diving pattern characterized
by (1) continuous diving during all periods at sea; (2) deep diving, relative
to other pinnipeds and cetacea; (3) long-duration dives, relative to other diving mammals, interspersed with brief surface intervals (about 1/10 the

mean dive duration) and a few unpredictable surface intervals longer than 10 minutes; and (4) submergence for the majority of the time (83–92%) spent at sea. Data supporting these generalizations are summarized below.

VARIATION IN DIVE PATTERN

Early Development

First Trip to Sea.   On the very first trip to sea when the animals are only 3½ months old, both the depth and duration of dives are great, relative to other diving mammals. Two seals, a male and a female, exhibited mean dive durations of 9.5 and 10.5 minutes with maximum dive durations of 18 and 22 minutes, respectively. The one animal, on whom we measured dive depth over a 30-day period, exhibited a mean dive depth of 206 m and a maximum dive depth to 553 m. These figures are impressive, for they exceed the adult diving performance of most other pinnipeds that have been studied. Further details on the first diving records of these young seals are found in P. H. Thorson and B. J. Le Boeuf (this volume).

Juvenile Diving.   Two-year-olds of both sexes dived to the same mean depths, in excess of 400 m, and remained submerged as long as adult males and females, a mean of 18 minutes or more (fig. 13.2). That is, by the end of the fourth trip to sea, when juveniles average 270 ± 26 kg, the adult pattern is essentially set, and 2-year-olds are accomplished divers. This has important implications for conducting studies of diving in the laboratory. It is feasible to transport juveniles but considerably more difficult to transport adult females whose mean nonpregnant weight is 395 ± 19 kg.

Age/Mass Effects

Dive Duration.   One expects that dive duration will scale approximately linearly with mass because dive duration is limited by oxygen stores (Scholander 1940; Calder 1984). In seals, oxygen stores are determined mainly by blood volume and hemoglobin and myoglobin concentrations. Blood volume (V_b) is linearly proportional to body mass (M_b), $V_b = M_b^{1.0}$, and the relationship between mass and metabolic rate (MR), which determines how quickly the oxygen stores are used, is $MR = M_b^{.75}$. In elephant seals, hemoglobin and myoglobin concentrations increase with age (Thorson and Le Boeuf, this volume), and there is a high and positive correlation between age and mass (fig. 13.3; Deutsch et al., this volume; Clinton, this volume). Indeed, one expects dive duration to scale to both age and mass in postbreeding females because of the close relationship between these two variables (fig. 13.3).

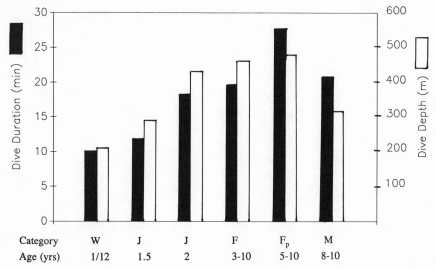

Fig. 13.2. Mean dive duration and mean dive depth of free-ranging northern elephant seals as a function of age.

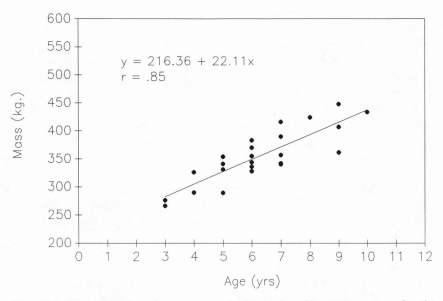

Fig. 13.3. The regression of mass on age in 23 postbreeding, nonpregnant female northern elephant seals.

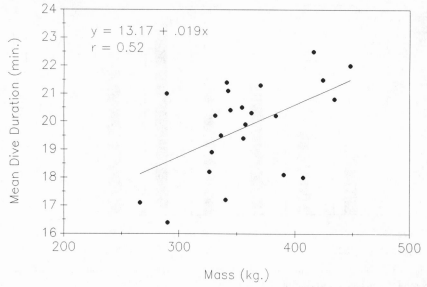

Fig. 13.4. Mean dive duration as a function of mass in 23 postbreeding, nonpregnant female northern elephant seals.

The relationship between dive duration and mass is positive in postbreeding females, but the association is not strong (fig. 13.4). If one adds
weanlings, juveniles, pregnant females, and males to the picture (fig. 13.5),
predictability worsens at the upper end. Indeed, it is not clear whether it is
males or pregnant females that are outstanding. The long dives of pregnant
females contradict the logic that the fetus parasitizes the stores of the
mother, causing a shortening of dives. Whatever the explanation, the data
indicate that mass predicts dive duration up to a point, but other factors
besides mass and oxygen carrying capacity affect the duration of dives.
Dive duration does not scale in a simple linear way with mass.

Dive Depth.   Mean dive depth does not vary systematically with mass
(fig. 13.6), and by association, with age (fig. 13.3), in nonpregnant females.
Similarly, there is no relationship between mean dive depth and age across
animals from all age categories (fig. 13.7).

Reproductive Condition
Pregnant females, whose diving behavior was recorded during the third
trimester of pregnancy, had mean dive durations 39% longer than nonpregnant females (table 13.2). Each one of them had maximum dive durations
in excess of 1 hour. Dives lasting more than 1 hour were never observed in

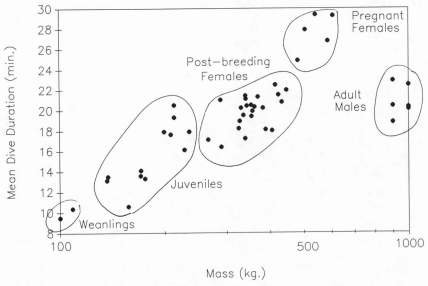

Fig. 13.5. Mean dive duration as a function of mass in northern elephant seals ranging in age from weanlings to adults of both sexes.

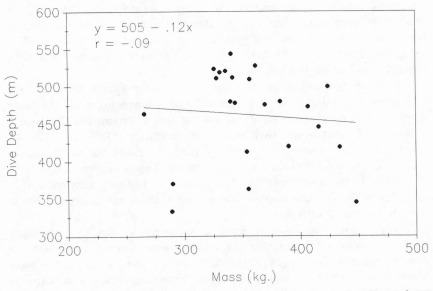

Fig. 13.6. Mean dive depth of 23 postbreeding, nonpregnant female northern elephant seals as a function of their mass.

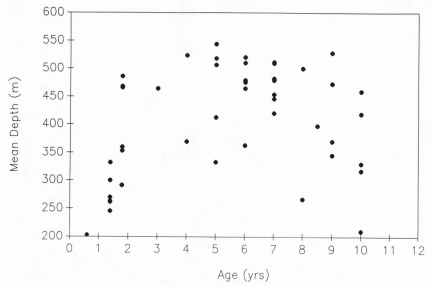

Fig. 13.7. Mean dive depth as a function of age in northern elephant seals.

the 26 nonpregnant females that were recorded. Despite the longer dives, the surface intervals of pregnant females were no longer than those of non-pregnant females. Moreover, pregnant females exhibited slower dive rates and spent less time at the surface than nonpregnant females. These differences were even more prominent in three females that were recorded in both conditions. Differences of similar magnitude have been observed in southern elephant seals, *M. leonina* (Hindell 1990; Hindell et al. 1992; Slip, Hindell, and Burton, this volume).

Why do pregnant females dive longer than nonpregnant females? One reason might be that it takes pregnant females longer to reach the same depths, the deep waters where prey are located. Pregnant females have a greater lipid composition than nonpregnant females (36% vs. 24%, on average), which makes them more buoyant; increased buoyancy implies greater diving effort and time to reach depths. This is suggested by M. A. Hindell's (1990) observation that pregnant and nonpregnant females do not differ in the duration spent at the bottom of dives; the differences are in descent and ascent duration.

The proximate mechanisms enabling pregnant females to dive longer than nonpregnant females must be either greater oxygen availability or reduced demand for oxygen. Pregnant female mammals are said to have about 40% more blood than nonpregnant females; however, it is not clear how much of this extra blood is involved in nurturing the developing fetus. Alternatively, pregnant females may have lower metabolic rates than non-

TABLE 13.2 Key statistics of the dive patterns of postbreeding females (nonpregnant), pregnant females, and males, 5 to 10 years of age.

	Mean dives/ hour	Mean % time on surface	Maximum depth (m)	Mean depth (m)	Maximum duration (min)	Mean duration (min)	Mean SI (min)
Postbreeding females (N = 10)[a]	2.6	9.7	1273	509 ± 147	47.5	20.8 ± 4.1	2.1 ± 0.5
Pregnant females (N = 5)	2.0	7.7	1181	473 ± 151	67.9	27.7 ± 6.4	2.1 ± 0.6
Adult males (N = 5)	2.5	13.5	1503	330 ± 127	66.7	21.2 ± 4.6	2.7 ± 0.9

[a] For comparability, only the records obtained with Hill TDRs from Wildlife Computers are included.

pregnant females. A preliminary observation consistent with this statement is that the horizontal swim speed of females during migration is significantly slower during the third trimester of pregnancy than when they are not pregnant (51.5 km vs. 75.0 km per day; t = 2.62, df = 2, p < .05). D. Renouf et al. (1991) report that gray seals, *Halichoerus grypus*, have lower resting metabolic rates when pregnant than when nonpregnant.

It is also possible that the differences in dive durations of pregnant and nonpregnant females reflect a change in prey or its availability (Lyons 1991). Pregnant females migrate farther away from the rookery than nonpregnant females, but much of the same route is traveled (see below). The diving records under both conditions are similar and offer no obvious evidence for this hypothesis.

Sex

Do the sexes differ with respect to diving pattern and migratory path? The great sexual dimorphism in this species makes this an interesting question: What are the bioenergetic consequences of extreme sexual dimorphism in size, especially with respect to foraging behavior?

Dive Pattern.   Males and females have similar dive patterns, except that males dive approximately 180 m less deep than nonpregnant females, on average (table 13.2, fig. 13.2; see also DeLong and Stewart 1991). The maximum depths we have recorded are 1,503 m for a 5-year-old subadult male and 1,273 m for a 4-year-old female.

The shallower dive pattern of males relative to females is due mainly to a single sex difference in diving pattern (Le Boeuf et al. 1993). The most common dive type in the records of females appears to serve pelagic foraging (Le Boeuf et al. 1988, 1992); these dives are characterized by direct descent to depth, with several vertical excursions at the bottom of the dive, followed by direct ascent to the surface. Males exhibit these dives, too, but their most frequent dives are flat bottomed with a mean depth of 331 ± 243 m. Females do not exhibit these dives or do so only rarely. The dive shapes and the location in which they occur suggest that males are pursuing benthic prey on the edge of the continental shelf or on seamounts or guyots. Because males are much larger than females (Deutsch et al., this volume), their daily energy requirements are approximately three times greater than those of females. Benthic prey might provide the additional energy a male requires, or it might be that it is less costly for males to catch these prey. This hypothesis is consistent with reports that males but not females feed on cyclostomes and elasmobranchs (Condit and Le Boeuf 1984; Antonelis et al., this volume). What prevents females from doing the same thing? Two reasons come to mind. They may not be large enough to capture benthic prey, or it may be more costly for females to migrate to the places where these benthic prey are located.

Migratory Routes. To determine migratory routes and possible foraging areas, we attached geographic location, time-depth recorders (GLTDRs) to 21 animals: 9 postbreeding, nonpregnant females; 5 postmolt, pregnant females (3 of which were also recorded as postbreeding females), 6 males (3 adults of age 10 or over and 3 subadult males, 5–6 years of age), and 1 juvenile female, 1.4 years of age, on its third trip to sea. A photocell in each GLTDR recorded light levels, providing estimates of the time of sunrise and sunset as the animal traveled across latitudes and longitudes; from these data, an algorithm calculated geolocation (DeLong, Stewart, and Hill 1992; Hill, this volume). Geolocation estimates of latitude were adjusted in accordance with surface temperatures recorded by the GLTDR matched to mean sea surface temperature locations compiled semimonthly from satellites by the National Meteorological Center of the National Weather Service (Ashville, N.C.); that is, no geolocation estimate was accepted unless it was consistent with the surface temperature in the area. Recent validation experiments of geolocation from GLTDRs against a ship's Global Positioning System (GPS) off the coast of British Columbia, Canada, in June and July 1992 indicate that light-level estimates have a southerly bias (S. Blackwell and B. Le Boeuf, unpubl. data).

Despite the error associated with light-level estimates of location (Hill, this volume), preliminary data reveal that both adult males and postbreeding females migrated long distances and the migratory paths of the sexes overlapped in the eastern Pacific, especially along the coast up to about 58°N Lat. (fig. 13.8). The most northerly and most westerly migrations were undertaken by males. One adult male traveled as far as the eastern Aleutian Islands in southern Alaska, a round-trip migration of approximately 7,500 km; two others remained near the coast and went no farther north than about 49 degrees, near the state of Washington. Postbreeding females moved in a broad expanse of the eastern Pacific, from near the coastline to as far west as 150°W Long. The longest round-trip migrations of postbreeding females were about 4,866 km. The juvenile female traveled far north into the Gulf of Alaska, exceeding the distances traveled by most postbreeding, adult females.

Pregnant females had minimum round-trip migrations of about 3,900 to 6,800 km. Three of them, recorded when pregnant as well as nonpregnant, took similar routes under both conditions (fig. 13.9). The distance traveled during these biannual migrations was always less during the nonpregnant period.

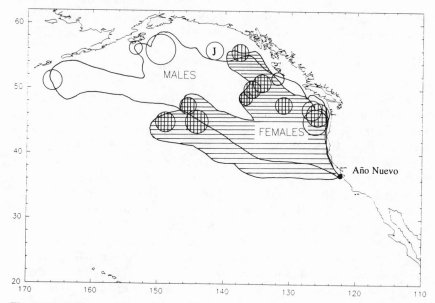

Fig. 13.8. Migratory destination or location at the end of recording of 6 breeding-age males (open circles), 9 postbreeding females (vertically hatched circles), and 1 juvenile female (J) that went to sea at Año Nuevo, California. The migratory paths of males are shown in the unhatched area; those of females are shown by horizontal hatching. The ultimate locations were derived from tracts plotted every two days as illustrated in figure 13.9.

The data suggest several points that merit further study:

1. Geolocation by light levels—corrected with sea surface temperature—is not accurate enough to place a seal in a specific location, such as correlating location with bathymetry, but it is sufficient for showing general migration direction and distance.

2. The data presented here, along with more recent findings (Le Boeuf et al. 1993) and reports of other investigators (DeLong and Stewart 1991; DeLong, Stewart, and Hill 1992; Stewart and DeLong, this volume), provide evidence of sex differences in foraging behavior. Adult males migrate to the northern edge of the North Pacific from the state of Washington west to the eastern Aleutians; they move directly to a foraging area, as defined by concentrated diving in this vicinity for up to two months. The migrations of adult females are more removed from the continental margins and are characterized by steady movement in the open ocean in the general range of 44–52°N lat.; females do not forage in narrowly focused areas but forage steadily en route. Differences in the frequency of dive types suggest that

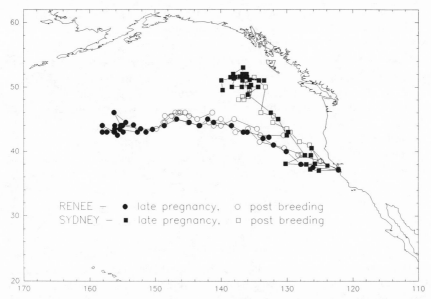

Fig. 13.9. Similarity in the migratory paths of females when pregnant (third trimester) and not pregnant (postbreeding period at sea). The pregnant period included the months November, December, and mid-January; the nonpregnant period included March, April, and mid-May. The records of pregnant females begin at roughly the furthest point from Año Nuevo; during the postbreeding period, the records begin at Año Nuevo and end at or near Año Nuevo. A third female, "B857," followed a similar route during both trips to sea; her record is not shown because it overlapped with that of the female "Renee."

       males are diving in areas where they can reach the bottom and that females are diving in the water column in deeper water (Le Boeuf et al. 1993); however, the geolocation data available are too imprecise to test this hypothesis.

3. Adult seals from Año Nuevo migrate to the same general areas as adult seals from San Miguel Island, located 450 km south of Año Nuevo (DeLong and Stewart 1991; DeLong, Stewart, and Hill 1992; Stewart and DeLong, this volume; Le Boeuf et al. 1993). This suggests that seals from other rookeries in Mexico and California might do the same. This would make for considerable mixing of animals during the foraging periods and would make reassortment on their return to their respective rookeries all the more interesting.

Homing and Translocation

The study of elephant seal diving has revealed a diving pattern that is unusual and difficult to explain with current theory that accounts well for the diving behavior of the shallower-diving sea lions and fur seals (Le Boeuf et al. 1988). How, for example, can elephant seals spend so little time on the surface following dives lasting over 1 hour? The diving pattern of elephant seals is evidently adapted to spending much of the year at sea and most of time at sea underwater. This regimen, however, is a problem for an investigator who wants to understand the physiology of diving by conducting short-term experiments of the type that have proven so successful with Weddell seals, *Leptonychotes weddelli* (Kooyman 1981; Guppy et al. 1986; Qvist et al. 1986; Hill et al. 1987). With this practical application in mind, my graduate students and I set out to determine if elephant seals translocated from Año Nuevo would home quickly and reliably.

Recent studies conducted with Guy Oliver, Pat Morris, and Phil Thorson showed that 43 of 45 juveniles, 1½- to 2-year-olds of both sexes, translocated from the Año Nuevo rookery in the spring when molting and in the fall when resting to release sites on a beach or at sea up to 70 km away returned "home" to the rookery. Most of the translocated animals wore radio transmitters that facilitated reidentification and determination of the time of return; the two exceptional animals bore no radios and may have returned but were overlooked. Nearly all returnees were back on the rookery within 1 to 7 days.

Sixteen of the translocated animals bore time-depth recorders. When these animals were released in a place where they had to traverse deep water to return to the rookery, a deep diving pattern indistinguishable from free-ranging diving was observed. We have subsequently used this paradigm to conduct doubly labeled water studies (D. Costa and B. J. Le Boeuf, unpubl. data), measure swim speeds (P. Thorson, G. Kooyman, and B. J. Le Boeuf, unpubl. data), and record heart rate and EKG during diving (R. Andrews, unpubl. data). This paradigm should facilitate short-term studies of diving, especially physiological studies of diving requiring completion in a day or two.

CONCLUSION

The study of free-ranging diving of marine mammals is developing fast. At the marine mammal conference in Chicago in December 1991, there were 24 talks dealing with 18 species that were instrumented with radio transmitters, satellite tags, or time-depth recorders or that were tracked with sonar: 11 pinnipeds, 6 cetacea, and the manatee. Increasingly sophisticated microcomputer time-depth recorders with more channels and larger memories are generating a wealth of detail about the diving behavior of pinnipeds. In the near future, it is anticipated that GPS devices will be attached to seals, permitting precise location of the animals during their migrations. This technology is now being transferred to cetacea.

The incoming data on elephant seals show that variation in dive pattern with age, sex, and reproductive condition as well as individual differences in foraging location can be substantial. Parametric studies taking this variation into consideration will be necessary to obtain a thorough understanding of the marine aspects of an animal's natural history. Satellite monitoring of the migratory path and diving pattern of juveniles during the first trip to sea may elucidate the causes of high at-sea mortality.

We are on the verge of learning much about the free-ranging dive pattern of elephant seals, as well as other diving mammals, and about diving biology in general. Studies of the underlying mechanics of the diving pattern of diving mammals is a fertile field for future investigation.

ACKNOWLEDGMENTS

Many collaborators played key roles in the studies summarized here, and they have graciously allowed me to refer to unpublished data. These include Dan Costa, Pat Morris, Phil Thorson, Dan Crocker, Yasuhiko Naito, Tomohiro Asaga, Maria Kretzman, Jeanine Williams, Guy Oliver, and many others. The research reported here was funded by grants from the National Science Foundation, the Minerals Management Service, and the G. MacGowan Trust Fund.

REFERENCES

Calder, William A., III. 1984. *Size, Function, and Life History*. Cambridge: Harvard University Press.

Condit, R., and B. J. Le Boeuf. 1984. Feeding habits and feeding grounds of the northern elephant seal. *Journal of Mammalogy* 65: 281–290.

DeLong, R. L., and B. S. Stewart. 1991. Diving patterns of northern elephant seal bulls. *Marine Mammal Science* 7: 369–384.

DeLong, R. L., B. S. Stewart, and R. D. Hill. 1992. Documenting migrations of northern elephant seals using day length. *Marine Mammal Science* 8: 155–159.

Guppy, M., R. D. Hill, R. C. Schneider, J. Qvist, G. G. Liggins, W. M. Zapol, and P. W. Hochachka. 1986. Microcomputer-assisted metabolic studies of voluntary diving of Weddell seals. *American Journal of Physiology* 250: R175–187.

Hill, R. D., R. C. Schneider, G. C. Liggins, A. H. Schuette, R. L. Elliott, M. Guppy, P. W. Hochachka, J. Qvist, K. J. Falke, and W. M. Zapol. 1987. Heart rate and body temperature during free diving of Weddell seals. *American Journal of Physiology* 253: R344–351.

Hindell, M. 1990. Population dynamics and diving behaviour of southern elephant seals. Ph.D. dissertation, University of Queensland, Australia.

Hindell, M. A., D. J. Slip, H. R. Burton, and M. M. Bryden. 1992. Physiological implications of continuous, prolonged, and deep dives of the southern elephant seal (*Mirounga leonina*). *Canadian Journal of Zoology* 70: 370–379.

Kooyman, G. L. 1981. *Weddell Seal: Consummate Diver.* Cambridge: Cambridge University Press.

Le Boeuf, B. J., D. P. Costa, A. C. Huntley, and S. D. Feldkamp. 1988. Continuous, deep diving in female northern elephant seals, *Mirounga angustirostris. Canadian Journal of Zoology* 66: 446–458.

Le Boeuf, B. J., D. P. Costa, A. C. Huntley, G. L. Kooyman, and R. W. Davis. 1986. Pattern and depth of dives in northern elephant seals, *Mirounga angustirostris. Journal of Zoology* 208: 1–7.

Le Boeuf, B. J., D. E. Crocker, S. B. Blackwell, P. A. Morris, and P. H. Thorson. 1993. Sex differences in diving and foraging behavior of northern elephant seals. In *Marine Mammals: Advances in Behavioural and Population Biology*, ed. I. Boyd, 149–178. Symposia of the Zoological Society of London no. 66. London: Oxford University Press.

Le Boeuf, B. J., Y. Naito, T. Asaga, D. Crocker, and D. P. Costa. 1992. Swim speed in a female northern elephant seal: Metabolic and foraging implications. *Canadian Journal of Zoology* 70: 786–795.

Le Boeuf, B. J., Y. Naito, A. C. Huntley, and T. Asaga. 1989. Prolonged, continuous, deep diving by northern elephant seals. *Canadian Journal of Zoology* 67: 2514–2519.

Lyons, K. J. 1991. Variation in feeding behavior of female sea otters, *Enhydra lutris*, between individuals and with reproductive conditions. Ph.D. dissertation, University of California, Santa Cruz.

Qvist, J., R. D. Hill, R. C. Schneider, K. J. Falke, G. C. Liggins, M. Guppy, R. C. Elliot, P. W. Hochachka, and W. M. Zapol. 1986. Hemoglobin concentrations and blood gas tensions of free-diving Weddell seals. *Journal of Applied Physiology* 61: 1560–1569.

Renouf, D., R. P. Gales, E. Noseworthy, and D. Rosen. 1991. Longitudinal studies of seasonal variations in phocid bioenergetics. Ninth Biennial Conference on the Biology of Marine Mammals, December 5–9, 1991, Chicago, Ill.

Scholander, P. F. 1940. Experimental investigations on the respiratory function in diving mammals and birds. *Hvalradets Skrifter Norske Videnskaps-Academie, Oslo* 22: 1–131.

FOURTEEN

Diving Behavior of Southern Elephant Seals from Macquarie Island: An Overview

David J. Slip, Mark A. Hindell, and Harry R. Burton

ABSTRACT. Results from 84 deployments of time-depth recorders on southern elephant seals, *Mirounga leonina*, at Macquarie Island are discussed. With loss and failure of instruments, data were collected from 31 seals: 14 postmolt females, 9 postbreeding females, 3 postmolt males, and 5 postbreeding males.

Water temperature data, also collected by the diving instruments, indicated that the major foraging areas of adult elephant seals from Macquarie Island were located in Antarctic waters and that males used areas over the Antarctic continental shelf, while females tended to inhabit deeper, offshore waters. Some individuals of either sex, however, spent the majority of their time in warmer waters associated with the Antarctic Polar Front or over the northern Campbell Plateau. Approximately 90% of the time at sea was spent submerged, with times on the surface generally less than 10 minutes.

Five characteristic dive types were identified, the most common of which were type 1 and type 2 dives. Both types are characterized by rapid descent to a depth followed by a protracted time at that depth interspersed with a number of small "wiggles," in turn followed by a rapid ascent to the surface. Type 1 dives exhibited marked diurnal variation, while type 2 dives showed no such variation and were usually in long sequences of very similar depths. We interpret type 1 as pelagic foraging dives and type 2 as benthic foraging dives. Type 2 dives were almost exclusively made by males while over the continental shelf. Type 1 dives were made by all animals.

The majority of dives made by southern elephant seals are characteristically deep (maximum depth = 1,430 m) and long (maximum duration = 120 min.). Estimated aerobic dive limits (ADL) were rarely exceeded by males or postbreeding females, but 44% of all dives made by postmolt (and, therefore, gestating) females exceeded ADL. Females in the last two trimesters of gestation performed the longest dives and most often exceeded ADL, suggesting that the estimated ADL is being underestimated in these animals and that these animals are making physiological adjustments that increase oxygen stores or reduce oxygen consumption.

Although much is known of the onshore behavior of the southern elephant seal (e.g., Carrick Csordas, and Ingham 1962; Ling and Bryden 1981; Hindell and Burton 1988), it is only recently that their aquatic behavior has come under investigation. Studies of diving in free-ranging seals are becoming increasingly common with the development of more sophisticated technology (e.g., Gentry and Kooyman 1986; Le Boeuf et al. 1988, 1989; Boyd and Arnbom 1991; Hindell, Burton, and Slip 1991; Hindell, Slip, and Burton 1991; Hindell et al. 1992). Studies such as these describe diving in relation to potential prey distribution, foraging range and movement patterns, and the diving performance in terms of physiological implications.

The southern elephant seal is a major predator of squid and fish in the Southern Ocean ecosystem (Laws 1977). Nevertheless, the diet of the southern elephant seal is one of the least known of the marine mammals of the Southern Ocean, with studies based on stomach samples collected while seals were ashore for their annual breeding or molting haul-outs (Laws 1977; Clarke and MacLeod 1982; Green and Williams 1986). Difficulties in collecting dietary samples from southern elephant seals near their main foraging areas prevent us from constructing a complete picture of the feeding ecology of one of the major vertebrate consumers of the Southern Ocean.

The annual cycle of adult southern elephant seals has two distinct aquatic phases: one postbreeding and one postmolt. Females come ashore to breed on subantarctic islands between September and November, whereas breeding males are ashore between August and December (Ling and Bryden 1981; Hindell and Burton 1988). After giving birth and weaning a pup, females spend about 10 weeks at sea feeding before returning to shore to molt in January and February. Adult males spend 12 to 14 weeks at sea feeding before returning to molt in mid-March. The molting process takes about 4 weeks, after which the seals return to sea until the following breeding season. By deploying time-depth recorders (TDRs) on adult male and female southern elephant seals as they leave Macquarie Island (lat. 54°35'S, long. 158°55'E) for the postbreeding and postmolt period at sea, we have been able to gain insight into their diving behavior that has added to out undertanding of their potential diet, foraging behavior, and movement patterns, all of which are crucial to understanding the role of the southern elephant seal in the Southern Ocean ecosystem. These studies have also raised some interesting questions regarding mammalian diving physiology.

In this chapter, we present an overview of our current knowledge of the diving behavior of the southern elephant seal at Macquarie Island with respect to their movement patterns, diving patterns, and physiology.

DEPLOYMENT AND RECOVERY OF RECORDERS

Time-depth recorders from Wildlife Computers were deployed on adult male and female southern elephant seals at Macquarie Island in 1988 and 1990. These units were microprocessor-controlled recording units equipped with pressure transducer, temperature probe, and internal clock. In addition, 50% of the units deployed in 1990 were equipped with a light-sensing geolocation option (DeLong Stewart, and Hill 1989). The movements of two adult males were determined in 1990 and 1991 using platform transmitter terminals and the Argos satellite-based location and data collection system. Seals were sedated using a combination of ketamine hydrochloride and either diazepam or xylazine (see Woods, Hindell, and Slip 1989). The units were attached to the seals using quick-setting epoxy (Araldite K268 road-marker adhesive, Ciba-Geigy), and a radio transmitter was attached to the TDR to facilitate the recovery of the units. Each complete package weighed approximately 3 kg (see fig. 14.1).

Eighty-four TDRs have been deployed on southern elephant seals at Macquarie Island. In 1988, 6 and 13 units were deployed on postmolt males and females, respectively; 10 units were deployed on postbreeding males; and 10 were deployed on postbreeding females. In 1990, 13 and 12 units were deployed on postmolt males and females, respectively; 10 units were deployed on postbreeding males; and 10 were deployed on postbreeding females. The TDRs were programmed to record depth every 30 seconds and temperature every 300 seconds, a sampling protocol that allowed for up to 80 days of continuous recording. The units that were deployed on postmolt animals were programmed to begin recording at different intervals to cover the entire period the seals were at sea. Recorders began sampling in February, April, or June. Postbreeding females were at sea for about 70 days and postbreeding males for about 100 days, so recording began as the animals left Macquarie Island.

The recovery rate of recorders varied between sexes and time of year. The recovery rate was slightly higher for females than for males. Return rates were 68% for postmolt females and 55% for postbreeding females. For males, return rates were 47% and 40% for the postmolt and postbreeding periods, respectively. In addition, electronic failures reduced the number of successful recoveries to 16% and 56% for postmolt males and females and 25% and 45% for postbreeding males and females.

LOCATION OF ANIMALS

The location of the foraging areas of southern elephant seals was initially estimated by examining the sea surface temperatures recorded for each

Fig. 14.1. Male elephant seal with time-depth recorder inshore at Macquarie Island. Photograph by David J. Slip.

animal. These temperatures provided sufficient resolution to distinguish three general regions: cold Antarctic waters with sea surface temperatures less than 0°C, warmer subantarctic waters with sea surface temperatures greater than 4°C, and a midregion that we term "convergence" waters which corresponds to the region around the Antarctic Polar Front (APF), where sea surface temperatures are above 0°C but generally do not exceed 4°C. ·

To estimate the areas of the ocean where the seals were located with greater resolution, the daily sea temperature/depth profiles were examined for each animal and matched with detailed oceanographic data for the Southern Ocean (Gordon and Molinelli 1982). Full details of this technique for estimating foraging areas are given in M. A. Hindell, H. R. Burton, and D. J. Slip (1991). The accuracy of this technique is entirely dependent on the accuracy of the oceanographic data and thus may be subject to some error.

Although the estimated foraging areas were different for males and females, the major foraging areas of adult southern elephant seals were located in the cold Antarctic waters, with other areas located along the APF, and in warmer subantarctic waters north to about 50°S. All five postbreeding males showed temperature/depth profiles consistent with being in Antarctic waters. Comparisons with oceanographic data identified three regions of the Southern Ocean, all close to the Antarctic continent (fig. 14.2a).

Postmolt males appear to have three distinct foraging areas: two close to the Antarctic continent, and one in the warmer subantarctic waters close to Campbell Island (fig. 14.2b). The estimated foraging areas of adult males close to the Antarctic coast and the area close to Campbell Island were over shelf waters with depths of between 500 and 1,000 m.

The foraging areas of postbreeding females were either in Antarctic waters or close to the APF. The seven females that moved south to colder waters had temperature/depth profiles consistent with three discrete areas located close to the Antarctic coast but farther north than the areas estimated for adult males (fig. 14.2a). One of these females remained in Antarctic waters for about four weeks before it moved north and spent about three weeks in the warmer waters around the APF. Only two postbreeding females did not move into Antarctic waters, and the temperature/depth profiles for these animals were consistent with the oceanographic profiles of an area around the APF (fig. 14.2a). The temperature/depth profiles for postmolt females were consistent with three areas in Antarctic waters and one area around the APF (fig. 14.2b). The three areas in Antarctic waters were to the north of the estimated locations of the postmolt males, in water of depths greater than 1,000 m. Nine postmolt females had profiles consistent with these areas of Antarctic waters. Two postmolt females had temperature/depth profiles consistent with being north of the APF, one had

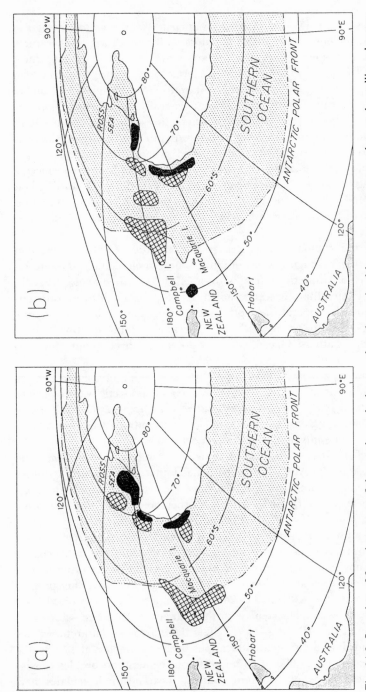

Fig. 14.2. Location of foraging areas of the southern elephant seal as determined by water temperature data and satellite telemetry (Slip, unpubl. data) for (A) postbreeding males, December to March (dark hatching), and postbreeding females, November to January (light hatching); and (B) postmolt males, March to August (dark hatching), and postmolt females, February to September (light hatching).

profiles consistent with waters to the south of the APF, and two had profiles consistent with being close to the APF, at times crossing it.

Subsequent satellite telemetry data from two male southern elephant seals are consistent with the estimations made from seawater temperatures, with one animal located in the Ross Sea and another located to the north-west of Campbell Island (unpubl. data). Preliminary analyses of geo-location data for males and females are also consistent with the estimations made from seawater temperature.

DIVE PATTERNS

All seals began diving as soon as they left Macquarie Island and dived almost continually throughout the recording period. The shelf around Macquarie Island extends only a few kilometers out to sea, so seals could begin deep diving within a few hours of leaving the island.

Approximately 90% of total time at sea was spent diving, and a dive rate of about 2.5 dives per hour was maintained throughout the entire period at sea. Approximately 99% of surface times between dives were of less than 10 minutes duration, with means ranging from 2.1 to 4.1 minutes for individual seals. Surface intervals longer than 10 minutes were defined as extended surface intervals (Le Boeuf et al. 1988) and generally occurred at higher frequencies at about the time the seals reached their foraging grounds. Extended surface intervals were more common at higher latitudes.

All seals that moved into Antarctic waters, both males and females, were recorded to haul out of the water for intervals of up to six hours two to three weeks after leaving Macquarie Island. It seems likely that these haul-outs were made onto ice floes, as few beaches are available in the areas indicated by the temperature profiles. Most of these seals made three to eight short haul-outs (usually less than three hours) over the recording period.

CHARACTERIZATION OF DIVE TYPES

Six characteristic dive types and the general form of the dive profile have been identified by principal component analysis (see Hindell, Slip, and Burton 1991). Typical examples of five of these dive types are illustrated in figure 14.3 (type 5 dives are rare; see below). Within these six dive types, there are two discrete groups of dives: those with more than one minute at the maximum depth, which were interpreted as foraging dives (types 1 and 2) and were the most common, constituting about 75% and 78% of total time at sea, respectively, for males and females, and those with less than one minute at the maximum depth, which were interpreted as nonforaging dives (types 3, 4, 5, and 6). The high proportion of type 1 and type 2 dives

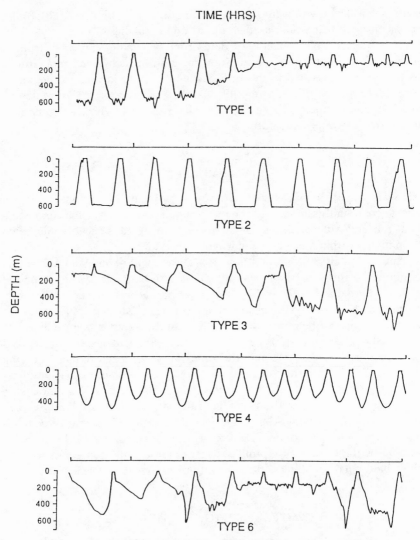

Fig. 14.3. Examples of dive types of the southern elephant seal.

is consistent with foraging dives, as elephant seals build up fat reserves quickly while at sea for reproduction and molting purposes.

Type 1 Dives

These dives were characterized by a rapid descent, followed by a period at the base of the dive containing shorter ascents and descents or "wiggles," prior to a rapid ascent to the surface (fig. 14.3). The depth and duration of these dives were variable, but they were generally of depths ranging from 300 m to 700 m (mean = 426 ± 177 m), with a mean of 11.3 ± 4.7 minutes between the end of descent and the beginning of ascent. Males and females spent different proportions of total time at sea on type 1 dives (35.9% for males, 77.6% for females). Females showed a marked diurnal pattern in the maximum depth, with type 1 dives becoming 200 to 400 m shallower at night. In males, a less pronounced diurnal pattern of depth occurred when the seals were close to Macquarie Island, but there was little evidence of any pattern of depth in these dives as the seals moved to higher latitudes or when they were over the Campbell Plateau.

We interpret these dives as pelagic foraging dives while the seals were foraging for prey characterized by diurnal vertical migrations. These dives are analagous to the Type D dives described for the northern elephant seal (Le Boeuf et al. 1988).

Type 2 Dives

These are characterized by a rapid descent followed by an almost flat period at depth with relatively few, short "wiggles" and a rapid ascent to the surface. These dives generally occur in bouts lasting about 18 hours, with little more than 10 to 20 m variation in depth between consecutive dives. Males spent 39% of total time at sea undertaking type 2 dives, but this type of dive was rarely seen in females, constituting only 0.5% of their total time at sea.

We interpret these dives as benthic foraging dives. In support of this interpretation, these dives were generally of depths ranging from 450 to 550 m (mean = 486 ± 83 m), with a mean of 13.9 ± 5.1 minutes spent at the bottom of the dive, and occurred over the Antarctic continental shelf and over the Campbell Plateau where water depths are around 500 m. When type 1 and type 2 dives occurred in the same area, type 1 dives were never deeper than type 2 dives, and when a series of type 2 dives occurred, no deeper dives were ever seen within the series, suggesting that the ocean floor limited the depth to which seals could dive.

Type 3 Dives

These are characterized by a short, rapid period of descent to a depth usually greater than 200 m, followed by a longer period of slow descent that occupied most of the dive, then a rapid ascent to the surface. These dives generally occurred in bouts of at least two dives and accounted for 4% of time at sea. The shape of these dives is consistent with the seals conserving energy by resting or sleeping, with the slower phase of descent resulting from a cessation of active swimming. Northern elephant seals are also believed to sleep underwater (Le Boeuf et al. 1988).

When seals were located in the more northerly areas of their range, a distinct pattern emerged: type 3 dives occurred in the early hours of the morning. This pattern was less pronounced at higher latitudes. There was considerable individual variation in the pattern of these dives, with some seals exhibiting a fairly regular 24-hour cycle over several weeks, while others were more erratic and did not undertake type 3 dives every day.

Type 4 Dives

These dives are a simple parabolic shape. They occurred predominantly in the first two to three weeks after the animals left Macquarie Island and appeared again as the animals approached Macquarie Island. This suggests that the function of these dives may simply be travel, as apparent velocities of between 80 and 240 km per day have been recorded by satellite for male southern elephant seals over the first three weeks after leaving Macquarie Island (unpubl. data). This type of dive had a mean depth of 343 ± 133 m and a mean duration of 20.4 ± 6.8 minutes. Seals spent 1.9% of total time at sea undertaking these dives. Why elephant seals would dive so deeply with no purpose other than travel is unclear. It may be that it is a predator avoidance strategy. Killer whales are the major predator of elephant seals (although sharks may also be important in the northern part of their range), and it may be that by maximizing the time they spend at depth, elephant seals reduce the risk of encountering these animals. There may also be an exploratory nature to these dives, and if prey are encountered, the animal may take advantage of this and forage. This interpretation is supported by the occurrence of, generally, type 4 dives in series that were sometimes punctuated by type 1 foraging dives. The occurrence of composite dives (see type 6 dives below) suggests that seals can change from one type of dive to another in mid-dive.

Type 5 Dives

These are short, shallow dives of less than 1.5 minute duration and are rare, accounting for 0.4% of total time at sea. The purpose of these dives is unclear. It may be that they are an artifact of the sampling protocol.

Type 6 Dives

These dives are quite variable in form, including simple spike dives and dives that may be composite dives, and do not fit into the above categories. They are the most common group of nonforaging dives, accounting for 7.2% of total time at sea. Type 6 dives had a mean depth of 485 ± 249 m and a mean duration of 19.5 ± 6.7 minutes. The large variance associated with mean depth was accounted for by the very deep dives that were usually simple spikes.

These dives probably have several functions, including exploration and escape from predators. A composite dive might occur if a foraging dive is interrupted by the threat of a predator or if prey is discovered during a travel or exploratory dive.

MAXIMUM DEPTHS AND DURATIONS

There was considerable individual variation in mean depth of dive, ranging from 269 ± 159 m to 589 ± 175 m, and in mean duration of dive, ranging from 16.0 ± 7.9 minutes to 36.9 ± 11.5 minutes (table 14.1).

The maximum dive depth recorded was 1,430 m by a postbreeding female (table 14.1). Although dives to such depths are relatively uncommon, constituting less than 1% of all dives, most seals undertook some very deep dives; 87% of females and 63% of males recorded dives of over 900 m in depth. This difference between the sexes might be a result of males foraging in shallower waters than females. The very deep dives were type 6 dives, generally a simple deep spike. These can occur in the middle of a set of type 1 or type 4 dives and are generally single deep dives. However, deep dives of over 1,000 m can occasionally occur in bouts of 3 to 10 dives over several hours. The functions of these deep dives are unclear, although possible explanations include avoidance of predators, exploration, or chasing prey.

Most seals undertook at least some dives with duration time greater than 50 minutes, and the longest dive duration recorded was 120 minutes by a postbreeding female (table 14.1). The longest male dive was 88.5 minutes. A full description of the 120-minute dive is given in Hindell et al. (1992). Long duration dives were generally type 6 composite dives. Their function might have been, for example, to escape an attack by a predator following a bout of foraging. These long dives do not appear in consistent patterns, occurring at any time during the course of a bout of more moderate dives. The only other phocid known to dive for as long or as deep is the northern elephant seal (Le Boeuf et al. 1988, 1989).

Given the similarities in morphology and life cycles of the two species of elephant seal, it is hardly surprising that there are many similarities in diving behavior between the southern elephant seal and its northern cogenitor. Adult southern elephant seals show a general pattern of continuous, prolonged deep diving. On leaving Macquarie Island, type 4 or traveling dives were common, interspersed with some type 1 pelagic foraging dives and occasional type 3 resting dives. After two to three weeks, the animals appeared to reach their foraging grounds, and females settled into a pattern

TABLE 14.1 Mean depth and duration of dives for each seal.

Seal	Sex/ Deployment	Mass (kg)	Depth (m)		Duration (min)		N
			M ± SD	Max.	M ± SD	Max.	
857	PMF	402	476 ± 197	1056	29.1 ± 8.4	58.6	1,881
1419	PMF	362	552 ± 207	1022	30.9 ± 8.0	78.5	2,836
1423	PMF	342	432 ± 151	1256	33.7 ± 13.6	120.0	1,362
1432	PMF	320	269 ± 159	856	16.0 ± 7.9	54.5	4,179
1440	PMF	422	396 ± 212	968	36.9 ± 11.5	78.5	2,386
3178	PMF	333	573 ± 195	1134	25.7 ± 6.6	52.0	3,927
3184	PMF	387	355 ± 157	758	28.8 ± 9.0	76.0	3,512
3186	PMF	278	446 ± 143	1044	20.1 ± 5.5	68.0	4,920
3187	PMF	304	380 ± 147	1158	21.7 ± 6.7	53.5	4,431
3189	PMF	282	296 ± 150	812	18.3 ± 7.7	55.5	5,071
3191	PMF	339	425 ± 170	1430	19.9 ± 4.9	56.5	2,577
3193	PMF	400	391 ± 176	928	28.9 ± 9.8	79.5	2,455
3196	PMF	430	456 ± 174	938	32.2 ± 9.1	98.5	2,554
3200	PMF	392	357 ± 119	1046	35.4 ± 8.2	66.0	2,353
905	PBF	295	382 ± 166	991	21.4 ± 6.6	66.0	4,458
1918	PBF	422	589 ± 166	1152	21.4 ± 6.6	53.0	4,554
1930	PBF	462	384 ± 166	911	21.3 ± 5.3	50.0	4,232
1938	PBF	366	490 ± 169	1100	22.3 ± 5.0	42.0	4,514
1948	PBF	425	347 ± 154	918	19.5 ± 6.8	52.5	4,568
6020	PBF	390	521 ± 167	978	21.1 ± 5.1	57.0	3,542
6010	PBF	344	547 ± 188	1004	20.7 ± 4.3	60.0	3,693
6012	PBF	367	420 ± 136	1118	17.0 ± 3.5	40.5	5,234
6002	PBF	298	414 ± 151	1218	19.0 ± 4.5	54.0	3,809
1475	PBM	1711	438 ± 137	1130	25.2 ± 6.4	63.0	3,836
1963	PBM	2122	426 ± 154	798	21.6 ± 6.3	88.5	4,462
1969	PBM	1657	401 ± 87	842	27.6 ± 6.1	62.5	3,654
6018	PBM	2143	526 ± 248	1210	26.7 ± 6.4	73.5	3,400
6035	PBM	1733	463 ± 144	1018	22.3 ± 5.5	67.5	4,407
4004	PMM	1275	390 ± 173	1282	22.0 ± 7.7	78.5	3,213
4017	PMM	2008	560 ± 145	846	31.9 ± 5.1	59.5	3,150
1453	PMM	3600	313 ± 224	1129	22.1 ± 1.1	78.5	3,453

NOTE: PMF = postmolt female, PBF = postbreeding female, PMM = postmolt male, PBM = postbreeding male, N = number of dives.

of type 1 pelagic foraging dives for most of the day, with a few hours of type 3 resting dives occurring in the early morning. The type 1 pelagic foraging dives became shallower at night, probably following the vertical migration of prey species. After males reached their foraging grounds, the general diving pattern was composed of up to 21 hours each day of type 2 benthic foraging, broken by a few hours of type 3 resting dives in the early hours of the morning. As the males and females returned to Macquarie Island, the most common dives were type 1 pelagic foraging dives and type 4 traveling dives.

PHYSIOLOGICAL IMPLICATIONS

The diving behavior of the southern elephant seal has some interesting physiological implications, as the animals dive continuously and deeply and remain submerged for long periods. An examination of the duration and depths of dives showed that postmolt females had significantly longer dive durations than postbreeding females, but there were no significant differences in dive depth between postbreeding and postmolt females (see fig. 14.4). When the data were pooled over years, it was possible to compare mean dive duration among four recording intervals—the three trimesters of pregnancy and the postbreeding period.

Seals recorded over the first trimester of pregnancy (February/April) and those recorded during the postbreeding period (November/January) had mean dive durations that were shorter than for those seals recorded over the second (April/June) and third (June/August) trimesters (fig. 14.5). However, mean dive depth and mean time spent at the bottom of the dive were not different among periods, suggesting that the longer dive durations were a result of slower ascents and descents.

When the theoretical aerobic dive limit (ADL) was calculated for each female using the equation ADL = (Lean Mass × TO_2)/RMR, where TO_2 = 0.079 1 O_2/kg (Kooyman 1989), and RMR = 0.0113 (Lean mass$^{0.75}$) 1 O_2/ minute (Schmidt-Neilsen 1983), it was possible to estimate the proportion of dives that exceeded this limit. Lean body mass was used to reduce bias introduced by differing amounts of metabolically inert blubber between individuals. Lean body mass was estimated from mean pre- and posthaul-out blubber mass of similar aged southern elephant seals (unpubl. data). Initial analysis of data from the 1988 deployments showed that 44% of dives made by postmolt females exceeded the calculated ADL, whereas only 7% of dives made by postbreeding females and less than 1% made by adult males exceeded this limit (Hindell et al. 1992). When data from 1990 were included, it was possible to make comparisons among different periods in a yearly cycle. Although there were considerable individual differences, animals from the postbreeding period (November/January) and from the first trimester (February/April) exceeded the ADL on a much smaller proportion of dives than did females from the second (April/June) and third (June/August) trimesters (table 14.2).

The small proportion of dives that exceeded the ADL in the postbreeding group suggests that the calculation of ADL for this group is relatively accurate and supports the assertion that the diving metabolic rate approxi-

MEAN DIVE DEPTH: FEMALES

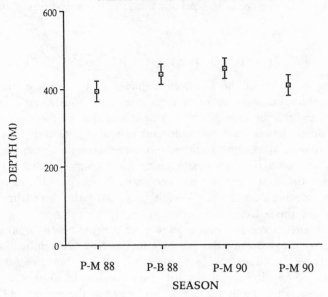

MEAN DIVE DURATION: FEMALES

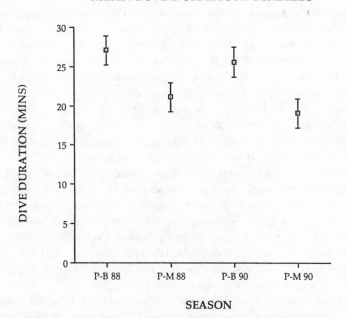

Fig. 14.4. Mean dive duration and mean dive depth for adult female southern elephant seals during the postmolt (PM) and postbreeding (PB) periods at sea for 1988 and 1990. Error bars are 95% confidence limits.

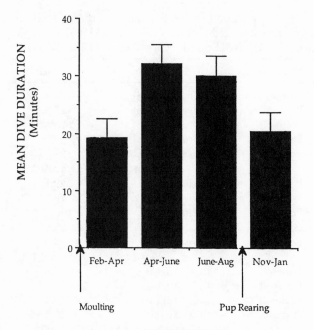

TIME OF YEAR

Fig. 14.5. Mean dive duration for female elephant seals during the three trimesters of pregnancy (Feb.–Apr., Apr.–June, June–Aug.) and the postbreeding period (Nov.–Jan.). Error bars are 95% confidence limits. Nested analysis of variance showed significant differences among groups ($F_{3,22} = 28.9$, $p < .0001$, number of dives = 83,008).

mates RMR for these animals. However, it seems unlikely that the ADL was exceeded by such a high proportion of dives in the last two trimester groups as the animals would have had to regularly cope with the buildup of toxic anaerobic metabolites by extending the duration of the surface intervals. Also, the absence of any extended surface intervals following dives that exceeded the calculated ADL suggests that the animals are, for the most part, diving aerobically. It has been suggested that northern elephant seals rarely exceed their ADL, even with dives as long as 50 minutes (Le Boeuf et al. 1988). Thus, the estimation of ADL is probably an underestimation for seals during the second and third trimesters. This would be accentuated by the presence of the fetus, which must be placing additional oxygen demands on the mother's oxygen stores and, therefore, lowering her effective aerobic limits. These seals may be making physiological adjustments to increase their ADL. There are several ways in which this might be achieved. A change in blood volume, possibly associated with pregnancy, would increase ADL, as would a reduction in metabolic rate. Although the seals would be gaining mass over this time, most of this mass gain is

TABLE 14.2   Proportion of dives of female southern elephant seals that exceed the estimated aerobic dive limit (ADL), where ADL = Lean mass $\times$ TO$_2$/RMR, and TO$_2$ = 0.0791 O$_2$/kg (Kooyman 1989).

Female	Deployment	Time of year	ADL (min)	% > ADL
1432	Postmolt	February–April	28.9	3.3
3191	Postmolt	February–April	27.9	3.8
3186	Postmolt	February–April	26.6	12.5
3189	Postmolt	February–April	26.7	13.4
3187	Postmolt	February–April	27.2	16.7
		Mean ± SD	27.5 ± 0.9	9.9 ± 5.4
3193	Postmolt	April–June	29.1	48.2
3184	Postmolt	April–June	28.9	54.3
1440	Postmolt	April–June	29.5	63.4
1419	Postmolt	April–June	28.4	68.4
3200	Postmolt	April–June	28.9	80.1
		Mean ± SD	29.0 ± 0.4	62.9 ± 11.1
3178	Postmolt	June–August	27.8	36.0
857	Postmolt	June–August	27.3	48.6
3196	Postmolt	June–August	29.6	63.5
1423	Postmolt	June–August	27.5	72.0
		Mean ± SD	28.1 ± 0.9	55.0 ± 13.8
6012	Postbreeding	November–January	28.8	0.3
1930	Postbreeding	November–January	30.2	2.2
6010	Postbreeding	November–January	28.0	2.4
1918	Postbreeding	November–January	29.5	2.9
6002	Postbreeding	November–January	27.0	3.8
1948	Postbreeding	November–January	29.5	4.1
6020	Postbreeding	November–January	28.9	5.5
1938	Postbreeding	November–January	28.5	8.0
905	Postbreeding	November–January	30.0	17.1
		Mean ± SD	28.9 ± 1.0	5.1 ± 4.7

through the addition of blubber reserves rather than an increase in lean mass. It may be that the seals reduce RMR through physiological adjustments, combined with behavioral changes that minimize energy consumption. It is suggested that the shorter dive durations immediately following the two periods of concentrated energy loss (molting and pup rearing, fig. 14.5) represent more concentrated and active periods of foraging to replenish the depleted fat stores. At these times, RMR would be elevated with respect to the less active foraging type 1 dives from April to August. Captive harbor seals, *Phoca vitulina*, continue to gain mass over the winter months in spite of reduced food intake and increased thermal energy demands, suggesting some form of reduction in metabolic rate (Renouf and Noseworthy 1990).

Although seals from the later trimesters exhibited longer dive durations than the other two groups, there were no differences in mean depth of dives and no differences in time spent at the bottom of the dive. Thus, the increase in dive duration must be accounted for by slower ascents and descents. Less strenuous swimming during the ascent and descent would conserve energy, reduce metabolic rate, and, hence, increase the ADL. Weddell seals, *Leptonychotes weddelli*, may reduce their metabolic requirements by 75% during the descent phase of a dive and by 55% during the ascent phase (Qvist et al. 1986). If the southern elephant seal can lower its metabolism by approximately 40% of resting metabolic rate, then aerobic capacity would be exceeded on very few dives (see Hindell et al. 1992).

Thus, it may be that southern elephant seals make behavioral and physiological changes during the long postmolt period at sea which reduce RMR and increase ADL. To test this more fully, we need to examine diving behavior of individual seals over the entire postmolt period, a task that now seems possible with the development of larger memory TDRs.

ACKNOWLEDGMENTS

We thank R. Woods, G. Bedford, K. Lawton, and all members of the 1988, 1989, 1990, and 1991 ANARE to Macquarie Island for their assistance in carrying out the fieldwork. K. Green also assisted in the field and commented on the manuscript, and J. Cox assisted with the figures. We thank the Tasmanian Department of Parks, Wildlife, and Heritage for permission to work at Macquarie Island.

REFERENCES

Boyd, I. L., and T. Arnbom. 1991. Diving behaviour in relation to water temperature in the southern elephant seal: Foraging implications. *Polar Biology* 11: 259–266.

Carrick, R., S. E. Csordas, and S. E. Ingham. 1962. Studies on the southern elephant seal, *Mirounga leonina* (L.). III. The annual cycle in relation to age and sex. *CSIRO Wildlife Research* 7: 119–160.

Clarke, M. R., and N. MacLeod. 1982. Cephalopods in the diet of elephant seals at Signy Island, South Orkney Islands. *British Antarctic Survey Bulletin* 57: 27–31.

DeLong, R. L., B. S. Stewart, and R. D. Hill. 1989. Tracking marine mammals at sea with an archival data recorder. In *Proceedings of the Eighth Biennial Conference on the Biology of Marine Mammals*, December 7–11, Pacific Grove, Calif.

Gentry, R. L., and G. L. Kooyman. 1986. *Fur Seals: Maternal Strategies on Land and at Sea*. Princeton: Princeton University Press.

Gordon, A. L., and E. J. Molinelli. 1982. *Southern Ocean Atlas*. Surrey: Columbia University Press.

Green, K., and R. Williams. 1986. Observations of the food remains in faeces of elephant, leopard, and crabeater seals. *Polar Biology* 6: 43–45.

Hindell, M. A., and H. R. Burton. 1988. Seasonal haulout patterns of the southern elephant seal (*Mirounga leonina*) at Macquarie Island. *Journal of Mammalogy* 69: 81–88.

Hindell, M. A., H. R. Burton, and D. J. Slip. 1991. Foraging areas of southern elephant seals, *Mirounga leonina*, as inferred from water temperature data. *Australian Journal of Marine and Freshwater Research* 42: 115–128.

Hindell, M. A., D. J. Slip, and H. R. Burton. 1991. The diving behaviour of adult male and female southern elephant seals, *Mirounga leonina*. *Australian Journal of Zoology* 39: 595–619.

Hindell, M. A., D. J. Slip, H. R. Burton, and M. M. Bryden. 1992. Physiological implications of continuous, prolonged and deep dives of the southern elephant seal (*Mirounga leonina*). *Canadian Journal of Zoology* 70: 370–379.

Kooyman, G. L. 1989. *Diverse Divers: Physiology and Behavior*. Berlin: Springer Verlag.

Laws, R. M. 1960. The southern elephant seal (*Mirounga leonina* Linn.) at South Georgia. *Norsk Hvalfangst-Tidende* 49: 466–476, 520–542.

———. 1977. Seals and whales of the Southern Ocean. *Philosophical Transactions of the Royal Society, London, Series B* 279: 81–96.

Le Boeuf, B. J., D. P. Costa, A. C. Huntley, and S. D. Feldkamp. 1988. Continuous, deep diving in female northern elephant seals, *Mirounga angustirostris*. *Canadian Journal of Zoology* 66: 446–458.

Le Boeuf, B. J., D. P. Costa, A. C. Huntley, G. L. Kooyman, and R. W. Davis. 1986. Pattern and depth of dives in northern elephant seals, *Mirounga angustirostris*. *Journal of Zoology, London* 208: 1–7.

Le Boeuf, B. J., Y. Naito, A. C. Huntley, and T. Asaga. 1989. Prolonged, continuous, deep diving by northern elephant seals, *Mirounga angustirostris*. *Canadian Journal of Zoology* 67: 2514–2519.

Ling, J. K., and M. M. Bryden. 1981. Southern elephant seal, *Mirounga leonina*. In *Handbook of Marine Mammals. 2. Seals*, ed. S. H. Ridgeway and R. J. Harrison, 297–327. London: Academic Press.

Naito, Y., B. J. Le Boeuf, T. Asaga, and A. C. Huntley. 1989. Long-term diving records of an adult female northern elephant seal. *Antarctic Record* 33: 1–9.

Qvist, J., R. D. Hill, R. C. Schneider, K. J. Falk, G. C. Liggins, M. Guppy, R. L. Elliot, P. W. Hochachka, and W. M. Zapol. 1986. Hemoglobin concentrations and blood gas tensions of free-diving Weddell seals. *Journal of Applied Physiology* 61: 1560–1569.

Renouf, D., and E. Noseworthy. 1990. Feeding cycles in captive harbor seals (*Phoca vitulina*): Weight gain in spite of reduced food intake and increased thermal demands. *Marine Behavioral Physiology* 17: 203–212.

Schmidt-Neilsen, K. 1983. *Animal Physiology: Adaptation and Environment*. Cambridge: Cambridge University Press.

Woods, R., M. A. Hindell, and D. J. Slip. 1989. Effects of physiological state on duration of sedation in southern elephant seals. *Journal of Wildlife Diseases* 25: 586–590.

FIFTEEN

Developmental Aspects of Diving in Northern Elephant Seal Pups

Philip H. Thorson and Burney J. Le Boeuf

ABSTRACT. The aim of this study was to describe the development of diving and measure concomitant changes in physiological correlates that enable this behavior in northern elephant seals, *Mirounga angustirostris*, during the first nine months of life.

Fifty-seven known-age juvenile seals born at Año Nuevo, California, were studied during the period 1988–1990. We measured (1) time spent in the water and changes in dive depth near the natal rookery during the 2½-month period following weaning, before the seals went to sea for the first time; (2) changes in blood volume, hematocrit, hemoglobin, and myoglobin in seals from near birth to eight months of age; (3) metabolic rate, from oxygen consumption, in 1½- to 3½- month-old juveniles diving in a hooded saltwater tank; and (4) the free-ranging dive pattern of 4-month-old juveniles during part of the first trip to sea.

Diving performance improved quickly during the 10-week period between weaning and going to sea, as reflected by increases in time spent in the water to 12½ hours per day, mean dive duration to 5.9 minutes, and mean dive depth to 16 m. Concurrently, blood and muscle oxygen stores increased, leading to a 46.7% elevation in mass specific oxygen stores, and diving metabolic rate decreased by approximately 50%. Metabolic rate also declined with increasing length of dives and total time submerged. The diving behavior of two 4-month-old seals during the first 12 to 26 days at sea resembled the continuous, deep, and long diving pattern of adults. Mean dive duration was 10 minutes (maximum = 22.3 min), mean surface interval was in the range 1.4 to 1.8 minutes, and mean dive depth was 206 m (maximum = 553 m); approximately 85% of the time at sea was spent submerged.

In only 10 weeks, newly weaned elephant seals undergo profound changes in mass specific blood volume, oxygen stores, and diving metabolic rate while learning to swim and dive near the rookery. These developments prepare them for long-duration deep diving over several months at sea.

Phocid seals suckle out of the water on the substrate where they are born. Consequently, in most species, pups get little or no experience swimming or diving prior to weaning. Newly weaned pups must develop these skills before they can forage on their own, often over long periods in the open ocean. Despite the critical importance for survival of the transition from a terrestrial to a marine existence, this period has not been studied in depth in any diving mammal.

Our aim was to study behavioral and physiological changes that accompany the development of diving in the northern elephant seal, *M. angustirostris*, a seal in which the move from land to sea is especially abrupt and demanding, requiring extreme adaptations for breath holding and withstanding high pressure. Adults lead a pelagic existence characterized by continuous, long duration, deep diving (Le Boeuf et al. 1986, 1988, 1989; DeLong and Stewart 1991; Hindell, Slip, and Burton 1991).

Northern elephant seal pups are nursed daily for up to 28 days before being weaned when the mother goes to sea (Le Boeuf, Whiting, and Gantt 1972). They remain on the natal rookery for the next 2½ months, fasting from food and water while learning to swim and dive. Within 2 weeks after weaning, the weanling enters the water for the first time, usually standing freshwater ponds, tide pools, or shallow water in protected coves or beaches (Rasa 1971; Reiter, Stinson, and Le Boeuf 1978). Initial attempts at swimming and diving are awkward and uncoordinated, but improvement is rapid. During this time, mean sleep apnea duration on land doubles from 4 to 8 minutes (Blackwell and Le Boeuf 1993). Within 8 to 10 weeks of weaning, these juveniles make the first pelagic trip to sea to forage, a journey that lasts approximately 4 months (Reiter et al. 1978).

The first trip to sea is a critical period in the life of elephant seals, as only a mean $46.0 \pm 7.7\%$ that depart in the late spring survive and return to the rookery in the fall (Le Boeuf, Morris, and Reiter, this volume). Little is known of the behavior of elephant pups during this period. Tag resight records reveal a general dispersal to the north, with pups from the most northerly rookeries in central California being observed as far north as northern California and Vancouver Island, British Columbia (Bonnell et al. 1979; Condit and Le Boeuf 1984). Examination of the stomach contents of juveniles reveals the remains of species common at depths of 200 m or more (Condit and Le Boeuf 1984; Antonelis et al. 1987).

The development of diving ability during the postweaning fast is important both for initial foraging success and for avoidance of white sharks, *Carcharodon carcharias*, a near-surface predator (Ainley et al. 1981; Le Boeuf, Riedman, and Keyes 1982). Critical to understanding diving performance is knowledge of the amount of oxygen stored, the rate that oxygen is utilized, and its effect on diving behavior. Changes in hemoglobin concentration, blood volume, and myoglobin concentration directly affect oxygen storage capacity (Snyder 1983; Kooyman 1989). Changes in metabolic rate with age affect the rate that oxygen stores are used. Oxygen storage capacity and diving metabolic rate determine the aerobic dive limit (ADL), the amount of time that a seal can remain submerged while diving aerobically (Kooyman et al. 1983).

The specific objectives of this study were to do the following: (1) describe the development of swimming and diving behavior of free-ranging juveniles near the natal rookery during the 2½-month period from weaning to departure from the rookery; (2) measure changes in blood volume, hematocrit, hemoglobin, and myoglobin in developing seals, from near birth to 8 months of age; (3) determine the metabolic rate of juveniles, 1½ to 3½ months old, during the course of diving, swimming, sleeping, and resting at the surface of a seawater tank; and (4) record the free-ranging dive pattern of juveniles, 3½ to 9 months of age, during the first pelagic trip to sea.

METHODS

Diving Behavior Near the Rookery during the Postweaning Fast

Diving behavior of juveniles, 1 to 3½ months of age, was studied in the waters surrounding their natal rookery at Año Nuevo, California, from February to May during the years 1988 to 1990. This encompasses the period from initial water entry to departure from the rookery on the first pelagic foraging trip to sea. All seals were known-age, having been marked with cattle ear tags (Dalton Jumbo Rototags, Oxon, England) in the interdigital webbing of the hind flippers a few days after weaning.

Three types of instruments were used to record changes in time per day spent in the water, dive duration, and dive depth. The amount of time spent in the water was recorded with modified digital watches (Cairns et al. 1987), which were attached to 15 juveniles at 1½ months of age. The watches were glued to the hair on the back of a sleeping juvenile, using 5-minute epoxy (Devcon, Danvers, Mass.). When the seal entered the water, the watch shorted out, and time on the watch did not advance during the period in the water. When the pup exited, the watch began to operate. Watches were read at least twice a week.

Dive duration data were obtained with radio transmitters (Titley Microelectronics, Blenheim, New Zealand) glued to the hair on the heads of 7 juveniles with 5-minute epoxy as they slept. Dive duration was measured using a stopwatch to record the time between when the radio signal was lost (submerged) and when the signal was recovered (surface). The range of the radio signal was approximately 4 km and was received with a Telonics TR-4 receiver (Mesa, Ariz.).

Maximum dive depth of 17 juveniles was measured with capillary tube depth recorders (Burger and Wilson 1988). Teflon tubing, 100 cm in length, was sealed at one end and dusted with blue dye powder. As the seal dived, pressure forced water part way up the tube, washing out the dye. The distance the water traveled through the tube was determined by pressure that is proportional to the maximum depth the seal attained. Maximum depth recorders (MDRs) were glued to the hair on the back of sleeping juveniles with 5-minute epoxy. After several days, the MDRs were recovered and maximum depth calculated from the equation of A. E. Burger and R. P. Wilson (1988).

Oxygen Storage Capacity

Total oxygen storage capacity was calculated as the sum of the blood, muscle, and lung oxygen stores, based on the equations of G. L. Kooyman et al. (1983).

Mass determinations, blood samples, and estimates of blood volume were obtained from 25 juveniles (11 males and 14 females) ranging in age from 2 days to 8 months old. The seals were restrained using a mixture of ketamine hydrochloride and diazepam at a dose of 4 mg/kg of body weight (Briggs, Hendrickson, and Le Boeuf 1975). Mass was determined for all seals by hoisting them in a modified canvas bag (Pernia, Hill, and Ortiz 1980) suspended from a $450 \pm .5$ kg spring scale (Chatillon, New York, N.Y.) attached to a tripod.

Blood samples were drawn from the extradural intravertebral vein using an 8.0-cm, 18-ga spinal needle (Geraci and Smith 1975). Samples were placed into sodium heparin Vacutainers (Becton-Dickson, Rutherford, N.J.). Blood volume was measured by first taking a blood sample, then injecting Evans blue dye into the extradural vein (4 mg/kg of body weight) (Linden and Mary 1983). A final blood sample was taken at least 20 minutes later, after allowing for equilibration.

Hematocrit was determined in duplicate from aliquots of whole blood that had been centrifuged at 11,500 rpm (IEC Micro Hematocrit, Needham Heights, Mass.). Hemoglobin (Hb) determinations were made using the cyanomethohemoglobin conversion method (Sigma Chemical Co., Assay Kit 525, St. Louis, Mo.). Blood volume was determined using the protocol of R. J. Linden and D. A. S. G. Mary (1983).

Myoglobin Assay

Muscle samples were obtained from 13 seals (8 females and 5 males) varying in age from stillborn pups to 8-month-old juveniles. For both fresh carcasses and live animals, muscle samples were taken from the latissimus dorsi just lateral to the pelvis. For biopsies of live seals, the site was cleaned with Betadyne solution, and a local anesthetic, lidocaine hydrochloride, was injected. A sterile 8 mm suction biopsy needle was inserted to obtain a tissue sample weighing approximately 75 mg (Dubowitz and Brooke 1973;

Evans, Phinney, and Young 1982). The assay for myoglobin concentration was performed using the protocol of B. Reynafarje (1963).

Metabolic Studies

Twelve juveniles ranging in age from 1½ to 3½ months old were transported from Año Nuevo to the Long Marine Laboratory where they were held in outdoor seawater tanks. Metabolic rate was determined using an open-circuit respirometry system to measure oxygen consumption in a manner similar to the method of T. M. Williams (1987).

This system consisted of a metabolic hood (a plexiglass dome measuring $2 \times 1 \times 0.5$ m) with an intake and exhaust port placed over a seawater tank which measured 2.5 by 1.8 by 2 m, the only area from which the seal could breathe. Ambient air was pulled through a dry gas meter (Singer, American Meter Division) and then through the metabolic hood by vacuum pump. An aliquot of the air exiting the dome was continually driven through a Baralyme column to remove carbon dioxide, and a Drierite column removed water before passing through an S-3A oxygen analyzer (Ametek, Sunnyvale, Calif.) to measure the fractional oxygen content. The analog output of the oxygen analyzer was converted to a digital signal (Sable Systems, Los Angeles, Calif.) and transferred to a computer. Equation 4b from P. C. Withers (1977) was used to calculate metabolic rate from the fractional change of oxygen.

Seals were weighed in the laboratory prior to experiments using a load cell platform scale (Senstek 2000, Canada). Water temperature of the metabolic tank ranged between 12 and 16°C, which is within the thermoneutral zone of weaned elephant seal pups (P. Thorson, unpubl. data).

Free-Ranging Dive Pattern at Sea

Time-depth recorders (TDRs) were attached to 5 juveniles (4 females and 1 male) during 1989 and 3 juveniles (2 males and 1 female) in 1990 at Año Nuevo just prior to departing on their first foraging trip to sea when they were 3½ months old. Two types of TDRs were used, one mechanical and the other a microprocessor system (Wildlife Computers, Woodinville, Wash.). The mechanical TDR measured 2.5 cm in diameter by 8.5 cm long with a mass of 70 (Naito, Asaga, and Ohyama 1990). The TDR used a mechanical pressure transducing system with a timing circuit to record time and depth on pressure-sensitive paper. The pressure transducing system had a threshold limit of 227 m. When recovered, the record was enlarged and digitized for analysis. The microprocessor TDR measured 15 cm long by 2.5 cm wide with a mass of 100 g. The TDR had 256 kilobytes of memory and was programmed to sample depth every 10 seconds (maximum depth limit = 2,000 m) and temperature every 10 minutes. At recovery, the data in the TDR were downloaded to a computer and analyzed.

The seals were chemically restrained, weighed, and blood sampled as mentioned above. TDRs were attached to the hair with 10-minute epoxy (Fibre Glass Evercoat, Cincinnati, Ohio), using the method of Le Boeuf et al. (1988, 1989).

RESULTS

Diving Behavior Near the Rookery during the Postweaning Fast

Newly weaned pups began entering the water at approximately 2 weeks postweaning at 1½ months of age. Time per day spent in the water was less than 2% at first and was concentrated at dawn and dusk. It increased to about 52% per day by 10 weeks postweaning, being concentrated at night, and remained at this level until departure on the first foraging trip to sea (fig. 15.1a).

The mean duration of dives in the waters surrounding the rookery increased from 1.9 minutes at initial water entry to 6.1 minutes at the end of the postweaning fast (fig. 15.1b). During this time, the seals did not venture far from the rookery and remained in water less than 12 m deep (except for one 16 m dive by seal H113 who was carrying a TDR; see below); mean dive depth increased with age (fig. 15.1c).

Changes in Physiological Variables over the First Eight Months of Life

Table 15.1 shows that as mass decreased during the postweaning fast, hematocrit, hemoglobin concentration, mass specific blood volume, and myoglobin concentration increased significantly (t-tests, all significant at $p < .005$). Consequently, mass specific oxygen stores increased by 46.7% over the postweaning fast, or 69.4% from the suckling period to the time when the seals were ready to go to sea. The highest levels of myoglobin concentration, mass specific blood volume, and total oxygen stores were reached in seals returning from their first trip to sea.

Metabolic Studies in Seawater Tanks during the Postweaning Fast

Diving metabolic rate decreased significantly (t-test = 5.43, df = 234, $p <$ .001), by about 50%, over the course of the postweaning fast (fig. 15.2a). Metabolic rate also declined as a function of increasing dive duration (fig. 15.2b) and increasing percentage of time spent submerged, when observed in 30-minute blocks (fig. 15.2c). This trend was seen throughout the postweaning fast, although it was more pronounced in the later period.

Diving Behavior during the First Foraging Trip to Sea

Two of the eight TDRs deployed on juveniles before their first trip to sea were recovered and contained diving data; both of them were carried by

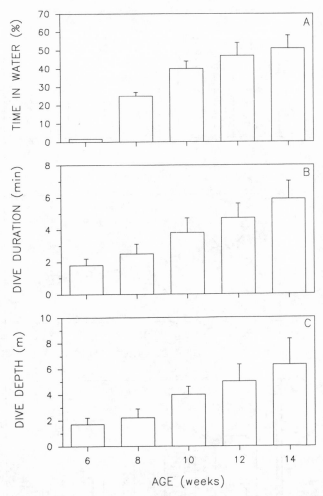

Fig. 15.1. *(A)* Percentage of time spent in water during the postweaning fast as a function of age based on 57 observations, from 15 individuals (9 females, 6 males). *(B)* Dive duration as a function of age during the postweaning fast, based on 121 observations of 7 individuals (4 males, 3 females). *(C)* Changes in dive depth during the postweaning fast as a function of age, based on 110 observations of 17 individuals (10 females, 7 males).

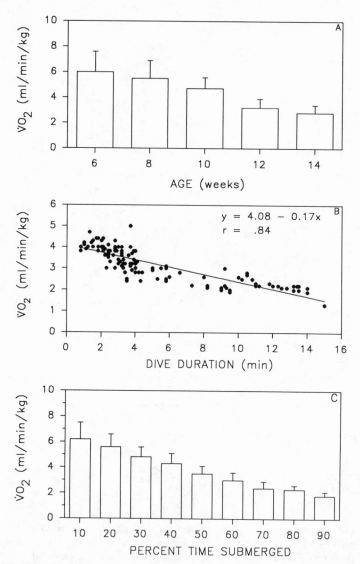

Fig. 15.2. *(A)* Changes in diving metabolic rate, measured as oxygen consumption, during the postweaning fast, based on 235 observations from 11 individuals (7 males, 4 females). *(B)* Change in metabolic rate, measured as oxygen consumption, as a function of dive duration. Data based on 137 observations from 6 individuals (4 males, 2 females). *(C)* Relation between metabolic rate, measured as oxygen consumption, and the percentage of time spent underwater. Data based on 115 30-min periods from 6 individuals (4 males, 2 females). Numbers in bars represent the number of observations for that interval.

TABLE 15.1 Mean hematocrit (Hct), hemoglobin concentration (Hb), mass specific blood volume (V_b), myoglobin concentration (Mb), and mass specific oxygen stores of northern elephant seals during nursing and the postweaning fast and after the first foraging trip to sea.

Age	Mass (kg)	Hct (%)	Hb (g/dL)	V_b (ml/kg)	Mb (g/100g)	O_2 Stores (ml/kg)
Nursing						
2–21 days	102	50.2	18.9	100	2.1	35.6
(n = 12)	(22.6)	(2.3)	(1.0)	(8.9)	(1.2)	
Postweaning						
4 weeks (n = 20)	134	52.4	20.0	111	3.4	41.1
	(19.2)	(2.9)	(1.8)	(14.6)	(1.1)	
8 weeks (n = 15)	103	59.6	22.6	127	4.5	52.3
	(20.2)	(6.4)	(2.3)	(12.9)	(1.2)	
12 weeks (n = 12)	96.1	61.6	23.6	144	5.1	60.3
	(13.3)	(4.2)	(3.2)	(18.2)	(1.3)	
After first foraging trip to sea						
8 months (n = 17)	90.1	58.1	22.8	175	5.7	68.8
	(23.2)	(6.5)	(2.4)	(10.4)	(1.1)	

NOTE: Numbers in parentheses represent ± 1 SD. Sample size is the same for all variables except myoglobin concentration (Mb), where n = 4 for all age classes.

females. The TDR of seal G372 was recovered in December 1990 at Año Nuevo, 7½ months after deployment; it recorded the first 12 days at sea, May 12–13, 1990. The TDR of seal H113 was recovered in March 1991 at Half Moon Bay, 45 km north of Año Nuevo, 10 months after deployment; it recorded the first 26 days at sea, May 14 to June 10, 1990.

A summary of the diving data for these two seals is shown in table 15.2. Both seals exhibited similar dive durations with a mean of about 10 minutes; seal H113 exhibited the longest dives, with the maximum lasting 22.3 minutes (fig. 15.3a). Approximately 90% of the surface intervals of both seals were between 0.5 and 1.75 minutes; only 5% were longer than 3 minutes (fig. 15.3b). Both seals had one extended surface interval longer than one hour, and both spent the majority of their time at sea underwater.

The mean depth of the dives of seal H113 was 206 m (median depth = 160 m), and its maximum dive depth was 553 m (table 15.2). Twenty-nine percent of the dives of seal G372 exceeded the depth limit of its TDR, 227 m. By comparison, 42% of the dives of the other seal exceeded 227 m.

Mean dive duration increased as a function of the time at sea for the first

TABLE 15.2 Summary of the diving behavior of seal G372 (12 days at sea) and seal H113 (26 days at sea).

Seal	No. of dives	Mean duration (min)	Max. duration (min)	Mean depth (m)	Max. depth (m)	Mean SI (min)	Max. SI (min)	% Time underwater
G372	1,408	9.5	17.4	157[a]	227[a]	1.8	66.1	84
		(1.2)		(66)		(3.2)		(3.7)
H113	3,124	10.6	22.3	206	533	1.4	217.1	88
		(3.5)		(120)		(4.4)		(4.1)

NOTE: Numbers in parentheses represent ± 1 SD.
[a] The TDR deployed on G372 had a depth limitation of 227 m, which the seal exceeded on 29% of its dives; therefore, the values given for mean and maximum dive depths are biased.

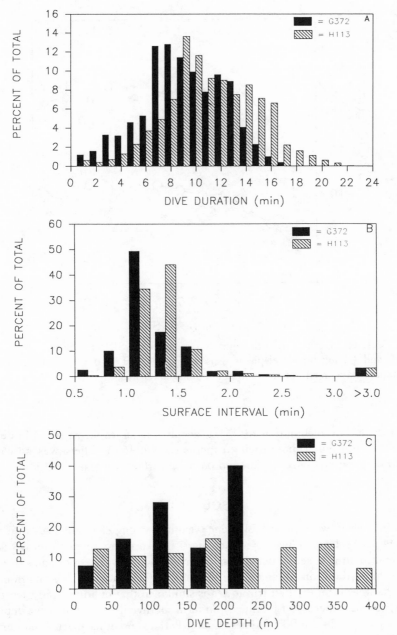

Fig. 15.3. *(A)* Frequency distributions of dive durations of 2 juvenile northern elephant seals during the first 12 to 26 days at sea. *(B)* Frequency distributions of surface intervals of 2 juvenile elephant seals during the first trip to sea. *(C)* Frequency distributions of dive depth of 2 juvenile northern elephant seals during the first trip to sea. The TDR of seal G372 had a depth limitation of 227 m.

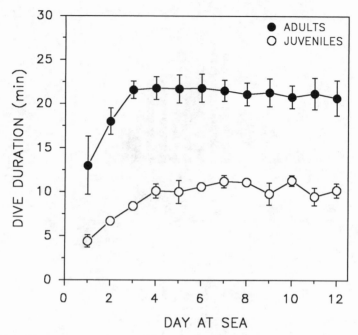

Fig. 15.4. Changes in mean dive duration per day (±1 SD) of 2 juvenile elephant seals and 4 adult female elephant seals during the first 12 days at sea following lactation. Data on adult females from unpublished records.

four days, but acclimation to diving deeply was as rapid as that of adult females with much more diving experience (fig. 15.4). There was no correlation between dive duration and postdive surface interval ($r = .01$).

DISCUSSION

During the 10-week period following weaning, young elephant seals undergo behavioral and physiological changes that prepare them to forage for several months at sea, a period characterized by virtually continuous, deep, and long duration diving. The diving performance of 3½-month-old elephant seals on the first trip to sea is similar to that of adults and exceeds that of most other adult pinnipeds, in terms of dive duration and dive depth (Kooyman 1989; Le Boeuf, this volume). Perhaps no other marine mammal must make such a drastic transition in such a short time.

We summarize and discuss the principal changes during this critical period of development that prepare the animal to be a deep and long duration diver. We emphasize that concomitant with improvements in swimming and diving performance, there occur significant increases in oxygen storage capacity and decreases in diving metabolic rate.

Oxygen Storage Capacity

Blood is the most important storage site of oxygen in phocid seals, as it contains approximately 65% of the total oxygen stores, followed by the muscle (30%) and the lungs (5%) (Kooyman 1985, 1989). Blood oxygen storage capacity is determined by hemoglobin concentration and blood volume (Snyder 1983). Diving animals tend to have higher hemoglobin concentrations and blood volumes than terrestrial animals. Among marine mammals, the deeper and longer duration divers have the highest hemoglobin concentrations and blood volumes (Ridgway and Johnston 1966; Sleet, Sumich, and Weber 1981; Duffield, Ridgway, and Cornell 1983; Snyder 1983; Kooyman 1985, 1989), and as we would expect from adult diving performance, these values have been reported to be high for northern elephant seal pups (Simpson, Gilmartin, and Ridgway 1970; Castellini, Costa, and Huntley 1986; Hedrick, Duffield, and Cornell 1986; Wickham 1989). Changes in these values with development have received little attention. A. M. Kodama, R. Elsner, and N. Pace (1977) reported that mass specific blood and hemoglobin concentrations increased during the first year in the harbor seal, *Phoca vitulina*. M. M. Bryden and G. H. K. Lim (1972) reported that southern elephant seals, *M. leonina*, increased their mass specific blood volume during the postweaning fast and during the first trip to sea, an increment that is similar to what we observed in northern elephant seal pups (table 15.1). By the time juvenile elephant seals are ready to go to sea for the first time, their body is 14.4% blood, as compared to 21.2% blood for an adult female (P. Thorson, unpubl. data).

Total oxygen stores increase rapidly to a high level in developing elephant seals, due to increases in blood and muscle oxygen stores (table 15.1). When it is time to make the first trip to sea, these young juveniles have already amassed mass specific oxygen stores of 60.3 ml/kg, 73.5% of the mass specific oxygen stores recorded in adult females (82.1 ml/kg) (P. Thorson, unpubl. data). The oxygen stores in juvenile elephant seals are similar to those found in adult phocids of other species (Kooyman 1985, 1989).

Metabolic Rate

By 3 months of age, elephant seals had significantly lower metabolic rates and longer dive durations than they exhibited at weaning (figs. 15.1B, 15.2A). This is not surprising, as it is known that metabolic rate decreases with age (Brody 1945; Miller and Irving 1975; Ashwell-Erickson and Elsner 1981) and with increasing time spent fasting (Brody 1945; Kleiber 1975; Ashwell-Erickson and Elsner 1981; Worthy and Lavigne 1987; Rea and

Costa 1992). Age and fasting effects alone, however, may not be responsible for the metabolic decrease we observed. During the postweaning fast, juvenile elephant seals spent an increasing amount of time swimming and diving, in the process utilizing protein and redistributing muscle mass (fig. 15.1; Bryden 1969). Using protein increases the metabolic rate, while diving decreases it.

The oxygen consumption pattern of juveniles was less consistent early in the postweaning fast than later on. This suggests that developing seals gained increasing control over their metabolic rate as their mass specific oxygen stores and diving experience increased. In addition, metabolic rate was inversely correlated with dive duration and the total amount of time spent diving (fig. 15.2), an effect that was most pronounced late in the postweaning fast. M. A. Fedak (1986) reported a similar trend for gray seals, *Halichoerus grypus*, swimming and diving in a water flume.

Diving Behavior on the First Trip to Sea

The diving pattern of juveniles during the first part of their first trip to sea suggests that they were diving aerobically. Postdive surface intervals are brief regardless of the duration of the previous dive (table 15.2). The percentage of dives that exceeded the ADL for the two juveniles in this study and for two juvenile Weddell seals is shown in table 15.3. With a diving metabolic rate of 1.5 times resting (Kooyman et al. 1973, 1983), 30.5 to 46.9% of the dives of the two juvenile elephant seals exceeded the ADL. This stands in marked contrast to the performance of immature Weddell seals who exceeded the ADL on only 4% of their dives, despite a 28% weight advantage. Moreover, for the immature Weddell seals, lactic acid and postdive surface intervals increased after dives above 10 minutes (Kooyman et al. 1983).

If the swim velocities of juveniles are relatively low on long duration dives, as preliminary data indicate (P. Thorson, unpubl. data) and as is the case in adult females (Le Boeuf et al. 1992; Crocker et al., this volume), the diving metabolic rate should be relatively low. Metabolic rate is highly correlated with swim velocity (Davis, Williams, and Kooyman, 1985; Ponganis et al. 1990; Williams, Kooyman, and Croll 1991); therefore, a decrease in swim velocity would decrease the diving metabolic rate and the ADL would increase. A diving metabolic rate of 0.36 to 0.39 l/min is 0.8 times the predicted basal metabolic rate (BMR) and would include even the longest dives of each pup at sea (table 15.3). Because diving metabolic rate in the laboratory decreases with increasing dive duration to levels below the predicted BMR (fig. 15.2b), it is possible for a seal to have a diving metabolic rate near resting when swim speed is low. Direct measurements of diving metabolic rate of Weddell seals have been recorded at near resting levels (Kooyman 1973; Castellini, Kooyman, and Ponganis 1992).

TABLE 15.3 Comparison of the estimated aerobic dive limit (ADL) of 2 juvenile Weddell seals (combined) and 2 juvenile elephant seals during the first foraging trip to sea.

Seal	Mass (kg)	O_2 stores (1 O_2)	VO_2 1 (O_2/min)	ADL (min)	% Dives (>ADL)
Weddell seal					
1.5·BMR 2.0 m/sec	140	8.2	0.69	11.9	3.5
Elephant seal					
1.5·BMR 2.0 m/sec	96	5.8	0.51–0.56	11.4	30.5
	109	6.6		11.8	46.9
1.0·BMR 0.8 m/sec	96	5.8	0.36	16.1	1.1
	109	6.6	0.39	16.9	4.7

NOTE: Estimates of ADL for the juvenile Weddell seals are from Kooyman et al. (1983). Estimates of the ADL for the elephant seals are from this study. Estimates of swim velocity for Weddell seals are from Castellini, Kooyman, and Ponganis (1992) and for elephant seals from P. Thorson (unpubl. data).

The depth of dives attained by juvenile elephant seals during the first 2 to 3 weeks at sea is remarkable in two respects. The dives are deeper than those of most other adult pinnipeds (see Le Boeuf et al. 1988, for review). Second, great depths are reached despite a body composition that averages 48% lipids, double the mean lipid composition of nonpregnant adult females and adult males (Kretzmann 1990; Rea and Costa 1992; unpubl. data). Because of the high ratio of fat to lean body mass, juveniles going to sea for the first time are positively buoyant. Thus, it would seem that juveniles would require greater effort to reach depths than adults.

In conclusion, the 10-week period following weaning that northern elephant seal juveniles spend swimming and diving near the natal rookery provides critical preparation for life at sea. Increases in blood volume, hematocrit, hemoglobin concentration, and myoglobin concentration result in large oxygen storage capacity, which, combined with an increased ability to decrease metabolic rate while diving, enables them to maximize time underwater for travel, foraging, rest, and predator avoidance.

ACKNOWLEDGMENTS

We thank V. Kirby, M. Kretzmann, P. Morris, L. Starke, and E. Theiss for assistance in the laboratory and in the field. M. A. Castellini, D. P. Costa, G. L. Kooyman, P. J. Ponganis, and G. A. J. Worthy provided invaluable advice on many aspects of this project. We are indebted to Y. Naito for providing four of the mechanical time-depth recorders as well as for digitizing the record of seal G372. We thank G. Strachan and the rangers at Año

Nuevo State Reserve for their cooperation. This work was supported by grants from the Earl and Ethyl Myers Oceanographic and Marine Biology Trust, Friends of the Long Marine Laboratory, Biology and Marine Sciences boards of the University of California, Santa Cruz, the G. MacGowan estate, and the National Science Foundation. This work was conducted under federal permit number 496 issued by the National Marine Fisheries Service.

REFERENCES

Ainley, D. G., C. S. Strong, H. P. Huber, T. J. Lewis, and S. H. Morrell. 1981. Predation by sharks on pinnipeds at the Farallon Islands. *Fishery Bulletin* 78: 941–945.

Antonelis, G. A., M. S. Lowry, D. P. De Master, and C. H. Ficus. 1987. Assessing northern elephant seal feeding habits by stomach lavage. *Marine Mammal Science* 3: 308–322.

Ashwell-Erickson, S., and R. Elsner. 1981. The energy cost of free existence for Bering Sea harbor and spotted seals. In *The Eastern Bering Sea Shelf: Oceanography and Resources*, ed. D. W. Hood and J. A. Calder, 869–899. Washington, D.C.: U.S. Department of Commerce.

Blackwell, S. B., and B. J. Le Boeuf. 1993. Developmental aspects of sleep apnoea in northern elephant seals. *Journal of Zoology, London* 231: 437–447.

Bonnell, M. L., B. J. Le Boeuf, M. O. Pierson, D. H. Dettman, and G. D. Farrens. 1979. *Summary Report 1975–1978. Marine Mammal and Seabird Surveys of the Southern California Bight Area. III. Pinnipeds*. Bureau of Land Management, Department of the Interior, Contract AA550-CT7-36 to Principal Investigators: Kenneth S. Norris (Cetacea), Burney J. Le Boeuf (Pinnipeds), and George L. Hunt, Jr. (Birds).

Briggs, G. D., R. V. Hendrickson, and B. J. Le Boeuf. 1975. Ketamine immobilization of northern elephant seals. *Journal of the American Veterinary Medicine Association* 167: 546–548.

Brody, S. 1945. *Bioenergetics and Growth with Special Reference to the Efficiency Complex in Domestic Animals*. New York: Hafner Press.

Bryden, M. M. 1969. Relative growth of the major body components of the southern elephant seal, *Mirounga leonina* (L.). *Australian Journal of Zoology* 17: 153–177.

Bryden, M. M., and G. H. K. Lim. 1972. Blood parameters of the southern elephant seal (*Mirounga angustirostris*, Linn.) in relation to diving. *Comparative Biochemistry and Physiology* 28: 139–148.

Burger, A. E., and R. P. Wilson. 1988. Capillary-tube depth gauges for diving animals: An assessment of their accuracy and capability. *Journal of Field Ornithology* 59: 345–354.

Cairns, D. K., K. A. Bredin, V. L. Bert, and W. A. Montevecchi. 1987. Electronic activity recorders for aquatic wildlife. *Journal of Wildlife Management* 51: 395–399.

Castellini, M. A., D. P. Costa, and A. C. Huntley. 1986. Hematocrit variation during sleep apnea in elephant seal pups. *American Journal of Physiology* 251: R429–R431.

Castellini, M. A., G. L. Kooyman, and P. J. Ponganis. 1992. Metabolic rates of freely diving Weddell seals: Correlations with oxygen stores, swim velocity, and diving duration. *Journal of Experimental Biology* 165: 181–194.

Condit, R., and B. J. Le Boeuf. 1984. Feeding habits and feeding grounds of the northern elephant seal. *Journal of Mammalogy* 65: 281–290.

Davis, R. W., T. M. Williams, and G. L. Kooyman. 1985. Swimming metabolism of yearling and adult harbor seals, *Phoca vitulina*. *Physiological Zoology* 58: 590–596.

DeLong, R. L., and B. S. Stewart. 1991. Diving patterns of northern elephant seal bulls. *Marine Mammal Science* 7: 369–384.

Dubowitz, V., and M. Brooke. 1973. *Muscle Biopsy: A Modern Approach.* Philadelphia: W. B. Saunders Co.

Duffield, D. A., S. H. Ridgway, and L. H. Cornell. 1983. Hematology distinguishes coastal and offshore forms of dolphins (*Tursiops*). *Canadian Journal of Zoology* 61: 930–933.

Evans, W. J., S. D. Phinney, and V. R. Young. 1982. Suction applied to a muscle biopsy maximizes sample size. *Medicine and Science in Sports and Exercise* 14: 101–102.

Fedak, M. A. 1986. Diving and exercise in seals: A benthic perspective. In *Diving in Animals and Man*, ed. A. Brubakk, J. W. Kanwisher, and G. Sundnes, 11–32. Tronheim: Tapir.

Fedak, M. A., L. Rome, and H. J. Seeherman. 1981. One-step-N_2 dilution technique for calibrating open-circuit VO_2 measuring systems. *Journal of Applied Physiology* 51: 772–776.

Geraci, J. R., and T. G. Smith. 1975. Functional hematology of ringed seals (*Phoca hispida*) in the Canadian arctic. *Journal of the Fisheries Research Board of Canada* 32: 2559–2564.

Hedrick, M. S., D. A. Duffield, and L. H. Cornell. 1986. Blood viscosity and optimal hematocrit in a deep-diving mammal, the northern elephant seal (*Mirounga angustirostris*). *Canadian Journal of Zoology* 64: 2081–2085.

Hindell, M. A., D. J. Slip, and H. R. Burton. 1991. The diving behaviour of adult male and female southern elephant seals, *Mirounga leonina* (Pinnipedia: Phocidae). *Australian Journal of Zoology* 39: 595–619.

Kleiber, M. 1975. *The Fire of Life: An Introduction to Animal Bioenergetics.* Huntington, N.Y.: Robert E. Kreiger Publishing.

Kodama, A. M., R. Elsner, and N. Pace. 1977. Effects of growth, diving history, and high altitude on blood oxygen capacity in harbor seals. *Journal of Applied Physiology* 42: 852–858.

Kooyman, G. L. 1973. Respiratory adaptation in marine mammals. *American Zoologist* 13: 457–178.

———. 1985. Physiology without restraint in diving mammals. *Marine Mammal Science* 1: 166–178.

———. 1989. *Diverse Divers: Physiology and Behavior.* Berlin: Springer Verlag.

Kooyman, G. L., M. A. Castellini, R. W. Davis, and R. A. Maue. 1983. Aerobic diving limits of immature Weddell seals. *Journal of Comparative Physiology* 151: 171–174.

Kooyman, G. L., D. H. Karem, W. B. Campbell, and J. J. Wright. 1973. Pulmonary gas exchange in freely diving Weddell seals, *Leptonychotes weddelli*. *Respiratory Physiology* 17: 283–290.

Kooyman, G. L., E. A. Wahrenbrock, M. A. Castellini, R. W. Davis, and E. E. Sinnett. 1980. Aerobic and anaerobic metabolism during voluntary diving in Weddell seals: Evidence of preferred pathways from blood chemistry and behavior. *Journal of Comparative Physiology* 138: 335–346.

Kretzmann, M. B. 1990. Maternal investment and the post-weaning fast in northern elephant seals: Evidence for sexual equality. Master's thesis, University of California, Santa Cruz.

Lane, R. A. B., R. J. H. Morris, and J. W. Sheedy. 1972. A haematological study of the southern elephant seal, *Mirounga leonina* (Linn.). *Comparative Biochemistry and Physiology* 42A: 841–850.

Le Boeuf, B. J., D. P. Costa, A. C. Huntley, and S. D. Feldkamp. 1988. Continuous, deep diving in female northern elephant seals, *Mirounga angustirostris. Canadian Journal of Zoology* 66: 446–458.

Le Boeuf, B. J., D. P. Costa, A. C. Huntley, G. L. Kooyman, and R. W. Davis. 1986. Pattern and depth of dives in northern elephant seals, *Mirounga angustirostris. Journal of Zoology, London* 208: 1–7.

Le Boeuf, B. J., Y. Naito, T. Asaga, D. Crocker, and D. P. Costa. 1992. Swim speed in a female elephant seal: Metabolic and foraging implications. *Canadian Journal of Zoology* 70: 786–795.

Le Boeuf, B. J., Y. Naito, A. C. Huntley, and T. Asaga. 1989. Prolonged, continuous, deep diving by northern elephant seals. *Canadian Journal of Zoology* 67: 2514–2519.

Le Boeuf, B. J., M. Riedman, and R. S. Keyes. 1982. White shark predation on pinnipeds in California waters. *Fishery Bulletin* 80: 891–895.

Le Boeuf, B. J., R. J. Whiting, and R. F. Gantt. 1972. Perinatal behaviour of northern elephant seal females and their young. *Behaviour* 43: 121–156.

Lenfant, C., K. Johansen, and J. D. Torrance. 1970. Gas transport and oxygen storage capacity in some pinnipeds and the sea otter. *Respiratory Physiology* 9: 277–286.

Linden, R. J., and D. A. S. G. Mary. 1983. The measurement of blood volume. *Techniques in the Life Sciences: Cardiovascular Physiology* P305: 25.

Miller, K., and L. Irving. 1975. Metabolism and temperature regulation in young harbor seals, *Phoca vitulina richardsi* in water. *American Journal of Physiology* 229: 506–511.

Naito, Y., T. Asaga, and Y. Ohyama. 1990. Diving behavior of Adélie penguins determined by time-depth recorder. *Condor* 92: 582–586.

Pernia, S., A. Hill, and C. L. Ortiz. 1980. Urea turnover during prolonged fasting in the northern elephant seal. *Comparative Biochemistry and Physiology* 65B: 731–734.

Ponganis, P. J., G. L. Kooyman, M. H. Zornow, M. A. Castellini, and D. A. Croll. 1990. Cardiac output and stroke volume in swimming harbor seals. *Journal of Comparative Physiology* 160B: 473–482.

Rasa, O. A. 1971. Social interaction and object manipulation in weaned pups of the northern elephant seal *Mirounga angustirostris. Zeitschrift für Tierpsychologie* 29: 82–102.

Rea, L. D., and D. P. Costa. 1992. Changes in standard metabolism during long-term fasting in northern elephant seal pups (*Mirounga angustirostris*). *Physiological Zoology* 65: 97–111.

Reiter, J., N. L. Stinson, and B. J. Le Boeuf. 1978. Northern elephant seal development: The transition from weaning to nutritional independence. *Behavioral Ecology and Sociobiology* 3: 337–367.

Reynafarje, B. 1963. Simplified method for the determination of myoglobin. *Journal of Laboratory and Clinical Medicine* 61: 139–145.

Ridgway, S. H., and D. G. Johnston. 1966. Blood oxygen and ecology of porpoises. *Science* 151: 456–457.

Simpson, J. G., W. G. Gilmartin, and S. H. Ridgway. 1970. Blood volume and other hematological values in young elephant seals (*Mirounga angustirostris*). *American Journal of Veterinary Research* 31: 1449–1452.

Sleet, R. B., J. L. Sumich, and L. J. Weber. 1981. Estimates of total blood volume and total body weight of a sperm whale (*Physeter catodon*). *Canadian Journal of Zoology* 59: 567–570.

Snyder, G. K. 1983. Respiratory adaptations in diving mammals. *Respiratory Physiology* 54: 269–294.

Wickham, L. L. 1989. Blood viscosity in phocid seals: Possible adaptations to diving. *Journal of Comparative Physiology* 159B: 153–158.

Williams, T. M. 1987. Approaches for the study of exercise physiology and hydrodynamics in marine mammals. In *Approaches to Marine Mammal Energetics*, ed. A. C. Huntley, D. P. Costa, G. A. J. Worthy, and M. A. Castellini, 127–145. Lawrence, Kan.: Allen Press.

Williams, T. M., G. L. Kooyman, and D. A. Croll. 1991. The effect of submergence on heart rate and oxygen consumption of swimming seals and sea lions. *Journal of Comparative Physiology* 160B: 637–644.

Withers, P. C. 1977. Measurement of VO_2, VCO_2, and evaporative water loss with a flow-through mask. *Journal of Applied Physiology* 42: 120–123.

Worthy, G. A. J., and D. M. Lavigne. 1987. Mass loss, metabolic rate, and energy utilization by harp and gray seal pups during the postweaning fast. *Physiological Zoology* 60: 352–364.

SIXTEEN

Postbreeding Foraging Migrations of Northern Elephant Seals

Brent S. Stewart and Robert L. DeLong

ABSTRACT. Adult northern elephant seals depart southern California Channel Islands rookeries in February and early March to forage and replenish body reserves that were depleted during intensive breeding season fasts. Females remain at sea for around 66 days and males for around 120 days before they return to the Channel Islands to molt. During that period seals dive—and presumably forage—deeply and continually while migrating between southern California rookeries and haulouts and offshore, northern foraging areas between 40° and 48°N latitude (females) and the Gulf of Alaska and eastern Aleutian Islands (males); females cover over 5,500 km and males over 11,100 km during these round-trip postbreeding migrations. Males and females differ in their vertical and geographic distributions during these migrations, but the reasons for that segregation are unknown.

Adult northern elephant seals depart terrestrial rookeries from late January through early March and remain at sea, diving continually, for around 2 (females) to 4 (males) months before returning to land to molt (Le Boeuf et al. 1989; DeLong and Stewart 1991). Their behaviors while at sea during those postbreeding periods have been documented in extraordinary detail in recent years (e.g., Le Boeuf et al. 1988, 1989; DeLong and Stewart 1991), though the geographic locations of the seals have been unknown. In 1987, we began, in collaboration with colleague R. Hill, to develop and test a microprocessor-based event recorder that would allow simultaneous documentation of the locations and diving patterns of foraging northern elephant seals throughout their long periods at sea (DeLong, Stewart, and Hill 1992). Using data collected with those instruments, we describe here the postbreeding migrations of 8 adult male and 5 adult female northern elephant seals between San Miguel Island, in southern California, and pelagic foraging areas in the North Pacific, and we interpret intra- and intersexual variation in dive patterns in light of this new information on seal dispersion.

METHODS

We instrumented 8 lactating females in 1990 and 16 adult males in 1989 and 1990 with microprocessor-based, geographic-location-time-depth recorders (termed geolocation recorders, or GLTDRs; DeLong, Stewart, and Hill 1992) at San Miguel Island (34°02′N, 120°23′W) at the end of the breeding season (February–March). We recovered the instruments when the seals returned to land to molt several months later (females, April–May; males, June–July). The design and function of the GLTDRs are described in detail by R. L. DeLong and B. S. Stewart (1991) and DeLong, Stewart, and Hill (1992). Briefly, we programmed the instruments to sample hydrostatic pressure (= depth ± 2 m) at 30- or 60-second intervals for the entire periods the seals were at sea. Measurements of sea-surface temperature (SST) and ambient light levels were made and stored during the seals' brief interdive surface periods. We estimated each seal's latitudinal location each day by calculating day length and longitude from local apparent noon using daylight profiles stored in the GLTDRs and computer algorithms developed by DeLong, Stewart, and Hill (1992). Determination of location from daylight profiles and the factors that affect location accuracy are discussed in detail elsewhere (DeLong, Stewart, and Hill 1992; Hill, this volume). The most important influences on accurate calculation of latitude are the durations of seals' dives near twilight and the equinox, when day length does not vary substantially with latitude. Near the vernal equinox (March 22) we compared GLTDR sea-surface temperature measurements to latitudinal distributions of SST from other sources to determine seals' latitudinal locations (see DeLong, Stewart, and Hill 1992). We also used SST comparisons to validate all other locations. Our field calibration studies of these instruments indicate that locations that are calculated and corrected using this technique are accurate to around 60 nm or better (ibid.; Stewart and DeLong, unpubl. data). We determined the number of days that each seal was at sea by direct inspection of the dive records, which indicated departure and return dates and times.

Statistical analyses of dive parameters were performed using Systat. We report sample statistics (i.e., $\bar{X}$ = sample mean, SD = standard deviation of the sample mean) as summary statistics for all dives of each seal and population statistics (i.e., μ = population mean, SEM = standard error of the population mean) as summary statistics for all dives of all seals. We used one- or two-way analysis of variance (Zar 1984) to compare seals' dive records.

RESULTS

We recovered instruments from 5 females and 9 males. The light-level sensor in one male's instrument failed, but all other GLTDRs contained depth, light-level, and SST data for the seals' entire periods at sea.

Females were at sea for around 66 days and males for around 120 days before returning to San Miguel Island to molt (tables 16.1, 16.2). All seals dove continually while at sea (figs. 16.1, 16.2, 16.3). Females' dives averaged 520 m deep (SEM = 15 m) and 22.3 minutes long (SEM = 1.3 min.), and males' dives averaged 367 m deep (SEM = 34 m) and 22.6 minutes long (SEM = 1 min.); interdive periods at the surface were routinely brief (females, $\mu = 2.1$ min., SEM = 0.05 min.; males, $\mu = 3.2$ min., SEM = 0.1 min.; fig. 16.4).

Seals spent little time at depths shallower than 200 m, except when rapidly descending or ascending to preferred depths, or greater than 800 m (figs. 16.1, 16.2, 16.3). As 20 to 50% of most dives were spent within a range of 30 m of maximum depth (DeLong and Stewart 1991; Stewart and DeLong, unpubl. data), we use maximum depth of each dive as an index of seals' water depth preference and presumably foraging habitat. Females dove deeper, on average, than males did (p < .01; tables 16.1, 16.2), although the greatest depths reached during their migrations were similar (females, 983–1,567 m; males, 831–1,581 m).

All seals began traveling north immediately upon entering the water in late February and March, covering about 90 to 100 km/day for approximately 16 (females, SD = 7.6 days) to 38 days (males, SD = 5.7 days) before travel speeds slowed (figs. 16.5, 16.6, 16.7). Seals then remained in somewhat more defined geographic areas for periods of around 36 (females, SD = 5.2 days) to 51 days (males, SD = 6.4 days). We refer to those areas as foraging areas and define them according to periods when distances covered between days during three or more consecutive days were less than 32 km, to distinguish them from rapid northward movements away from San Miguel Island in March and similar southward movements when seals were returning to the island to molt. We refer to the latter as north and south transits, respectively. South transits to molting beaches from foraging areas took females around 15 days and males around 31 days (table 16.2), traveling at minimum speeds of 90 to 100 km/day (= 1.04–1.15 m/sec.). Foraging areas of females (figs. 16.5, 16.6) were less obvious than those of males, whose day-to-day movements in northern areas were small and highly concentrated (fig. 16.7). Female foraging areas therefore appear to be series of high-density clusters of daily locations rather than the single clusters characteristic of males (figs. 16.5, 16.6, 16.7).

Females covered at least 5,500 km during their postbreeding migrations and males at least 11,100 km. Seals remained in deep water (from one to

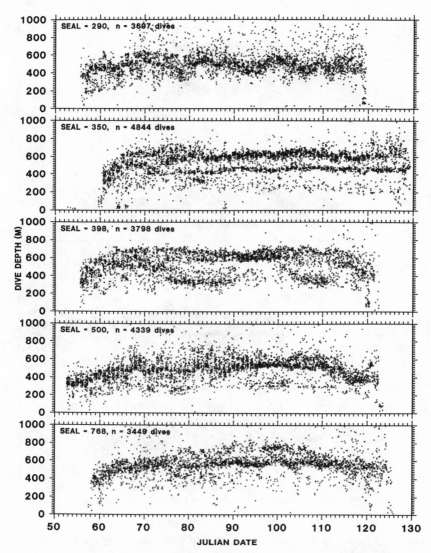

Fig. 16.1. Depths of all dives made each day for 5 northern elephant seal females during their postbreeding migrations in 1990.

several thousand or more meters) throughout their postbreeding migrations. Although males traveled through female foraging areas, between 40° and 48°N latitude, they did not linger there but continued rapidly north to the Gulf of Alaska and the eastern Aleutian Islands. Although the 1989 and 1990 foraging areas of some males overlapped, dives made in those areas

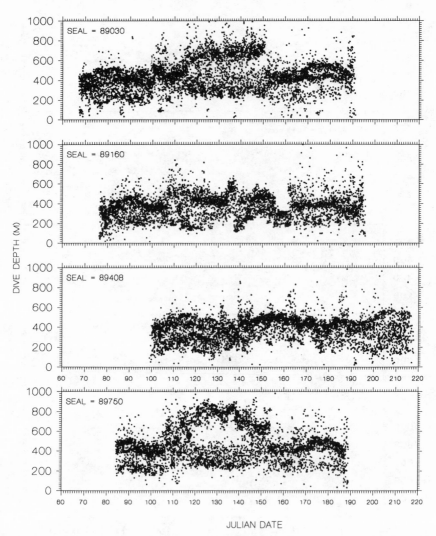

Fig. 16.2. Depths of all dives made each day for 4 adult northern elephant seal
males during their postbreeding migrations in 1989.

were shallower in 1990 ($\mu = 318$ m, SEM $= 56.2$) than they were in 1989
($\mu = 415.7$ m, SEM $= 24.4$; p $< .01$).

Dive depths were similar among all females, and dives made in foraging
areas were deeper than those made during north or south transits (table
16.1). Similarly, depths of foraging area dives of males were usually deeper
than transit area dives (table 16.2). Dives of 3 males in 1990 were notable
exceptions; seals 90640 and particularly 90480 spent substantial periods at

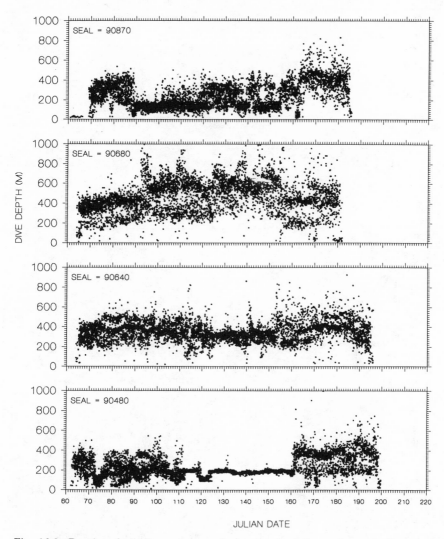

Fig. 16.3. Depths of all dives made each day for 4 adult northern elephant seal males during their postbreeding migrations in 1990.

shallow depths, and there was much less variation around maximum depth during those periods compared to the records of other males (figs. 16.2, 16.3). The patterns of these 2 males while in foraging areas differed fundamentally in this way from those of females. But the dive depths of 3 other males (89030, 89750, 90680) in their foraging areas south of the eastern Aleutian Islands were similar to dive depths of females that foraged in areas far to the south (tables 16.1, 16.2; figs. 16.2, 16.3, 16.4).

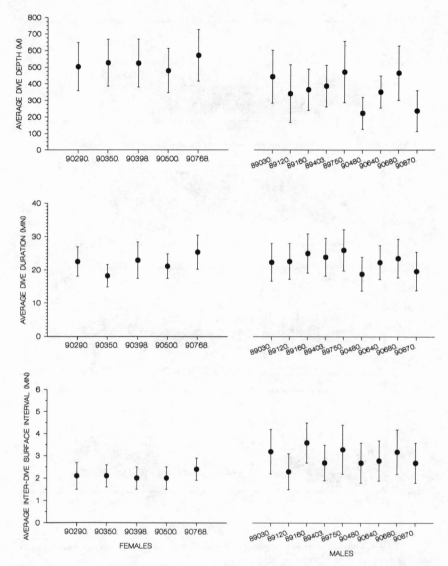

Fig. 16.4. Postbreeding season dive statistics for adult female and male northern elephant seals ranging from San Miguel Island, California (individual seal identification numbers are along the x axis; bars are standard deviations; there are no location records for male 89120).

TABLE 16.1   Depths of dives of adult female northern elephant seals during different stages of their postbreeding migrations.

Seal		North transit	Foraging area	South transit	Total
90290	Mean	402.5	520.1	508.2	502.6
	SD	112.0	135.5	153.1	143.8
	n_1	427	2056	1214	3697
	n_2	7	36	20	63
90350	Mean	492.4	553.5	541.1	526.3
	SD	151.5	117.1	146.7	140.6
	n_1	1957	1936	951	4844
	n_2	28	27	14	69
90398	Mean	495.6	538.9	504.3	523.7
	SD	127.7	133.7	180.9	143.2
	n_1	808	2324	666	3798
	n_2	14	40	13	67
90500	Mean	402.4	514.2	468.8	478.7
	SD	116.3	125.5	133.9	133.2
	n_1	1006	2425	908	4339
	n_2	16	39	16	68
90768	Mean	499.6	593.2	578.2	570.8
	SD	138.8	133.8	199.3	155.5
	n_1	705	2006	738	3449
	n_2	14	38	13	65

NOTE: SD = standard deviation; n_1 = number of dives; n_2 = number of days; mean and SD in m.

DISCUSSION

DeLong and Stewart (1991) reported significant diurnal and seasonal variation in dive parameters of adult northern elephant seal males. By simultaneously documenting dive patterns and geographic locations of seals, we have shown that location and migratory behavior (transiting vs. stationary) can explain such apparent changes. The dive patterns that we report here for adult males in 1989 and 1990 are similar to those reported for postbreeding males in 1988 (DeLong and Stewart 1991). Similarly, the dive patterns of females from San Miguel Island in 1990 are similar to those reported by Le Boeuf et al. (1988, 1989) and Le Boeuf (this volume) for females that breed at Año Nuevo. Females appear to be more consistent in their patterns of dive depths, as shown in figure 16.1, than are males, who show greater seasonal and interindividual variability in preferred dive depths (figs. 16.2, 16.3). We are uncertain about the reasons for this difference, but we suspect that the greater physical oceanographic complexity

TABLE 16.2 Depths of dives of adult male northern elephant seals during different stages of their postbreeding migrations.

Seal		North transit	Foraging area	South transit	Total
89030	Mean	384.2	518.9	428.9	442.4
	SD	117.5	187.2	122.5	158.1
	n_1	2767	2407	1697	6871
	n_2	38	46	39	123
89160	Mean	343.8	372.3	378.1	364.5
	SD	117.4	122.7	134.5	124.5
	n_1	1915	2636	1399	5950
	n_2	35	50	31	116
89408	Mean	348.4	432.8	366.6	386.2
	SD	111.7	120.3	131.1	125.3
	n_1	2530	2550	1171	6251
	n_2	43	44	27	114
89750	Mean	397.4	547	413	469.9
	SD	131.2	208	124	183.4
	n_1	1430	2369	1400	5200
	n_2	35	44	34	113
90480	Mean	208.0	184.4	323.7	222.7
	SD	88.1	37.5	118.9	96.7
	n_1	3553	3561	1870	8984
	n_2	48	50	36	134
90640	Mean	353.9	320.4	387.6	350.3
	SD	90.2	84.7	113.8	96.8
	n_1	3584	2312	1501	7397
	n_2	40	57	31	128
90680	Mean	386.4	524.5	410.2	463.3
	SD	133.7	149.6	170.7	162.5
	n_1	1857	3250	1064	6171
	n_2	31	58	30	119
90870	Mean	218.9	203.1	400.5	236.1
	SD	106.7	95.7	119.5	122.9
	n_1	2271	3982	1039	7292
	n_2	32	60	22	114

NOTE: SD = standard deviation; n_1 = number of dives; n_2 = number of days; mean and SD in m. The first 4 seals are 1989 records, and the latter 4 are from 1990.

near the Aleutian Islands compared with the central Pacific Subarctic Transition Zone (e.g., Favorite, Diomead, and Nasu 1976; Wetherall 1991) may result in greater regional variability in thermocline depth and vertical water mass discontinuities in the foraging areas of males.

The foraging area dives of seal 90480 are particularly interesting because of his shallow, narrow depth range of dives during a 50-day period. D. E. Crocker et al. (this volume) proposed that this type of diving is indicative of foraging in benthic or epibenthic habitats. However, our location records for seal 90480 indicate that he was feeding in an area where water depths exceeded several thousand meters and that the nearest shallow water areas (i.e., < 300 m) were 241 to 322 km to the north. Further, sea-surface temperatures from other sources compared to the seal's GLTDR SST data confirm the location records. Although some seamounts do rise from the Aleutian Trench near where the seal was foraging, the charted tops of those seamounts are at least several hundred meters below the seal's foraging depths. We attribute the prolonged shallow diving of this seal to his preference for feeding at that depth rather than epibenthic foraging in shallow coastal habitats or on near-surface seamounts or guyots. Similar depth preferences can be seen throughout his north transit period in the high-density depth bands in his dive records (fig. 16.3), although there is greater variability around the high-density bands during those times. Similar clustering of dive depths, which appear to represent interindividual differences in depth preferences, can be seen in the records of other males, although they are most obvious as shallow depth preferences in the records of seals 90870, 90640, and 89160 for portions of their migrations. We also attribute the interannual differences in dive depths among males that foraged in similar areas to differences in individual preferences, as the similarities in SSTs in those areas in 1989 and 1990 (compare figs. 16.6 and 16.7 with figs. 16.9 and 16.10) do not suggest any yearly differences in oceanographic conditions there. Canyon, seamount, or current divergence and convergence influences could cause sharp temperature discontinuities between vertical water masses, which would result in substantial local variation in thermocline depth. Such local physical variability would have strong influences on prey distributions. I. L. Boyd and T. Arnbom (1991) showed that the dive depths of one female southern elephant seal were closely linked to temperature characteristics of the water column that were likely influencing prey concentrations. Simultaneous collection of oceanographic data in areas where elephant seals are known to be foraging would be invaluable during future research on elephant seal foraging dynamics.

Adult female and male northern elephant seals that breed at San Miguel Island evidently migrate to different areas of the North Pacific to forage and recover the substantial body mass they lost during breeding season fasts. Females are at sea about half as long as males in spring before they must

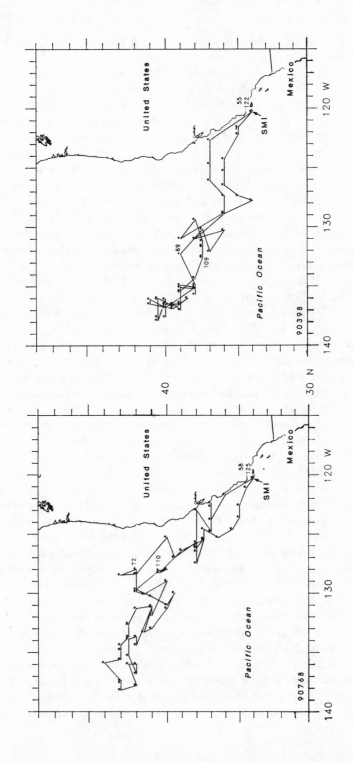

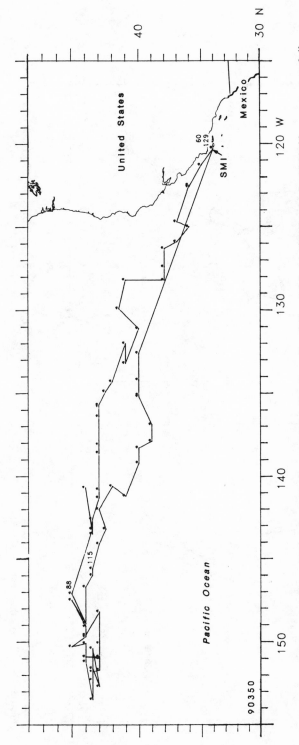

Fig. 16.5. Migratory routes of 3 northern elephant seal females from San Miguel Island, California, in 1990 (dots represent daily locations for all days that seals were at sea, except for seal 90350 whose instrument's memory filled up as it returned to San Miguel Island; numbers in lower left corners are seal identification numbers).

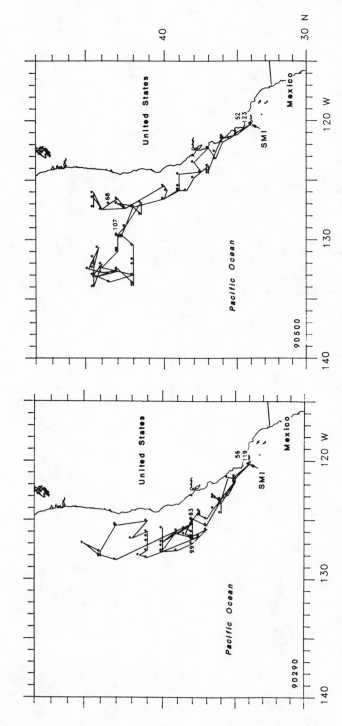

Fig. 16.6. Migratory routes of 2 northern elephant seal females from San Miguel Island, California, in 1990 (dots represent daily locations for all days that seals were at sea).

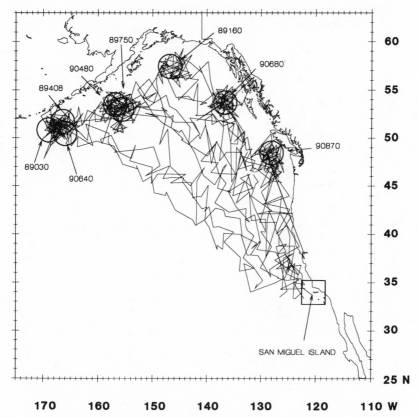

Fig. 16.7. Postbreeding migratory routes and foraging areas of 8 northern elephant seal males in 1989 and 1990 (all days that seals were at sea are plotted).

return to land to molt (tables 16.1, 16.2). This difference may explain why females do not continue to travel northward to the areas where males forage. A round-trip alone to those areas without lingering there would take around 60 days. It is intriguing, though, that males do not remain in the southern areas where females forage even though there appear to be adequate prey resources. Latitudinal differences in prey quality or size may constrain males to continue north in pursuit of more rewarding food resources. Sea-surface temperatures differ by 3 to 5° between the foraging areas of males and females (figs. 16.8, 16.9, 16.10), but we do not think that temperature alone would constrain the distribution of elephant seal adults in the North Pacific. Records from southern elephant seals (Fedak et al., this volume; Slip, Hindell, and Burton, this volume) show that adult males and females may occupy the same coastal Antarctic waters where SSTs are

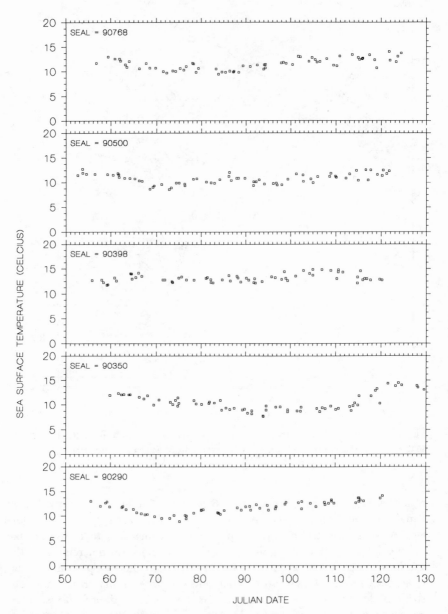

Fig. 16.8. Sea-surface temperature measurements recorded by GLTDRs attached to 5 adult northern elephant seal females during their postbreeding migrations in 1990.

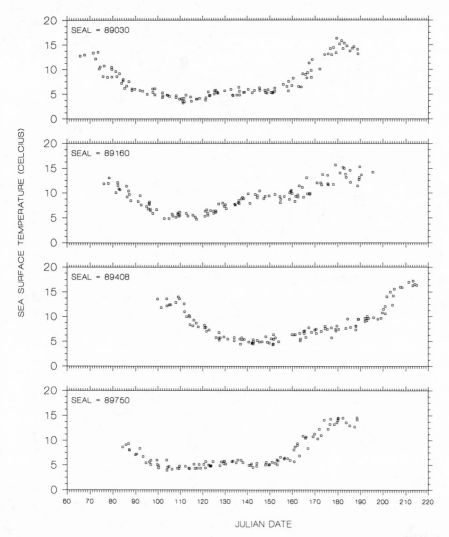

Fig. 16.9. Sea-surface temperature measurements recorded by GLTDRs attached to 4 adult northern elephant seal males during their postbreeding migrations in 1989.

several degrees colder than in the Gulf of Alaska and the eastern Aleutian Islands where male northern elephant seals forage.

Le Boeuf (this volume) reported that some females from Año Nuevo Island traveled into offshore British Columbia waters to feed, an area of overlap with adult males. We can think of no reason other than time constraints that females from the Channel Islands do not migrate that far north. Certainly, additional studies are needed to examine the biotic and abiotic

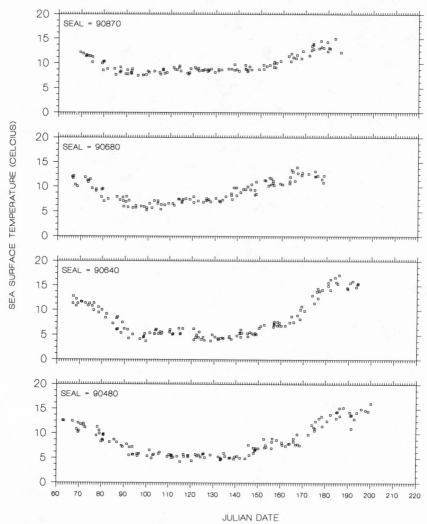

Fig. 16.10. Sea-surface temperature measurements recorded by GLTDRs attached to 4 adult northern elephant seal males during their postbreeding migrations in 1990.

factors that influence the differential distributions of foraging northern elephant seals during their seasonal migrations.

Our studies indicate that northern elephant seals that breed on San Miguel Island feed in offshore waters in the mesopelagic zone while pursuing vertically migrating prey. Studies of the diets of these seals (Condit and Le Boeuf 1984; Antonelis et al., this volume; Stewart and DeLong

1993) indicate that those prey are principally squid of several families that are known to be vertical migrators inhabiting mesopelagic habitats (Jefferts 1983; Roper and Young 1975). While in the California Current north of San Miguel Island, male and female elephant seals eat similar prey (Antonelis et al., this volume; Stewart and DeLong, unpubl. data). While transiting through that area, however, males and females are diving to different depths (tables 16.1, 16.2); thus they may be foraging on different size classes of those common prey, but we are unable to confirm this with the available data. Data on the diet of elephant seals in their offshore foraging areas are lacking, but we do know from the stomach contents of other predators (e.g., sperm whales, beaked whales, northern right whale dolphins, Pacific white-sided dolphins, Dall's porpoise) and midwater trawls (R. L. DeLong, unpubl. data) that squid species that are eaten by elephant seals in the southern part of the California Current are also present in the Subarctic Transition Zone, near the eastern Aleutian Islands and in the Gulf of Alaska. We presume that elephant seals are pursuing those prey during most of their periods at sea, but latitudinal differences in squid age and sex composition and behavior could explain the apparent differences in dive depths between male and female northern elephant seals.

Elephant seals depart breeding beaches in lean condition, having lost 40% or more of their body mass during breeding season fasts and lactation (Costa et al. 1986; Deutsch, Haley, and Le Boeuf 1990). We presume that they begin feeding immediately on entering the water, a hypothesis that is supported by diving records and the large increases in mass of males and females by the time they haul out to molt (Le Boeuf et al. 1988, 1989; De-Long and Stewart, unpubl. data). The continuity and characteristics (e.g., diurnal variation in dive depth to presumed depths of the deep scattering layer) of diving also suggest that the seals forage continuously while at sea. The identification of specific destinations where seals, particularly males, spend several weeks or more suggests that some areas of the North Pacific are more productive and energetically more rewarding to northern elephant seals than others. Satellite imagery has revealed dynamic primary productivity south of the eastern Aleutian Islands and in parts of the Gulf of Alaska, particularly in spring and summer when ocean warming and stabilization and longer days promote explosive phytoplankton blooms (Lewis 1989). Elephant seal males are evidently attracted to the deeper biological communities that respond to that heightened surface productivity. Southern elephant seals also appear to be attracted to similar areas of seasonally enhanced biological productivity along the Antarctic Peninsula and to other coastal Antarctic regions (Fedak et al., this volume; Slip, Hindell, and Burton, this volume). Elucidating the cues that attract, guide, and drive elephant seals to those areas will be challenging topics for future research.

The California Current and the North Pacific Transition Zone (NPTZ) and Subarctic Frontal Zone in the central North Pacific, where northern elephant seal females migrate to forage, are also known to be highly productive areas. The NPTZ, in particular, attracts large numbers of seabird, turtle, and other marine mammal predators during some seasons (e.g., Wetherall 1991). The near-surface communities of that area have been relatively well studied in recent years, especially since the explosive growth of the drift net squid fishery in the central North Pacific (ibid.). But little is known about the mesopelagic communities in that area which appear to play an important role in the foraging dynamics of female northern elephant seals.

The abilities of northern elephant seals to make long-distance, deepwater, foraging migrations appear to be important adaptations that permit them to range widely during most of the year to accumulate substantial resources that are essential to the energetically demanding terrestrial-bound activities of breeding and molt. Because elephant seals from the San Miguel Island and Año Nuevo rookeries evidently mix in foraging areas yet show strong fidelity to molting and breeding sites, continued study of intercolony differences and similarities in foraging patterns and migrations will be important to understanding the mechanisms that will limit and regulate the growth of each colony and of the still-increasing northern elephant seal population (see Stewart et al., this volume).

ACKNOWLEDGMENTS

We thank P. Yochem, B. DeLong, D. DeLong, S. Melin, G. Antonelis, S. Osmek, H. Huber, T. Ragen, and J. Francine for field assistance, the Channel Islands National Park Service for facilitating our research at San Miguel Island, and G. Antonelis, J. Baker, I. Boyd, R. Gentry, J. R. Jehl, B. Le Boeuf, and P. Yochem for comments on earlier drafts of the manuscript. We thank Clairol Research Laboratories for supplying hair dye and bleach products used for marking seals and R. Hill, S. Hill, and M. Braun for consultation during development and field testing of the GLTDRs. GLTDRs were manufactured by Wildlife Computers (Woodinville, Wash.), and VHF transmitters were manufactured by ATS (Isanti, Minn.). BSS was supported by a contract to the U.S. Air Force, Space Systems Division. The research was conducted under Marine Mammal Permit No. 579 to BSS.

REFERENCES

Boyd, I. L., and T. Arnbom. 1991. Diving behaviour in relation to water temperature in the southern elephant seal: Foraging implications. *Polar Biology* 11: 259–266.

Condit, R., and B. J. Le Boeuf. 1984. Feeding habits and feeding grounds of the northern elephant seal. *Journal of Mammalogy* 65: 281–290.

Costa, D. P., B. J. Le Boeuf, A. C. Huntley, and C. L. Ortiz. 1986. The energetics of lactation in the northern elephant seal. *Journal of Zoology* 209: 21–33.

DeLong, R. L., and B. S. Stewart. 1991. Diving patterns of northern elephant seal bulls. *Marine Mammal Science* 7: 369–384.

DeLong, R. L., B. S. Stewart, and R. D. Hill. 1992. Documenting migrations of northern elephant seals using day length. *Marine Mammal Science* 8: 155–159.

Deutsch, C. J., M. P. Haley, and B. J. Le Boeuf. 1990. Reproductive effort of male northern elephant seals: Estimates from mass loss. *Canadian Journal of Zoology* 12: 2580–2593.

Favorite, F., A. J. Diomead, and K. Nasu. 1976. Oceanography of the Subarctic Pacific Region, 1960–1971. *International North Pacific Fish Commission Bulletin* 33: 1–187.

Hindell, M. A., H. R. Burton, and D. J. Slip. 1991. Foraging areas of southern elephant seals, *Mirounga leonina*, as inferred from water temperature data. *Australian Journal of Marine and Freshwater Research* 42: 115–128.

Jefferts, K. 1983. Zoogeography and systematics of cephalopods of the northeastern Pacific Ocean. Ph.D. dissertation, Oregon State University, Corvallis.

Le Boeuf, B. J., D. P. Costa, A. C. Huntley, and S. D. Feldkamp. 1988. Continuous, deep diving in female northern elephant seals, *Mirounga angustirostris. Canadian Journal of Zoology* 66: 446–458.

Le Boeuf, B. J., Y. Naito, A. C. Huntley, and T. Asaga. 1989. Prolonged, continuous, deep diving by northern elephant seals. *Canadian Journal of Zoology* 67: 2514–2519.

Lewis, M. R. 1989. The variegated ocean: A view from space. *New Scientist* 1685: 1–4.

Roper, C. F. E., and R. E. Young. 1975. Vertical distribution of pelagic cephalopods. *Smithsonian Contributions to Zoology* 209: 1–51.

Stewart, B. S., and R. L. DeLong. 1990. Sexual differences in migrations and foraging behavior northern elephant seals. *American Zoologist* 30: 44A.

———. 1993. Seasonal dispersion and habitat use of foraging northern elephant seals. In *Marine Mammals: Advances in Behavioural and Population Biology*, ed. I. L. Boyd, 179–194. Symposia of the Zoological Society of London no. 66. London: Oxford University Press.

Wetherall, J. A., ed. 1991. Biology, oceanography, and fisheries in the North Pacific Transition Zone and Subarctic Frontal Zone. *NOAA Technical Report NMFS* 105: 1–110.

Zar, J. 1984. *Biostatistical Analysis*. Englewood Cliffs, N.J.: Prentice-Hall.

SEVENTEEN

Functional Analysis of Dive Types of Female Northern Elephant Seals

Tomohiro Asaga, Yasuhiko Naito, Burney J. Le Boeuf, and Haruo Sakurai

ABSTRACT. The aim of this study was to elucidate the function of individual dive types observed in the dive records of female elephant seals, *Mirounga angustirostris*. Free-ranging dive records spanning 29 to 81 days were obtained from three adult females from Año Nuevo, California, in 1990, using time-depth recorders glued to the pelage on their backs.

Type D dives, assumed to serve pelagic foraging, (1) accounted for 75 to 80% of all dives, (2) occurred in series with a mean length of 10.1 to 22.9 dives, (3) had a bottom of dive element that accounted for 28 to 44% of the total dive duration, and (4) exhibited longer durations and deeper depths during the day than at night. Type A, or "transit," dives were sparsely and widely distributed in the record, accounting for only 1.7 to 7.1% of all dives; they rarely occurred in a series and were the deepest dives observed in all records. Type C dives accounted for 2.6 to 6.8% of all dives, and they were the shallowest dives. They occurred in series with a mean of 3.6 to 4.7 dives. The second descent segment of these dives, which accounted for 52 to 58% of the total dive duration, showed a 30 to 83% reduction in descent rate over the preceding descent segment. There was an inverse relationship between Type D and Type C dive frequency of occurrence as a function of time of day; C dives peaked between 0400 and 1000 hours, the time interval when D dives were least frequent.

The results of this study are consistent with the hypothesis derived from swim speed analysis that Type A, D, and C dives serve transit, pelagic foraging, and physiological processing functions, respectively.

By providing a continuous record of the duration and depth of dives, time-depth recorders (TDRs) provide insights into the at-sea behavior of marine mammals. Data from TDRs allowed us to focus on the individual dives of northern elephant seals, *M. angustirostris*, with the aim of elucidating their function and role in foraging.

The diving pattern of northern elephant seals differs in many respects from that of other pinnipeds. After lactation and weaning their pups, adult females go to sea to feed for 2½ months. During this period, their dives are long and deep. Mean dive depth is 446 to 544 m, and mean dive duration is 17.1 to 22.5 minutes (Naito et al. 1989; Le Boeuf et al. 1989). Sea lions (*Zalophus californianus* and *Z. c. wollabaeki*) and fur seals (*Callorhinus ursinus, Arctocephalus gazella, A. pusillus, A. australis,* and *A. galapagoensis*) dive to mean depths of less than 100 m, and mean dive durations are less than 3 minutes (Feldkamp, DeLong, and Antonelis 1989; Gentry, Kooyman, and Goebel 1986). Weddell seals, *Leptonychotes weddelli,* dive to less than 200 m most of the time, and more than half of their dives are less than 10 minutes in duration (Kooyman 1989). Elephant seals dive deeper than sperm whales, *Physeter catodon,* who dive to mean depths of 314 to 382 m, with most dives being less than 500 m (Papastavrou, Smith, and Whitehead 1989). These comparisons suggest that northern elephant seals dive longer and deeper than other pinnipeds and most whales.

Several studies have showed that many diving mammals dive in bouts, a series of dives over a certain period of time; diving bouts are followed by rest (Gentry, Kooyman, and Goebel 1986). Northern elephant seals do not dive in bouts but rather dive continuously for the duration that they are at sea. The dive bouts of Weddell seals, California sea lions, and fur seals last for several hours (Feldkamp, DeLong, and Antonelis 1989; Kooyman 1989; Kooyman et al. 1980), and so do those of blue-eyed shags, *Paracrocorax atriceps,* and Adélie penguins, *Pygoscelis adeliae* (Croxall et al. 1991; Naito et al. 1989). Female northern elephant seals dive continuously, sometimes up to 2½ months at a stretch during which the seal may exhibit more than 5,000 dives with only short surface intervals of less than 3 minutes between dives (Le Boeuf et al. 1989). Thus, an analysis of dive bouts, the usual approach in studying otariid diving behavior, is not applicable to the study of elephant seals. This discrepancy suggests that elephant seal diving differs fundamentally from the diving behavior of otariids.

Among California sea lions and northern fur seals, there may be rest at the surface, rest on land, or transit to foraging areas between dive bouts (Feldkamp, DeLong, and Antonelis 1989; Kooyman and Gentry 1986). These activities are also observed in gentoo penguins, *Pygoscelis papua,* and chinstrap penguins, *P. antarctica* (Trivelpiece et al. 1986). The diving behavior of northern elephant seals is devoid of swimming at the surface (Le Boeuf et al. 1992; see chap. 10), and it is not likely that they sleep at the surface (Le Boeuf et al. 1988). This is important for understanding their diving behavior, their migrations to foraging areas, and their rest or sleep activities at sea. There are temporal and frequency differences in dive types that we think elucidate the function of their unusual diving behavior.

In this study, we analyze the distribution of distinguishable dive types of free-ranging female northern elephant seals during the 2½-month period at sea after breeding. Our investigation is aimed at understanding the function of dives in relation to foraging and in elucidating their physiological basis.

METHODS

The field aspects of this study were conducted at Año Nuevo Point, California, in 1988. Three adult female northern elephant seals were immobilized with ketamine hydrochloride (Briggs et al. 1975) during mid-February near the end of their lactation periods. A TDR and a radio transmitter (Advanced Telemetry Systems, Bethel, Minn.) were attached to the pelage of the back above the shoulders of each female with marine epoxy (Evercoat Ten-Set, Fibre-Evercoat Co., Cincinnati, Ohio). Each seal was weighed at this time (see Le Boeuf et al. 1988 for method).

The TDR (Naito et al. 1989) was 52 mm in diameter and 193 mm in length and weighed 980 g in air. It was housed in an aluminum casing that withstood pressures to 3,000 m in depth. The instrument contained a diamond stylus that inscribed a line on aluminum-coated paper (20 μ thick) proportional to water pressure. The motor was capable of running for 130 days, being powered with two 1.5 v lithium batteries. The depth range was 0 to 900 m. Recording error was estimated at less than 2% of depth and duration.

When the seals returned to the rookery in May, each one was weighed and the TDRs recovered. The recording paper was subsequently enlarged 14½ times with a reader-printer (Minolta PR507). From the strip chart records, the dives for each female were classified into five dive types based on their time-depth profiles (fig. 17.1). This classification is based on that of B. J. Le Boeuf et al. (1988) with little modification and is the same classification used in Le Boeuf et al. (1992) and D. E. Crocker et al. (this vol.). Type A dives have a straight descent to a sharp point, then direct ascent to the surface. Type B dives are similar except that the bottom of the dive is rounded. Both A and B dives have no bottom time. Type C dives have direct descent to a depth, at which point the descent rate decreases dramatically until the bottom of the dive, then following a rather sharp inflection point, ascent to the surface is direct. Type D dives are characterized by direct descent to a depth, at which point there occur 2 to 12 vertical excursions or "wiggles," ending in direct ascent to the surface. Type E dives show direct descent to the bottom of the dive, which is flat, and end in direct ascent to the surface. Dives that could not be put into one of these categories were excluded from analysis. Type E dives in this study correspond to Type E and F dives in Le Boeuf et al. (1988).

A Hitachi H-F8844-65 was used to digitize the dives. Each classifiable dive was digitized with 3 to 5 points (fig. 17.1), which gave a measure of dive depth, duration, and descent and ascent rates. The minimum mea-

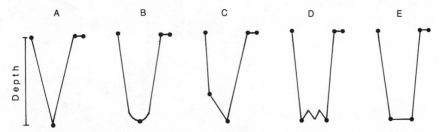

Fig. 17.1. Schematic profiles of five dive types observed in the dive records of northern elephant seals (adapted from Le Boeuf et al. 1992). The points of each dive type were digitized to determine dive elements.

sured value of surface intervals (SIs) was 0.4 minutes, set by the time resolution of the instrument. SIs greater than 10 minutes, extended surface intervals (ESIs), were excluded from statistical analysis.

In an attempt to elucidate the temporal patterning of dive types and their function, we analyzed the temporal patterning of each dive type. For analytical purposes, we defined a dive sequence as a series of dives of the same type bordered by a different dive type or an ESI. That is, we regarded a change in dive type or a surface interval of 10 minutes or more as an interruption signaling the end of a dive type sequence and the dives following a change in dive type or an ESI as the beginning of a new dive type sequence. An additional criterion was that a dive type sequence contain a minimum of three dives.

RESULTS

All females returned to the breeding site. Their TDRs were recovered, and they were weighed. All females gained a mean of 1.0 kg/day, an indication that they succeeded in foraging and were not impeded by the instruments.

Summary Statistics

The entire dive records of two females, Tuf (81 days at sea) and Tow (64 days at sea), were obtained. For the third subject, Vi, the first 48 days of 77 days at sea were recorded; the last 29 days of the record were lost due to a tear in the recording paper. Summary statistics for all females are shown in table 17.1. Diving performance was similar to that of eight females reported in Le Boeuf et al. (1988).

Dive Type Distribution and Temporal Pattern

To clarify the temporal frequency of each dive type, we divided each record into three periods: (1) beginning—the first five days; (2) end—the last five days; and (3) middle—the entire rest of the record. Half of the Type B

TABLE 17.1 Summary statistics for 3 adult females.

Seal	Recording duration (days)	Total no. of dives	Mean ± SD dive depth (m)	Max. dive depth (m)	Mean ± SD dive duration (min)	Max. dive duration (min)	Mean ± SD surface intervals (min)
Tuf	81	5657	524 ± 126	1093	18.2 ± 3.7	44.4	2.0 ± 0.7
Vi	48	3260	519 ± 136	1131	20.2 ± 4.3	45.0	1.6 ± 0.8
Tow	64	3640	446 ± 125	1250	22.5 ± 5.0	50.1	2.4 ± 0.8

TABLE 17.2   Percentage of the various dive types in each dive period: beginning (first 5 days in a dive record), end (last 5 days in a dive record), and middle (intervening days).

		Dive types				
Seal	Dive periods	A	B	C	D	E
Tuf	Beginning	0.9	1.9	0.2	1.8	0.4
	Middle	7.1	2.1	6.8	74.4	0
	End	0.1	0.1	0.2	3.6	0.6
	Total	8.1	4.1	7.2	79.8	1.0
Vi	Beginning	0	7.6	0.2	2.2	0
	Middle	7.1	2.7	7.2	78.8	0
	End	—	—	—	—	—
	Total	7.1	10.3	7.4	81.1	0
Tow	Beginning	1.5	1.9	0	3.7	0.6
	Middle	1.7	2.2	2.6	79.6	0
	End	0.2	0.7	0	4.6	0.7
	Total	3.4	4.8	2.6	87.9	1.3

dives and all Type E dives occurred during the beginning and end periods (table 17.2). The mean depth of dives during these periods was not stable compared to dives in the middle period. Because dive depth is constrained by the shallow ocean floor between the rookery and the continental slope (Le Boeuf et al. 1988; Le Boeuf et al. 1989), the dive records for the beginning and end periods of each record were excluded from analysis. That is, all analysis here excludes the first five days of all records and the last five days of the records of females Tuf and Tow.

D dives were the dominant dive type in the middle period of all records (fig. 17.2). These dives occurred in long series. Type C dives occurred more frequently than Type A dives, except for female Tuf, and Type C dives occurred several times every day, except for female Tow. Type C dives were the majority of dives (68%) following ESIs. Type A dives occurred unpredictably. There were no Type E dives, and Type B dives were rare (2.1–2.7% of all dives); consequently, these dive types are excluded from further analysis.

Characteristics of Dive Parameters

Dive Depth-Duration Ratio. We tested the dive depth-duration difference among dive types. The mean dive depth-duration ratio of A, C, and D dives was 29.5 to 35.9, 13.7 to 21.2, and 22.4 to 31.3, respectively. Differences in these ratios among dive types were significant for all females (Mann-Whitney $U = 13.3$, $p < .05$). This indicates that the classification by dive depth-duration profiles was suitable for dive type analysis.

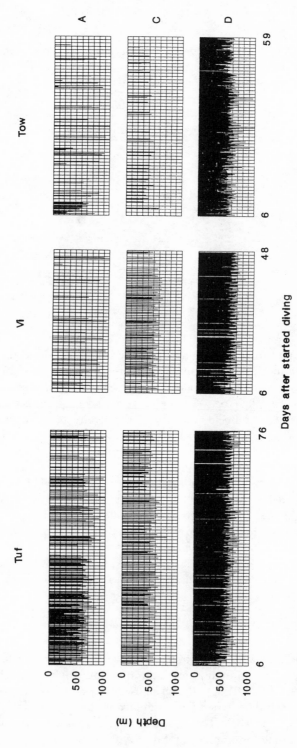

Fig. 17.2. The dive records of 3 females showing the distribution of A, C, and D dive types during the middle period at sea. This excludes the first five days of each record, when the animal is going from shallow to deep water, and the last five days, when the reverse occurred.

TABLE 17.3 Summary statistics of dive elements for three type dives.

Dive type	Seal	Mean ± SD dive depth (m)	Maximum dive depth (m)	Mean ± SD duration (min)	Maximum dive duration (min)	Mean ± SD surface intervals (min)	Mean ± SD bottom time[a] (min)	Mean ± SD descent rate[b] (m/sec)	Mean ± SD ascent rate (m/sec)
A	Tuf	594 ± 123	1093	16.7 ± 3.5	40.7	2.0 ± 0.7	—	0.80 ± 0.16	2.69 ± 1.06
	Vi	654 ± 239	1131	18.2 ± 6.0	36.8	1.8 ± 1.0	—	1.12 ± 0.28	1.35 ± 0.53
	Tow	547 ± 234	1250	18.8 ± 6.8	34.2	2.2 ± 0.9	—	0.82 ± 0.27	1.34 ± 0.52
C	Tuf	479 ± 92	791	22.7 ± 3.9	35.7	2.0 ± 0.8	13.1 ± 3.3	0.82 ± 0.16 (0.34 ± 0.06)	1.79 ± 0.90
	Vi	470 ± 79	662	25.5 ± 5.1	45.0	1.7 ± 0.8	13.3 ± 3.6	1.79 ± 0.82 (0.31 ± 0.11)	0.87 ± 0.24
	Tow	355 ± 85	583	26.4 ± 6.8	50.1	2.6 ± 1.0	14.0 ± 4.3	0.84 ± 0.26 (0.25 ± 0.04)	0.73 ± 0.32
D	Tuf	545 ± 90	834	17.7 ± 2.9	31.7	2.0 ± 0.6	4.9 ± 2.5	0.92 ± 0.16	3.24 ± 2.05
	Vi	558 ± 94	845	19.7 ± 3.0	31.4	1.6 ± 0.7	5.9 ± 2.5	1.33 ± 0.28	1.32 ± 0.69
	Tow	478 ± 105	1002	21.8 ± 3.9	45.9	2.5 ± 0.8	9.7 ± 3.8	0.99 ± 0.21	1.67 ± 0.43

[a] Mean bottom time of C-type dives correspond to mean duration in second descent segment.
[b] Mean descent rate of C-type dives correspond to that in first descent segment; mean descent rate in second descent segment is given in parentheses.

Dive Depth, Dive Duration, and Surface Intervals. Table 17.3 shows that the mean depths of Type A dives were significantly deeper in all records than other dive types (t = 3.4, p < .05). Indeed, most dives (64–92%) deeper than 800 m were Type A dives, and these were the maximum depth dives for all females. In contrast, Type C dives were significantly shallower than other dive types (t = 13.3, p < .05). The longest mean dive durations of all females were Type C dives (t = 6.5, p < .05).

There were no significant differences in SIs among dive types in all females except for SIs following Tow's Type A dives, which were significantly longer than those of other females (t = 2.5, p < .05).

Bottom Time and Descent and Ascent Rates. The mean bottom duration of Type D dives accounted for 28 to 44% of the total duration of dives. The mean duration of the second descent segment of Type C dives (fig. 17.1) took up 52 to 58% of the total duration of the dives.

The mean descent rate during the second descent segment of Type C dives was significantly slower than in any other dive type (t = 8.8, p < .05). The mean descent rate during this segment represents a 30 to 83% reduction over the descent during the first descent segment of Type C dives. In addition, the mean descent rates between A and D dives, except for Vi, were statistically significant (t = 5.8, p < .05). The mean ascent rate of Type C dives was slower than that of other dive types (t = 5.2, p < .05).

Interrelationships between Dive Elements. Correlations between dive depth, dive duration, bottom time, and descent and ascent rates are shown in table 17.4. Significant correlations were observed between dive depth and dive duration for all dive types and all females. The positive relationship between these two variables was highest for Type C dives and lowest for Type D dives. The relationship between transit time and depth of Type D dives was strong for all females.

For Type C dives, there were high and positive correlations between dive depth and the duration of the second descent segment, as well as between total dive duration and the second descent segment. For Type D dives, there were weak negative correlations between dive depth and bottom duration. There were positive correlations between total dive duration and duration at the bottom of dives. Weak relationships were in evidence between dive depth and descent rate, total duration and descent rate, and total duration and ascent rate.

Diel Pattern

Number of Dives. There was an inverse relationship in the daily frequency pattern of C and D dives in all females (fig. 17.3). Type C dives peaked

TABLE 17.4 Correlation coefficients between the dive elements of the major dive types for each female.

	Type	Seals	r*	Equation
Duration vs. depth	A	Tuf	0.32	y = 0.017x + 6.557
		Vi	0.70	y = 0.024x + 2.786
		Tow	0.36	y = 0.118x + 9.084
	C	Tuf	0.50	y = 0.028x + 9.089
		Vi	0.25	y = 0.023x + 14.573
		Tow	0.54	y = 0.065x + 3.298
	D	Tuf	0.35	y = 0.012x + 11.082
		Vi	0.47	y = 0.017x + 10.146
		Tow	0.13	y = 0.005x + 19.404
Transit vs. depth	D	Tuf	0.43	y = 0.019x + 2.191
		Vi	0.48	y = 0.019x + 3.251
		Tow	0.51	y = 0.019x + 3.03
Bottom[a] vs. depth	C	Tuf	0.47	y = 0.019x + 3.904
		Vi	0.20	y = 0.014x + 6.704
		Tow	0.61	y = 0.043x − 1.089
	D	Tuf	−0.05	y = 0.007x + 8.883
		Vi	−0.10	y = 0.002x + 6.897
		Tow	−0.24	y = 0.014x + 16.366
Bottom[a] vs. duration	C	Tuf	0.52	y = 0.602x − 0.569
		Vi	0.54	y = 0.546x − 0.597
		Tow	0.59	y = 0.513x + 0.478
	D	Tuf	0.25	y = 0.326x − 0.848
		Vi	0.38	y = 0.462x − 3.157
		Tow	0.45	y = 0.654x − 4.516
Descent rate vs. depth	D	Tuf	0.10	y = 0.001x + 0.664
		Vi	0.15	y = 0.001x + 0.894
		Tow	0.25	y = 0.001x + 0.577
Descent rate vs. duration	D	Tuf	−0.22	y = 0.016x + 1.208
		Vi	−0.10	y = 0.017x + 1.660
		Tow	−0.24	y = 0.018x + 1.373
Ascent rate vs. duration	D	Tuf	−0.15	y = 0.039x + 2.673
		Vi	−0.16	y = 0.010x + 1.116
		Tow	−0.25	y = 0.016x + 1.165

[a] Bottom duration of C-type dives are equal to the duration in second descent segment.
* All correlations significant at p < .10 level.

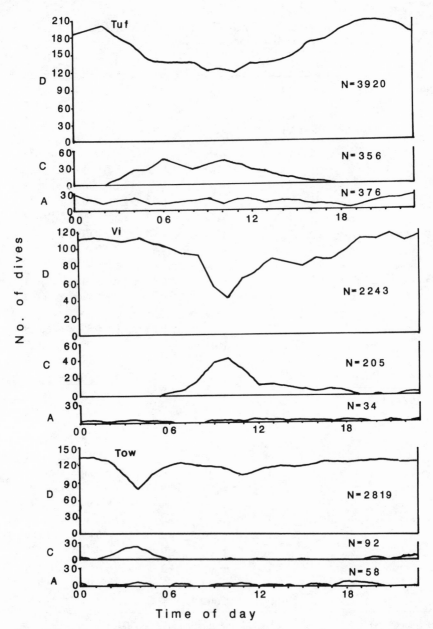

Fig. 17.3. A frequency distribution of three dive types as a function of time of day for each of 3 females.

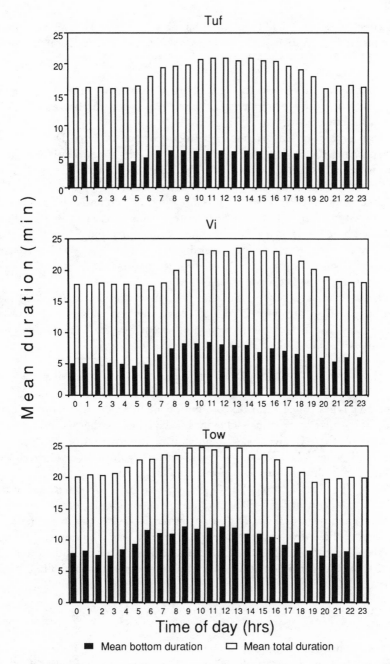

Fig. 17.4. Mean dive duration and mean bottom time of D-type dives as a function of time of day for 3 females.

in frequency between 0400 and 1000 hours, when Type D dives were most infrequent. Type A dives did not vary systematically as a function of time of day.

Dive Elements. We describe the daily rhythm of dive elements of Type D dives only because the frequency of other dive types was low and did not vary with time of day.

Mean dive duration and mean bottom duration of Type D dives were longer during the day than at night (fig. 17.4). The time when these durations were longest corresponded with the modal frequency of Type C dives. To further investigate this association, we selected 10 Type A, C, and D dives at random, each of which was preceded and followed by a series of 10 consecutive Type D dives. We then tested for differences in dive durations of the Type D dives as a function of the preceding dive type. The majority of dive durations of Type D dives, 70 to 100% of them, increased following Type C dives but showed little increase after Type A dives (0–10%) and Type D dives (0–30%).

The mean dive depths of Tuf and Vi during the day (0600–2000 hrs) were deeper than at night (2100–0500 hrs), but this was not the case for female Tow. Mean SI, descent rate, and ascent rate for Type D dives did not vary with hour of the day for all females.

Dive Type Sequence

Type A dives showed the lowest number of dives in a sequence among the three dive types (table 17.5). Except for female Tow, 80% or more of Type C dives occurred in series. The mean duration of the second descent segment of Type C dives was about half of the dive sequence duration. The depth where the second descent segment began in Type C dives was constant in a dive sequence regardless of its maximum depth (F = 3.41, p < .01). The dive duration of Type C dives gradually increased toward the end of a dive sequence (50–80% of total dive sequence number).

More than 90% of Type D dives occurred in dive sequences. The total duration of these dive sequences in a day was 15.5 to 20.5 hours. The bottom duration corresponded to 24 to 39% of dive sequence duration, indicating that females remained at the maximum depth of dives for 3.6 to 8 hours every day.

DISCUSSION

This study provides additional information on the putative function of dive types in elephant seals. Le Boeuf et al. (1988) hypothesized that Type D dives serve foraging. Using information provided by measurement of swim

TABLE 17.5 Dive sequence characteristics for 3 adult female northern elephant seals.

Dive type	Subjects	Dive no.	Dives incl. (%)	Sequence no. per day	Mean ± SD sequence duration (hr)	Mean dive no.	Mean no. range	% Sequence underwater ± SD	% Sequence bottom time[a] ± SD
A	Tuf	14	13	0.2	1.1 ± 0.2	3.4	3–4	92.8 ± 1.4	—
	Vi	—	—	—	—	—	—	—	—
	Tow	1	3	0.01	—	—	—	—	—
C	Tuf	60	81	0.9	1.9 ± 1.0	4.7	3–22	93.7 ± 2.4	55.5 ± 5.2
	Vi	39	82	0.9	1.8 ± 0.5	4.3	3–8	94.3 ± 2.4	52.0 ± 4.9
	Tow	10	39	0.2	1.6 ± 0.3	3.6	3–6	92.1 ± 2.9	50.8 ± 5.3
D	Tuf	357	93	5.0	3.1 ± 3.0	10.1	3–72	91.3 ± 1.6	23.5 ± 6.4
	Vi	97	99	2.3	8.7 ± 7.0	22.9	3–82	92.5 ± 1.8	27.8 ± 5.6
	Tow	147	99	2.7	7.6 ± 6.5	18.9	3–78	91.0 ± 2.3	39.2 ± 7.9

[a] Bottom time of C dives refers to the duration of the second descent segment.

speed, Le Boeuf et al. (1992) attributed a similar function to Type D dives and added that they served pelagic foraging. They further hypothesized that Type E dives served benthic foraging, especially in males, that Type A and B dives functioned primarily as transit dives during migrations, and that Type C dives were processing dives that facilitated digestion, were instrumental in the removal of anaerobic metabolites, or served rest or sleep. We discuss the relevance of the present data to the presumed function of dive types.

Type D dives made up 75% or more of the dive types in each of the dive records in this study, a percentage similar to that reported for females in an earlier study (Le Boeuf et al. 1988). The predominance of Type D dives—coupled with the large proportion of time spent at the dive bottom (34.0 ± 8.7% of the total dive duration), the up-and-down activity of the seal at the dive bottom, the rapid descent and ascent rates, and the length and duration of a series of these dives—is consistent with the hypothesis that they serve foraging (Le Boeuf et al. 1988; Le Boeuf et al. 1992).

This hypothesis is bolstered by a number of ancillary findings regarding Type D dives.

1. They occurred regularly throughout the period at sea with the exception of the first few days at sea and the last few days at sea when the animal was moving across the continental shelf. This suggests that feeding is not restricted to foraging areas but occurs throughout migration (Le Boeuf et al. 1992; see Le Boeuf, this volume).

2. The mean dive duration, mean bottom-of-dive duration, and mean dive depth of these dives are greater during the day than at night. This is consistent with the hypothesis that Type D dives reflect foraging in the deep scattering layer, which contains prey that come closer to the surface at night and retreat to depths at midday (Le Boeuf et al. 1988).

3. The frequency of Type D dives was lowest during the day when the frequency of Type C dives peaked. This is consistent with the idea that the latter type are processing dives and that the best time for the seal to reduce its dive rate and rest, digest food, or get rid of anaerobic metabolites is when it must dive deepest for prey. Data from ongoing studies show that Type D dives occur with great frequency in both sexes and in seals ranging in age from weanlings on their first trip to sea to adults (Le Boeuf, this volume).

Le Boeuf et al. (1992) hypothesized that A and B dives were primarily transit dives. This was based in part on the fact that the mean horizontal distance covered, 1.3 km, was greater for this dive than for other dive types (0.6–1.1 km), and they argued further that this dive shape was most efficient for horizontal travel. Our data show that Type A dives are on average deeper and of shorter duration than Type D dives. The shorter duration of Type A dives might be due to the greater energy expenditure required to go deeper, which puts a lower limit on dive duration. The swim speed data indicate that foraging may occur during transit. The consistency of Type A dives in females throughout the length of the periods at sea supports this idea. We suggest further that the dives might have an exploratory function. This is consistent with the dive's depth and its spiked bottom; at depth, the seal may decide to abort the dive and ascend when it does not find suitable prey.

With the exception of M. A. Hindell's dive analysis on southern elephant seals, *M. leonina* (Hindell 1990; Hindell, Slip, and Burton 1991), no other investigators have attempted detailed descriptions of dive types in other species. Dives resembling the A-type dives of elephant seals, however, have been observed in northern fur seals and South African fur seals (Gentry, Kooyman, and Goebel 1986; Kooyman and Gentry 1986); 90% of the dives of these animals had spiked bottoms resembling the shape of A-type dives of elephant seals. The investigators suggested that these dives served foraging. G. L. Kooyman (1981) came to the same conclusion about some of the dives of Weddell seals.

Several findings in this study are consistent with the hypothesis that Type C dives have a processing function and that the animal is drifting down during the second segment of descent (Le Boeuf et al. 1992). First, Type C dives are as long as the dive durations of the other dive types, even though reduced energy expenditure is expected due to the animal drifting down during the second segment of descent rather than powering down as in Type A and D dives. Second, the mean distance traveled during these dives (0.6 km) is less than for any other dive type (ibid.). Third, Type C dives were the most common dive type following an extended surface interval. Since we know from the swim speed study (ibid.) that the animal is not swimming, this is a strong argument that it is resting. The contiguity of ESIs and Type C dives suggests that the latter may have a "processing" function similar to that of ESIs. Fourth, Type D dives that followed Type C dives had a longer mean duration than normal. This suggests that the animal was now rested or in some way better prepared to dive long. Last, the peak frequency of Type C dives, coupled with the low frequency of Type D dives, suggests that these two dive types have opposite functions.

We conclude that our analysis of dive types is consistent with the hypotheses advanced regarding their function (ibid.). The ultimate tests of these hypotheses will be empirical and will require more sophisticated diving instruments than the ones currently in use.

ACKNOWLEDGMENTS

We thank D. Costa and P. Thorson for their support with fieldwork at Año Nuevo Point, California, and N. Satoh, head of the computer center of the National Institute of Polar Research in Tokyo, and his staff for their advice and help with data processing. This study was financed in part by a grant from the National Science Foundation.

REFERENCES

Briggs, G. D., R. V. Hendrickson, and B. J. Le Boeuf. 1975. Ketamine immobilization of northern elephant seals. *Journal of the American Veterinary Medicine Association* 167: 546–548.

Croxall, J. P., Y. Naito, A. Kato, P. Rothery, and D. R. Briggs. 1991. Diving patterns and performance in the Antarctic blue-eyed shag, *Phalacrocorax atriceps*. *Journal of Zoology, London* 225: 177–199.

Feldkamp, S. D., R. L. DeLong, and G. A. Antonelis. 1989. Diving patterns of California sea lions, *Zalophus californianus*. *Canadian Journal of Zoology* 67: 872–883.

Gentry, R. L., G. L. Kooyman, and M. E. Goebel. 1986. Feeding and diving behavior of northern fur seals. In *Fur Seals: Maternal Strategies on Land*, ed. R. L. Gentry and G. L. Kooyman, 61–78. Princeton: Princeton University Press.

Hindell, M. A. 1990. Population dynamics and diving behaviour of a declining population of southern elephant seals. Ph.D. dissertation, University of Queensland, Australia.

Hindell, M. A., D. J. Slip, and H. R. Burton. 1991. The diving behaviour of adult male and female southern elephant seals, *Mirounga leonina* (Pinnipedia: Phocidae). *Australian Journal of Zoology* 39: 595–619.

Kooyman, G. L. 1981. *Weddell Seal: Consummate Diver*. Cambridge: Cambridge University Press.

———. 1989. *Diverse Divers: Physiology and Behavior*. Berlin: Springer Verlag.

Kooyman, G. L., and R. L. Gentry. 1986. Diving behavior of South African fur seals. In *Fur Seals: Maternal Strategies on Land*, ed. R. L. Gentry and G. L. Kooyman, 142–152. Princeton: Princeton University Press.

Kooyman, G. L., E. A. Wahrenbrock, M. A. Castellini, R. W. Davis, and E. E. Sinnett. 1980. Aerobic and anaerobic metabolism during voluntary diving in Weddell seals: Evidence of preferred pathways from blood chemistry and behavior. *Journal of Comparative Physiology* 138: 335–346.

Le Boeuf, B. J., D. P. Costa, A. C. Huntley, and S. D. Feldkamp. 1988. Continuous, deep diving in female northern elephant seals, *Mirounga angustirostris*. *Canadian Journal of Zoology* 66: 446–458.

Le Boeuf, B. J., Y. Naito, T. Asaga, D. Crocker, and D. P. Costa. 1992. Swim speed in a female northern elephant seal: Metabolic and foraging implications. *Canadian Journal of Zoology* 70: 786–795.

Le Boeuf, B. J., Y. Naito, A. C. Huntley, and T. Asaga. 1989. Prolonged, continuous, deep diving by northern elephant seals. *Canadian Journal of Zoology* 67: 2514–2519.

Naito, Y., B. J. Le Boeuf, T. Asaga, and A. C. Huntley. 1989. Long-term diving records of an adult female northern elephant seal. *Nankyoku Shiryo (Antarctic Record)* 33(1): 1–9.

Papastavrou, V., S. C. Smith, and H. Whitehead. 1989. Diving behavior of sperm whale, *Physeter macrocephalus*, off the Galápagos Islands. *Canadian Journal of Zoology* 67: 839–846.

Trivelpiece, W. Z., J. L. Bengtson, S. G. Trivelpiece, and N. J. Volkman. 1986. Foraging behavior of gentoo and chinstrap penguins as determined by new radiotelemetry techniques. *Auk* 103: 777–781.

EIGHTEEN

Swim Speed and Dive Function in a Female Northern Elephant Seal

*Daniel E. Crocker, Burney J. Le Boeuf, Yasuhiko Naito, Tomohiro Asaga,
and Daniel P. Costa*

ABSTRACT. The objective of this chapter is to discuss the behavioral and metabolic implications of swim speed data recorded from a northern elephant seal during free-ranging diving. Variation in swim speed elucidates the observed diving pattern and aids in understanding foraging tactics.

The data were obtained from a swim speed-distance meter and a time-depth recorder attached to the shoulders of an 8-year-old female at Año Nuevo, California, in February 1990. Swim speed, distance, depth, and duration of dives were recorded continuously for 29 days.

Average swim speeds ranged from 0.91 to 1.66 m/sec for all dive segments save one. These numbers are similar to those obtained from diving otariids and fall within the predicted cruising speed range for an aquatic animal less than 3 m in length. Horizontal swim speed during transit is similar to indirect measurements of transit velocity made by other investigators. Based on swim speed and distance traveled for each of five dive types apparent in the record, the two-dimensional shape of the various dive types observed, and the temporal patterning of dive types, we hypothesize that the four most common dive types observed serve three general functions: transit, foraging (pelagic and benthic), and internal physiological processes.

The advent of the use of time-depth recorder (TDR) technology has opened up a new level of behavioral and physiological analysis of diving in marine mammals (Kooyman 1965, 1968; Kooyman, Billups, and Farwell 1983; Guppy et al. 1986; Hill et al. 1987; Le Boeuf et al. 1989). Changes in depth, however, represent only one dimension of the behavior exhibited by an animal as it moves through the water column. Crucial to the understanding and expansion of the ideas arising from the TDR data is the development and deployment of an instrument that measures swim velocity. Historically, these data have been hard to acquire. Recent development and deployment of a number of different velocity instruments have allowed researchers to begin to incorporate this measurement into their analyses of diving behavior and physiology (Ponganis et al. 1990, 1992; Le Boeuf et al. 1992).

Long duration divers like elephant seals must reconcile the conflicting metabolic demands of responses to hypoxia and exercise. The physiological adaptations with which the animal defends against hypoxia must be balanced against the rate of oxygen utilization required for active swimming. Much of what is known about the relationship between nondiving swimming velocity and metabolism comes from exercise studies in harbor seals, *Phoca vitulina* (Davis, Williams, and Kooyman 1985; Williams, Kooyman, and Croll 1990), gray seals, *Halichoerus grypus* (Fedak 1986), and California sea lions, *Zalophus californianus* (Feldkamp 1987; Williams, Kooyman, and Croll 1990). The sea lion study showed that oxygen consumption increased as an exponential function of swimming velocity. In contrast, investigations on freely diving Weddell seals, *Leptonychotes weddelli* (Kooyman et al. 1973), showed that reductions in metabolic rate together with extremely efficient swimming hydrodynamics allow the seal to swim while diving at only slightly higher than resting energy costs.

Swim velocity measured in a freely diving elephant seal offers an important clue to understanding the balance achieved between the opposing demands of exercise and apnea. The rarity of extended recovery periods in northern elephant seals, *Mirounga angustirostris*, and its implication of aerobic metabolism further underscores the importance of understanding this balance. Variations in swim velocity can offer significant insights toward understanding foraging tactics as well as other behaviors exhibited during diving.

Through simultaneous deployment of velocity and TDR instruments, the animal's swim velocity can be combined with the vertical component of its movements derived from the TDR data to map dives in two dimensions. This allows calculation of angles of descent and ascent as well as the horizontal component of the animal's movement. The shapes derived from this two-dimensional analysis and the energetic and behavioral implications derived from the velocity data allow us to begin to interpret the functions of the dive types classified as a result of the TDR data.

In this chapter, we discuss the behavioral and metabolic implications of swim speeds and the diving record obtained from an 8-year-old postpartum female northern elephant seal in 1990. We summarize the data from B. J. Le Boeuf et al. (1992) and offer further discussion of the results.

MEASURING SWIM SPEED

The swim velocity instrument or, more accurately, the swim speed-distance meter (SSDM) measured distance traveled by the seal as a function of time

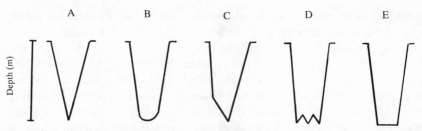

Fig. 18.1. Schematic representations of five dive types that appear in northern elephant seal diving records. The dive types depicted are idealized and not adjusted for depth or time. (Adapted from Le Boeuf et al. 1992.)

from which swim speed was calculated. The SSDM and TDR used in this analysis are described in detail in Le Boeuf et al. (1992).

DIVE TYPE CLASSIFICATION

From the TDR data, each dive was classified into one of five categories and a category of unclassifiable dives from shape on a strip chart representation of the time-depth profile (fig. 18.1). Most unclassifiable dives appeared to be composites of two of the five categories. Dive types were classified as follows: Type A dives, characterized by direct descent to a sharp inflection point, followed by direct ascent to the surface; Type B dives, characterized by direct descent ending in a rounded bottom, followed by direct ascent to the surface; Type C dives, characterized by direct descent to depth, at which point descent rate decreases noticeably but proceeds directly to the bottom of the dive, followed by direct ascent to the surface; Type D dives, characterized by direct descent to a depth, followed by 2 to 12 vertical excursions, followed by direct ascent to the surface; and Type E dives, characterized by direct descent to a flat bottom followed by direct ascent to the surface.

SWIM SPEED AND DISTANCE CALCULATIONS

Swim speeds were calculated for dive segments and paired with the TDR data to draw two-dimensional vector triangles, the missing dimension being the spatial orientation of the animal at any time. The horizontal distances calculated in this analysis do not equate to linear swimming distance as the animal could have been turning during the course of a dive. These distances calculated are simply the horizontal component of the animal's movement during the course of a dive segment or dive. As the maximum calibrated speed measured by the instrument was exceeded on some dives, all speeds and distances calculated should be considered minimum estimates. Similarly, angles of ascent and descent calculated are maxima. For a detailed explanation of the methods and potential sources of error, see Le Boeuf et al. (1992).

SWIM SPEED MEASUREMENTS

Excluding the second descent segment of Type C dives, mean swim speeds for dive segments ranged from 0.91 to 1.66 m/sec. The mean velocity for all segments of all dives was 1.24 ± 0.21 m/sec.

On average, mean descent speeds were 27% faster than ascent speeds. This difference was statistically significant for all dive types. Descent speeds were similar across all dive types except for the second descent segment of Type C dives, which exhibited a significant reduction in swim speed. The second descent segment of Type C dives averaged 0.59 m/sec. in swim speed. Velocity on descent averaged 1.5 m/sec., while ascent and bottom velocity averaged 1.0 m/sec. No swimming was recorded at the surface above the stall speed of the instrument (0.4 m/sec.). Energetic concerns might preclude swimming above this speed due to increased drag at the surface. No swim speeds below 0.4 m/sec. were recorded during diving.

COMPARISON OF SWIM SPEEDS

The swim speeds recorded by the SSDM fall predominantly within the range of 1 to 2 m/sec., suggested by G. L. Kooyman (1989) as the cruising swimming velocity for animals up to 3 m in length. Recent investigations of swimming velocities in otariids (Ponganis et al. 1990) found a similar range for four species (0.9–1.9 m/sec.). It might seem that the larger elephant seal should swim faster than the smaller otariids since drag increases with the surface area or L^2, while the power increases directly with muscle volume or L^3, where L = length. Y. U. G. Aleyev (1977) suggests that swim speed varies directly with the frequency of the propulsive movement, which decreases with increasing muscle length, but that the drag/power relationship holds true for animals with a length less than 4.5 m. This would include our experimental animal.

The shorter dive duration, trip duration, energetically costly surface swimming, and different foraging strategies seen in the otariids suggest that a different compromise must be reached between the metabolic requirements of exercise and apnea and may explain the similar swim velocities seen in the long duration diving elephant seals. Swimming at MCT velocities allows an animal to cover a given distance with the minimum total oxygen consumption. If, however, the animal's emphasis is on achieving a lower rate of oxygen consumption and thereby a longer duration dive, swimming at lower than MCT velocities might occur. A nonactive swimming or drift component to diving would contribute to this effect.

This idea is further supported by an energetic and physiological comparison of otariids and phocids. Studies have found that four species of lactating otariids expend energy at sea at five times the predicted basal level (Costa and Gentry 1986; Costa, Croxall, and Duck 1989; Costa, Thorson, and Kretzmann 1989; Costa, Antonellis, and DeLong 1990). Metabolic rates of diving elephant seals were estimated from dive durations and available oxygen stores to be only 1.3 times the predicted basal rate. Phocids have increased blood oxygen storage capacity due to a higher hematocrit (Lenfant, Johansen, and Torrance 1970). This increases blood viscosity and reduces optimal oxygen transport (Hedrick, Duffield, and Cornell 1986). Otariids, in contrast, have lower hematocrits that fall within a more optimal range for oxygen transport (Hedrick and Duffield 1991).

COMPARISON WITH OTHER ELEPHANT SEAL INVESTIGATIONS

A number of investigators in this volume present average daily transit velocities that allow comparison with the speeds measured by this instrument. We predict that these transit velocities would be somewhat lower than our recorded swim speeds as they ignore the vertical component of the animal's movement and include surface intervals for which our data indicate a lack of swimming. We can derive a comparable measure by dividing the calculated horizontal distance covered for each presumed transit dive by the dive duration. This yields a mean horizontal transit velocity of 1.00 ± 0.46 m/sec.

This speed is similar to the values reported for adult northern elephant seals, 0.89 to 1.03 m/sec. (Le Boeuf, this volume) and 1.04 to 1.15 m/sec. (Stewart and DeLong, this volume), and southern elephant seal females, 0.55 to 1.03 m/sec. (Fedak et al., this volume). D. J. Slip, M. A. Hindell, and H. R. Burton (this volume) present a range of transit speeds for southern elephant seals from 80 to 240 km/day or 0.9 to 2.8 m/sec. While the lower part of this range is similar to our measurements, a more complete accounting of which individuals exhibited higher transit speeds is required for meaningful comparison.

If we assume similar geometries to those obtained in this study, a transit speed of 2.8 m/sec. seems somewhat high. Using angles of ascent and descent derived from this study and assuming a nonswimming, surface interval of 1.5 minutes and an average dive duration of 20 minutes, a horizontal transit speed of 2.8 m/sec. yields a sustained swimming velocity of 3.4 m/sec. for the transit period. This is clearly beyond the capabilities exhibited by our female.

One possible explanation is that elephant seals might sometimes adopt a different mode of swimming that maximizes transit velocity at the expense of the total dive duration. This could represent flexibility in diving behavior that might be essential in a high predation environment where the depth of dives might be bathymetrically constrained. This alteration in swimming mode would be evident in the relationship between transit velocity and dive duration. A number of new records obtained from southern elephant seals show a dramatic shift from high transit velocity, short duration dives over the continental shelf to the more typical diving pattern as the animals reach deep water (C. Campagna, B. J. Le Boeuf, unpubl. data).

Recent deployment of velocity instruments on translocated yearling elephant seals has measured swim velocities similar to those presented in this study (R. Andrews, pers. comm.; P. Thorson, pers. comm.).

SPEED CHANGES DURING DIVE SEGMENTS

Since the sampling regimen of the SSDM was linear with respect to distance, not time, we were unable to assess instantaneous velocity or acceleration. However, numerous instances of what we termed "burst diving," rapid decreases and increases in speed within dive segments, were observed in the data. This level of detail was lost in the average speeds calculated for dive segments. To quantify this aspect of diving behavior, we looked at the variance among speed values within segments of dives. Burst diving occurred predominantly during descent and most often at the bottom of Type D dives. This behavior was significantly less frequent on the ascent segments of all dive types. Changes in swim speed occurred most infrequently during the second descent segment of Type C dives.

ANGLES OF ASCENT AND DESCENT AND DISTANCE TRAVELED

Figure 18.2 shows the average two-dimensional plots for each dive type. Angles of descent were on average shallower (30° to 56°) than angles of ascent (52° to 82°). Angles of descent and ascent could not be calculated for Type B dives, due to their rounded bottom. However, the great similarity between Type B and Type A dives in every other aspect of this analysis argued for combining them as Type AB dives for subsequent analysis.

FUNCTIONS OF DIVE TYPES

The data suggest that the different dive types serve three main functions. We hypothesize that Type AB dives may serve as transit dives. The average two-dimensional shape of Type A and B dives is notable for the great amount of horizontal distance covered: the seal averaged 1.2 km per dive.

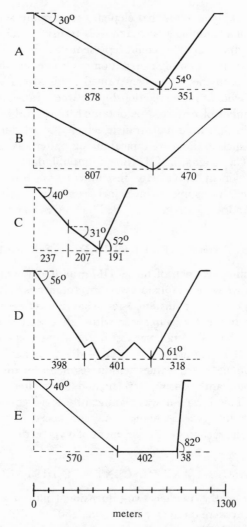

Fig. 18.2. Two-dimensional shapes for each of the five classified dive types, showing mean angles of ascent and descent and mean distance covered for segments of dives. (Adapted from Le Boeuf et al. 1992.)

Similarly, Type A and B dives had the shallowest angles of ascent and descent. Moreover, there is the strong degree of asymmetry seen in these dives. Horizontal distance covered was approximately three times greater on descent than ascent.

Returning to the central question of oxygen utilization and the balance between swimming effort and responses to hypoxia, we propose that this asymmetry represents a more energetically efficient way of covering horizontal distance. If the animal is negatively buoyant, it could drift down or swim aided by gravity and then swim actively to the surface. On an average Type AB dive, this would enable the animal to cover 1.2 km of horizontal distance, while actively swimming or swimming unaided by buoyancy for only half of that distance. The energy savings would have to be substantial, as actual distance covered is increased. However as metabolic expenditure is directly related to swimming effort, this is possible.

Type AB dives were interspersed throughout the record but occurred in greatest numbers in long uninterrupted series during the first 19 days at sea. The strong asymmetry in shape, great horizontal distance covered, and temporal patterning of Type AB dives are all consistent with their hypothesized function as transit dives.

The data suggest that Type C dives may be "processing" dives, where behavior and energy utilization are focused on internal processes such as digestion and clearing of metabolites rather than external processes. Type C dives showed the shortest horizontal distance covered. The second descent segment of Type C dives averaged about a third of the speed of other segments of all dive types. This reduction in swim speed tended to occur at the same depth within a series of Type C dives. On average, this slow swimming period accounted for almost half (46.9%) of a Type C dive's duration.

It has been hypothesized that diving mammals are trying to maximize their time underwater while foraging (Kooyman 1981; Fedak, Pullen, and Kanwisher 1988; Kramer 1988). While this idea is not necessarily applicable to other types of dives, we argue that the pattern of C dives exhibited by northern elephant seals suggests its application to Type C dives. These dives occur in distinct bouts, with little variance in the inflection depth or second segment descent rate. It is unreasonable that the seal would return to the surface with remaining oxygen reserves and then spend potential foraging time and energy to return to depth and repeat the same behavior. If this is true—and whatever happens metabolically during diving is a constant across dive types—a dive type with a significantly lower swim speed would be expected to have a significantly longer duration. However, the mean duration of Type C dives (21.4 min.) is not significantly different from that of Type AB dives (21.6 min.), where the animal swims faster and covers on average twice the horizontal distance. Analysis of TDR records from 23 elephant seals including yearlings, adult females, and adult males show that Type C dive durations are not significantly different from Type A dive durations (Paired t-test, $t = -0.05$, $p > .05$; B. Le Boeuf, P. Morris, and D. Crocker, unpubl. data). From this, we infer that something different is happening metabolically during Type C dives.

Oxygen savings from reduced swim speed might be used in the processing of food or metabolites incurred during diving. Oxygen savings could be used in perfusion of organs that incurred restricted blood flow on other types of dives. Type C dives did not appear in the record until day 17. This coincides with a relative decrease in the frequency of Type AB dives and a concomitant increase in the frequency of Type D dives.

Feeding studies on northern elephant seals have shown increases in metabolic rate averaging 65% 4 to 5 hours after feeding (P. Thorson, pers. comm.). If foraging occurs in long bouts, the increase in energy expenditure due to digestion might negatively impact the balance between the increased oxygen utilization required for active foraging and the hypoxia of diving. However, if the metabolic changes associated with food processing could be coupled temporally with decreased oxygen requirements due to decreased swimming effort, this impact could be significantly lessened.

A subsequent paper will test a number of predictions based on this idea in an attempt to demonstrate a more conclusive pattern of evidence supporting this hypothesis. The real test, however, will be to attempt to illicit this dive type in the homing paradigm (see Le Boeuf, this volume) and to directly measure perfusion of the digestive tract and swimming muscles. The metabolic issue raised here, the overall significantly lower swim speed, and the temporal patterning of this dive type in the record, are consistent with Type C dives serving a physiological process.

Type D and Type E dives appear to have characteristics consistent with foraging. The vertical excursions seen at the bottom of Type D dives are consistent with searching, pursuing, and capturing prey. Type D dives occurred in long series and were the most common dive type. Both of these characteristics are consistent with foraging. Type E dives might serve feeding on the bottom, or benthic foraging. These dives also occurred in long series. The depths of the bottom segments within a series of Type E dives were relatively invariant. Though the depth sometimes changed slightly during the course of a bottom segment, the succeeding bottoms in a series appeared to start close to the depth at which the preceding bottom ended. This gives the impression that the animal's movement was following the bottom topography, though geolocation and bottom depths are needed to be conclusive. Moreover, the descent and ascent angles were abrupt in both of these dive types, suggesting that the animal descended directly to a depth where prey were found, spent one-third of the dive's duration foraging, then returned directly to the surface. The increased incidence of speed changes seen in both dive types, especially during the bottom segment of Type D dives, is also consistent with foraging.

A recent investigation of sex differences in foraging patterns in northern elephant seals (Le Boeuf et al. 1993) offers strong support for this characterization of Type D and E dives. Diurnal variation in depth for Type D dives was found which is consistent with movements of the deep scattering layer and was not evident in any other dive type. This dive type might account for most of the diurnal variation in dive depth and duration reported by many investigators.

A strong association of Type E dives with coastal regions was also demonstrated. However, the geolocation of many of these dives might be sufficiently far away from the continental shelf to preclude benthic foraging. Stewart and DeLong (this volume) also argue that the geolocation of their Type E dives precludes benthic foraging. However, a recent shipboard validation study of a geolocation TDR revealed a strong, consistent, southernly bias to geolocation estimates in the area in question (S. Blackwell and B. J. Le Boeuf, unpubl. data). This bias might preclude the use of light-level geolocation to obtain precise correlations between diving behavior and bathymetric features.

This analysis represents a first effort at using velocity and TDR instruments together to investigate questions concerning diving behavior. New and better instruments that allow assessment of instantaneous velocity and acceleration will allow a much finer level of analysis than that presented here. Future generations of instruments will also have higher upper ranges so that a more accurate picture will emerge. However, the relationships between the numbers in this study allowed us to formulate hypotheses that will form the basis of future investigations.

REFERENCES

Aleyev, Y. U. G. 1977. *Nekton*. The Hague: W. Junk.

Briggs, G. D., R. V. Henrickson, and B. J. Le Boeuf. 1975. Ketamine immobilization of northern elephant seals. *Journal of the American Veterinary Medicine Association* 167: 546–548.

Costa, D. P., G. P. Antonelis, and R. DeLong. 1990. Effects of El Niño on the foraging energetics of the California sea lion. In *Effects of El Niño on Pinnipeds*, ed. F. Trillmich and K. Ono, 156–165. Berlin: Springer Verlag.

Costa, D. P., J. P. Croxall, and C. Duck. 1989. Foraging energetics of Antartic fur seals, *Arctocephalus gazella*, in relation to changes in prey availability. *Ecology* 70: 596–606.

Costa, D. P., and R. L. Gentry. 1986. Reproductive energetics of the northern fur seal. In *Fur Seals: Maternal Strategies on Land and at Sea*, ed. R. L. Gentry and G. L. Kooyman, 79–101. Princeton: Princeton University Press.

Costa, D. P., P. H. Thorson, and M. Kretzmann. 1989. Diving and foraging energetics of the Australian sea lion, *Neophoca cinera*. *American Zoology* 29: 71A.

Davis, R. W., T. M. Williams, and G. L. Kooyman. 1985. Swimming metabolism of yearling and adult harbor seals, *Phoca vitulina*. *Physiological Zoology* 58: 590–596.

Fedak, M. A. 1986. Diving and exercise in seals: A benthic perspective. In *Diving in Animals and Man*, ed. A. Brubakk, J. W. Kanwisher, and G. Sundnes, 11–32. Kongsvold Symposium, Royal Norwegian Society of Sciences and Letters, June 3–7, 1985. Trondheim: Tapir.

Fedak, M. A., M. R. Pullen, and J. Kanwisher. 1988. Circulatory responses of seals to periodic breathing: Heart rate and breathing during exercise and diving in the laboratory and open sea. *Canadian Journal of Zoology* 66: 63–69.

Feldkamp, S. D. 1987. Swimming in the California sea lion: Morphometrics, drag, and energetics. *Journal of Experimental Biology* 131: 117–136.

Guppy, M., R. D. Hill, R. C. Schneider, J. Qvist, G. C. Liggins, W. M. Zapol, and P. W. Hochachka. 1986. Micro-computer-assisted metabolic studies of voluntary diving of Weddell seals. *American Journal of Physiology* 250: R175–R187.

Hedrick, M. S., and D. A. Duffield. 1991. Haematological and rheological characteristics of blood in seven marine mammal species: Physiological implications for diving behavior. *Journal of Zoology, London* 225: 273–283.

Hedrick, M. S., D. A. Duffield, and L. H. Cornell. 1986. Blood viscosity and optimal hematocrit in a deep-diving mammal, the northern elephant seal (*Mirounga angustirostris*). *Canadian Journal of Zoology* 64: 2081–2085.

Hill, R. D., R. C. Schneider, G. C. Liggins, A. H. Schuette, R. L. Elliot, M. Guppy, P. W. Hochachka, J. Qvist, K. J. Falke, and W. Zapol. 1987. Heart rate and body temperature during free diving of Weddell seals. *American Journal of Physiology* 253: R344–R351.

Kooyman, G. L. 1965. Techniques used in measuring diving capacities of Weddell seals. *Polar Record* 12: 391–394.

———. 1968. An analysis of some behavioral and physiological characteristics related to diving in the Weddell seal. In *Antarctic Research Series*, vol. 11, *Biology of the Antarctic Seas III*, ed. W. L. Schmitt and G. A. Llano, 227–261. Washington, D.C.: American Geophysical Union.

———. 1981. *Weddell Seal: Consummate Diver*. Cambridge: Cambridge University Press.

———. 1989. *Diverse Divers: Physiology and Behavior*. Berlin: Springer Verlag.

Kooyman, G. L., J. O. Billups, and W. D. Farwell. 1983. Two recently developed recorders for monitoring diving activity of marine birds and mammals. In *Experimental Biology at Sea*, ed. A. G. Macdonald and I. G. Priede, 197–214. London: Academic Press.

Kooyman, G. L., R. L. Gentry, and D. L. Urquhart. 1976. Northern fur seal diving behavior: A new approach to its study. *Science* 193: 411–412.

Kooyman, G. L., D. H. Kerem, W. B. Campbell, and J. J. Wright. 1973. Pulmonary gas exchange in freely diving Weddell seals (*Leptonychotes weddelli*). *Respiration Physiology* 17: 283–290.

Kramer, D. L. 1988. The behavioral ecology of air breathing by aquatic animals. *Canadian Journal of Zoology* 66: 89–94.

Le Boeuf, B. J., D. E. Crocker, S. A. Blackwell, P. A. Morris, and P. Thorson. 1993. Sex differences in foraging in northern elephant seals. In *Marine Mammals: Advances in Behavioural and Population Biology*, ed. I. L. Boyd, 149–178. Symposia of the Zoological Society of London no. 66. London: Oxford University Press.

Le Boeuf, B. J., Y. Naito, T. Asaga, D. E. Crocker, and D. P. Costa. 1992. Swim speed in a female elephant seal: Metabolic and foraging implications. *Canadian Journal of Zoology* 70: 786–795.

Le Boeuf, B. J., Y. Naito, A. C. Huntley, and T. Asaga. 1989. Prolonged, continuous, deep diving in female northern elephant seals, *Mirounga angustirostris*. *Canadian Journal of Zoology* 67: 2514–2519.

Lenfant, C., K. Johansen, and J. D. Torrance. 1970. Gas transport and oxygen storage capacity in some pinnipeds and the sea otter. *Respiration Physiology* 9: 277–286.

Naito, Y., B. J. Le Boeuf, T. Asaga, and A. C. Huntley. 1989. Long-term diving records of an adult female northern elephant seal. *Antarctic Record* 33: 1–9.

Ponganis, P. J., R. L. Gentry, E. P. Ponganis, and K. V. Ponganis. 1992. Analysis of swim velocities during deep and shallow dives of two northern fur seals, *Callorhinus ursinus*. *Marine Mammal Science* 8: 69–75.

Ponganis, P. J., E. P. Ponganis, K. V. Ponganis, G. L. Kooyman, R. L. Gentry, and F. Trillmich. 1990. Swimming velocities in otariids. *Canadian Journal of Zoology* 68: 2105–2112.

Williams, T. M., G. L. Kooyman, and D. A. Croll. 1990. The effect of submergence on heart rate and oxygen consumption of swimming seals and sea lions. *Comparative Physiology* 160B: 637–644.

PART IV

Physiological Ecology

Apnea Tolerance in the Elephant Seal during Sleeping and Diving: Physiological Mechanisms and Correlations

Michael A. Castellini

ABSTRACT. To better understand the diving behavior of elephant seals, it is necessary to study how their diving physiology limits their diving behavior and how behavior fits into the window of physiological options. Unfortunately, the diving physiology of elephant seals is very difficult to study because the seals are at sea and inaccessible during most of the year. However, when on land, they exhibit long duration breath holds during sleep that can last over 20 minutes. By studying these periods of breath holding during sleep, we have found that the physiology of sleep apnea appears to be very similar to the physiology of diving apnea. We suggest that the control processes involved in both states may be similar enough to allow us to study some of the aspects of diving physiology by instead examining the animals while they hold their breath on land.

Diving behavior and physiology are integrally related components to the study of "diving biology" in any species. Given the physiological limits that may be placed on a species, diving behavior must fit into those constraints. For example, most natural dives of the Weddell seal, *Leptonychotes weddelli*, appear to fall within the physiological time window of the seal's aerobic diving limit (Kooyman et al. 1980). While a tremendous amount of knowledge has recently been gained about the diving behavior of the elephant seal, the seal's pelagic nature makes it almost impossible to study diving physiology. However, these seals also exhibit extremely long duration (up to 20 min.) breath holding while sleeping on land. This behavior allows us to study the physiology of apnea under more controlled conditions and extrapolate our findings to what may be occurring during diving apneas at sea. Using this method, we have been attempting to understand some of the components of diving physiology in elephant seals that may not be attainable by other methods. Unfortunately, the sleeping habits of seals are one of those behaviors that marine mammal biologists have often informally observed but not documented. Consequently, it is "common knowledge" that seals hold their breath while sleeping and that in captivity they often sleep on the bottom of their pools. For example, at the National Zoo in Washington, D.C., there is a sign in front of the gray seal exhibit telling the public that the seals lying motionless underwater on the bottom of the tanks are not dead, just sleeping. There have been very few formal studies of sleeping seals, although there are some data available from various projects in which sleeping seals were coincidentally observed as part of larger behavior or physiology programs. Sleep in seals, however, offers a window into the study of mammalian metabolic and physiological tolerance to apnea that is not easily modeled by any other system. The primary question in this review is whether sleep breath holding is analogous to diving.

The study of the physiology of naturally diving marine mammals is confounded by the fact that the animals are at sea, and it is very difficult to even track them, let alone obtain solid physiological data on basic parameters such as heart rate or body temperature. The vast majority of natural diving physiology data has come from work on Weddell seals in Antarctica. By working on the sea ice from experimental dive sites, scientists have been able to study this species in a relatively confined area and be reasonably sure that the seal will return to the experimental hole after each dive. This experimental protocol has provided most of the physiological data on natural diving in pinnipeds (Kooyman 1981, 1989). Under these conditions, however, natural diving is complicated by its two major components: underwater exercise and breath holding. These two processes compete with one another for the limited supply of oxygen that the seal carries with it from the surface. Exercise increases oxygen demand, while diving calls for reducing oxygen consumption. Seals balance such seemingly conflicting demands on each dive, but it is difficult for the scientist to distinguish how the physiological and metabolic data reflect that balance. For example, does the pattern of heart rate variability during diving reflect the demands of exercise, or does diving reduce the heart rate that would occur if the seal was simply exercising? This may at first seem like a mere semantic distinction, but it is not. To understand how seals survive extremely long periods underwater, it is critical to know how they balance their response to hypoxia and their response to exercise. Understanding this balance is why there has been such a concerted effort to obtain swimming velocities on diving seals. Given a constant supply of oxygen, a seal swimming quickly would presumably consume oxygen faster than if it was swimming slowly. This would make the aerobic dive time shorter and alter underwater efficiency. Such alterations have clear implications for foraging theory and for migration. No matter what aspect of natural diving behavior is being studied, there is always some question about the amount of oxygen carried and the rate at which it is utilized.

How does sleep physiology interact with this area of diving physiology? During the long periods of apnea in sleep, the seal experiences hypoxia but without the simultaneous demands of exercise. Thus, the seal in sleep apnea offers the opportunity to study how the animal reacts to breath holding without the additional energy requirements of swimming, foraging, or underwater traveling. It is a completely natural process in these animals, and as a group, the seals exhibit the longest duration normothermic sleep apnea of any mammal. Long duration sleep-associated apnea appears to be a strictly phocid attribute and has not been seen in otariids or cetaceans. Even so, it has been formally reported in less than half a dozen species of seals. By far, the greatest amount of information on sleep physiology in seals comes from studies of the northern elephant seal. G. A. Bartholomew (1954) first described the phenomenon by placing his hands on the chests of sleeping elephant seals and counting the heartbeats. R. C. Hubbard (1968) discussed the general cardiorespiratory pattern of sleeping elephant seals. A. C. Huntley (1984) cataloged some of the basic cardiorespiratory and sleep stage patterns in pups. S. B. Blackwell and B. J. Le Boeuf (1993) described developmental changes in sleep apnea from birth to adulthood. Since 1987, our laboratory has been involved in detailed studies of the metabolic implications of long duration sleep apnea in seals, mainly elephant seals (for review, see Castellini 1991).

PHYSIOLOGY OF SLEEP-ASSOCIATED APNEA

There are four primary areas to consider when discussing the physiology of sleep apnea in seals: respiratory, cardiac, circulatory, and metabolic alterations. The electrophysiology of sleep in pinnipeds is also of great interest but will only be referred to here when sleep staging is important to better understand the four areas noted above. In each of these areas, data from diving apnea will be compared and contrasted. The goal of this section is to ask how sleep and dive apnea may be related, keeping in mind that while at sea, seals may sleep while at depth and thus combine the two events. Most of this information comes from studies on elephant seal pups, with some comparative work on Weddell seals. This is all very recent work and is still in the process of being analyzed. Thus, a great deal of this discussion is based on personal observations, but references to work already published will be noted as available.

Respiratory Pattern

In elephant seal pups, the typical respiratory pattern while sleeping is to link together several periods of apnea and eupnea into one "bout" of sleep. Thus, the seal may sleep for 30 to 40 minutes and during that time go through a 10-minute period of apnea, 2 to 3 minutes of eupnea, another apnea, another short eupnea, and then a final apnea before waking. Figure

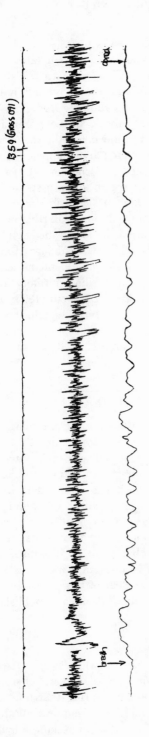

Fig. 19.1. Polygraph recording of a sleeping, 3-month-old northern elephant seal pup. The top line is a time trace with 5-second tick marks. Line 2 is the electroencephalogram (EEG) showing large-voltage, slow-frequency recordings typical of slow wave sleep (SWS). Line 3 is a respiratory trace showing the end of a 12-minute apnea, about 2.6 minutes of eupnea, and then the beginning of another apnea. Line 4 is the instantaneous heart rate showing the low heart rate during apnea, the postapnea tachycardia, the appearance of the normal sinus arrhythmia, and then the low heart rate of the next apnea. The heart rate calibration is on the left. Line 5 is the electrocardiogram (EKG), which varies depending on the respiratory cycle.

19.1 shows a short (2.5-min.) breathing interval between two longer apnea periods. There are many variations to this pattern, but it is critical to note that apnea is not a prerequisite for sleep (i.e., a seal can be ventilating and sleeping) and that sleep is not a prerequisite for apnea (an awake seal can breath hold). However, the longest apneas appear only during sleep, and awake apnea seems to have a limit of about 5 minutes. The longest period of apnea that has been reported on a 4-month-old elephant seal pup is just over 11 minutes (Castellini, Costa, and Huntley 1986), but recent recordings have exceeded 14 minutes. Average apnea duration ranges from about 6 to 8 minutes for a 3½- to 4-month-old pup. It is interesting to note that the pattern of long apnea relative to short eupnea is similar to the repetitive diving habits of both elephant seals (Le Boeuf et al. 1986) and Weddell seals (Castellini, Davis, and Kooyman 1988). This pattern implies that oxygen loading and carbon dioxide dumping are accomplished quickly during the respiratory period and that there is no metabolic processing of hypoxic end products. In fact, as will be discussed in detail later, there is no apparent change in plasma lactate or glucose during or after sleep apnea in elephant seals. Neither we nor Hubbard (1968) were able to find a clear correlation between the length of the eupnea and the length of the preceding apnea, although Huntley (1984) saw such a relationship in his study, which was conducted using different methods on restrained pups. Similarly, there is no relationship between the surface interval duration and preceding dive duration interval in elephant seals (Le Boeuf et al. 1988). In Weddell seals, there is no relationship between dive time and surface time during aerobic diving, but there is a longer surface recovery time correlated to increasing postdive lactate loads after anaerobic diving (Kooyman et al. 1980). Thus, both sleep apnea and diving apnea have the similar appearance of short eupnea periods between longer apnea periods, occur in bouts, and appear to be mostly aerobic.

Cardiac Pattern

The bradycardia associated with natural diving seals is well documented and has been shown to be related to the length of the dive (Kooyman and Campbell 1973) and to ascent and descent patterns (Hill et al. 1987) and therefore, presumably, to both effort and the need to stay underwater as long as possible. Similarly, Bartholomew (1954) noted that there was bradycardia in sleeping elephant seals with a reduction in heart rate from about 65 beats per minute (BPM) to 54 BPM in adults. G. L. Kooyman (1968) recorded the heartbeats of sleeping Weddell seals and also observed a decline in heart rate of about 30 to 40% in adults. S. H. Ridgway, R. J. Harrison, and P. L. Joyce (1975) demonstrated a slowing of heart rate in sleeping gray seals, and Huntley (1984) recorded the electrocardiogram (EKG) of restrained, sleeping elephant seal pups and showed that the average sleeping heart rate declined during apnea to about 70% of average, awake, breathing values. This pattern of bradycardia during the sleep apnea in seal pups appeared straightforward until we recently found two patterns that suggest the heart rate changes may be more complicated. First, the "bradycardia" in sleeping northern elephant seal pups seems to be age related in that 2-month-old pups neither drop their heart rate as low as 4-month-old pups nor seem to hold the heart rate steady during apnea. This same pattern has been seen in Weddell seal pups (Kooyman 1968). Second, there may not actually be a true bradycardia in the sense of the dramatic instantaneous decline seen in diving animals. By analyzing the EKG with an instantaneous beat-to-beat heart rate analyzer, we found that during eupnea, the older pups show a normal sinus arrhythmia. That is, as they inhale, heart rate increases, and as they exhale, heart rate declines. During eupnea, heart rate varies from about 80 BPM high to about 50 BPM low, for an "average" heart rate of about 65 BPM. As the pups enter into apnea with the last exhalation of their breathing period, the heart rate declines just as it had during the exhalation in eupnea and then stays at the 50 BPM rate during the long apnea. This pattern is clearly shown in figure 19.1. Thus, it seems that the "bradycardia" associated with sleep apnea is actually the low heart rate of a normal respiratory cycle and that the apnea is just a very long breath pause, at least in terms of cardiac control. The only time that the heart rate becomes very low during sleep is when the pup moves from slow wave sleep (SWS), which is the predominant type of sleep state, into rapid eye movement (REM) sleep. At this point the heart rate can get very low (around 20–25 BPM) and become quite variable.

The heart rate of freely diving elephant seals is being studied at this time (R. Andrews, pers. comm.). Preliminary evidence suggests that the most common heart rate of freely diving elephant seals may, in fact, be closer to the rate seen during sleep apnea than to low rates more often associated with forced diving conditions. However, very low heart rates have been observed and reinforces the point that the heart rate in these animals is not a reflex and is probably under higher-level control. Therefore, what at one time seemed to be clear sleep and dive apnea bradycardia may not remain as clear as more information is collected.

Circulatory Alterations

In 1980, Kooyman et al. found that the hemoglobin (Hb) levels in the blood of freely diving seals varied before and after dives. When the seals returned from a dive, the Hb levels would be high but declining. Later, J. Qvist et al. (1986) showed that Hb and hematocrit (Hct) in Weddell seals began to increase as soon as the dive began, leveled out at high values during the dive, and then declined afterward. M. A. Castellini, R. W. Davis, and G. L. Kooyman (1988) showed that the Hct in Weddell seals tended to stay high during an entire diving bout and only returned to resting levels during very long breaks in diving. This variation in Hct has been proposed to be caused by the sequestering and release of red blood cells (RBC) by the spleen during diving events (Zapol 1987). The maximization of Hct in the interdive surface interval would certainly facilitate the rapid loading of oxygen and also maximize the amount of oxygen that could be carried by the blood. Increased Hct is also known to occur in racing horses and dogs during sprint events (Harris et al. 1986). This is a perfect example of the difficulty of trying to separate diving into its exercise and apnea components. Does the change in Hct in diving seals come about because of exercise, or is it related to breath holding? As it turns out, sleeping seals may provide the answer.

In 1986, M. A. Castellini, D. P. Costa, and A. C. Huntley found that the Hct in sleeping elephant seal pups began to increase as soon as the apnea started and then declined as soon as breathing began. It has recently been found that when several apnea-eupnea-apnea cycles are linked together, the Hct stays elevated during the entire cycle and then drops to resting values when the pup is awake and breathing for a long period. Therefore, it seems reasonable to conclude that the change in Hct that occurs in diving seals most likely arises from the apnea response and not from exercise. If the spleen is the modulating organ for this phenomenon, then it must begin to contract on the initial apnea and sequester RBC during eupnea. However, if the next apnea follows soon after the short eupnea, there would not be enough time for all the cells to be gathered, and thus the Hct will stay somewhat elevated during the breathing period.

The striking similarity between the diving and sleeping apnea alterations in Hct suggests that the neurological mechanisms involved in initiating and maintaining the apnea are the same for both diving and sleeping. This is a critical point because sleep apnea in seals appears to be centrally controlled. That is, it is a neurologically influenced event and not obstructive apnea. In obstructive apnea, which is very common in humans, the upper airway becomes blocked during sleep as the tissues around the trachea relax and the patient begins, essentially, to suffocate. In obstructive apnea, the patient tries to breathe but cannot and must awaken to break the pattern (Strohl, Cherniak, and Gothe 1986). In seals, there is no attempt to breathe during the sleep apnea event (fig. 19.1). The apnea is controlled from higher central nervous system centers, and the same neural inputs that initiate the breath hold must also contract the spleen. The advantage of a high Hct, in both sleeping and diving seals, is that they can load oxygen quickly during the short eupnea period following the apnea.

There are additional changes that occur in the circulation of sleeping seals that strike parallels with the diving condition. It is well known that there are marked circulatory perfusion shifts that occur in diving seals. The classic dive response involves shunting blood flow away from the peripheral tissues and conserving the oxygen-rich blood for the more aerobic central organs, such as the brain and heart. This shunting has been visualized in a variety of methods, but one that is relevant here is a procedure that examines how plasma radioisotope tracers can show such shifts. When a radioactive metabolic tracer is injected into the circulation of a diving seal, the tracer slowly equilibrates into the blood pool and is only slowly utilized until the dive ends. At that point, the tagged tracer is metabolized at the normal resting rate. At the point of inflection, the specific activity of the tracer in the plasma falls dramatically and provides a qualitative method to visualize the transition. This process appears to occur in both laboratory dives (Castellini et al. 1985) and natural dives (Guppy et al. 1986). Similarly, such transition points have been observed at the apnea/eupnea transition at the end of sleep apnea in northern elephant seal pups (Castellini 1986). When a tracer is injected into a sleeping seal, it follows a distinct pattern during the apnea and is then altered as soon as breathing occurs. These data imply that the same type of circulatory shifts that have been so well documented in diving seals may also occur in the sleeping seal.

On the basis of these two different indications of circulatory modifications that occur in sleeping seals, it is tempting to suggest that many of the same control mechanisms that regulate circulation during diving also occur in the sleep apnea event.

Metabolic Changes

While diving, there are a variety of blood chemistry changes that can occur. The first and most obvious is that blood oxygen decreases as the dive progresses, and the animal becomes hypoxic. Carbon dioxide partial pressure increases, and there is a respiratory acidosis induced by the high CO_2 (Kooyman et al. 1980; Qvist et al. 1986). During long dives, beyond the aerobic diving limit (ADL), lactate accumulates in the periphery and is flushed into the circulation when the peripheral tissues are reperfused after the dive ends (Guppy et al. 1986; Kooyman et al. 1980). Despite the low oxygen levels that are reached in all dives, the majority of dives, at least in Weddell seals, are known to be aerobic and do not show the characteristic increase in lactate after the dive. Similarly, the concentration of plasma glucose, the ultimate substrate for the lactate, drops during anaerobic diving but does not change during aerobic diving (Castellini, Davis, and Kooyman 1988; Guppy et al. 1986; Kooyman et al. 1980). Finally, during bouts of dives, there appears to be very little change in glucose or lactate over hours of diving unless a long dive occurs (Castellini, Davis, and Kooyman 1988).

During sleep apnea, blood oxygen declines to very low levels, CO_2 increases, and there is a respiratory acidosis (Kooyman et al. 1980). However, plasma lactate and glucose remain stable and do not change before, during, or after any single apnea or bout of apnea (Castellini, Costa, and Huntley 1986; Castellini and Castellini 1989). Thus, sleep apnea would appear to be mostly aerobic. This makes sense, since the seal sleeping on the beach can simply breathe when it becomes necessary. However, this does raise an interesting problem for seals that may be sleeping underwater at sea. If a sleeping seal has dropped its blood oxygen and raised its carbon dioxide to the point where is it necessary to breathe, it cannot ventilate if it is at 500 m depth. It would seem to be necessary for the seal to either awaken and swim to the surface or to somehow stay asleep and get to the surface. We know that sleeping elephant seal pups can come to the surface from about 0.5 m in a tank and ventilate without having to awaken. But floating to the top of a 0.5 m tank is considerably different from swimming to the surface from 500 m. We are left with trying to construct a control mechanism that signals to the seal when it will be necessary to breathe and get the animal to the surface while it is sleeping. Perhaps, however, elephant seals do not sleep while at sea, although this is unlikely given that they are pelagic for months at a time.

There is one last area of metabolic alteration that is of importance, and this concerns the metabolic cost of diving or sleeping. In Weddell seals, it has been shown that diving is not very costly and only elevates metabolism by 1.5 to 2 times over resting (Kooyman et al. 1973). We have recently found that for dive events and sleep apnea events of the same duration, diving only costs about 1.5 times the cost of sleep in Weddell seals (Castellini, Kooyman, and Ponganis 1992). If this is an energy demand that is typical of phocids, then we might be able to predict the metabolic cost of diving in elephant seals from the oxygen requirements of sleeping. For elephant seal pups, this would not be a difficult calculation, because there is a considerable amount of data available on the oxygen consumption patterns of pups during both apnea and eupnea periods. For adult elephant seals, however, this would involve finding a way to measure the oxygen consumption rate of a large and intractable animal. However, measuring the oxygen consumption rate of a sleeping elephant seal on land is infinitely easier than obtaining the same information on one that is diving at sea.

CONCLUSION

Are diving and sleep apnea similar? Based on the information available, it would appear that many of the same responses seen in diving seals occur in seals that are breath holding on land. Given that it is much easier to study sleeping seals on land than diving seals, this approach could be worthwhile as a starting point for species that are simply too difficult for study while at sea. However, the study of sleep apnea in and of itself is also interesting and has implications for the study of sleep apnea syndrome and sudden infant death syndrome (SIDS) in humans. Sleep apnea in seals is perfectly normal; it is not a disease or a syndrome and instead is part of a natural breathing pattern and is adaptive for a diving life-style.

The goal here was to relate some of the physiological mechanisms involved in the phenomenon of breath holding during sleep to breath holding while diving. After years of work in this area, our conclusions are that the two events are extremely similar and that many of the same control processes are involved. In the future, it is our hope that when seals are sleeping during a biological study, the scientists involved will not just casually note that the seal is resting but will instead look a little closer at an event that is like no other in the mammalian order.

REFERENCES

Bartholomew, G. A. 1954. Body temperature and respiratory and heart rates in the northern elephant seal. *Journal of Mammalogy* 35: 211–218.

Blackwell, S. B., and B. J. Le Boeuf. 1993. Developmental aspects of sleep apnoea in northern elephant seals, *Mirounga angustirostris*. *Journal of Zoology, London* 231: 437–447.

Castellini, M. A. 1986. Visualizing metabolic transitions in aquatic mammals: Does apnea plus swimming equal "diving"? *Canadian Journal of Zoology* 66: 40–44.

———. 1991. The biology of diving mammals: Behavioral, physiological, and biochemical limits. In *Advances in Comparative and Environmental Physiology*, vol. 8, 105–134. Berlin: Springer Verlag.

Castellini, M. A., and J. M. Castellini. 1989. Influence of hematocrit of whole blood glucose levels: New evidence from marine mammals. *American Journal of Physiology* 256: R1220–R1224.

Castellini, M. A., D. P. Costa, and A. C. Huntley. 1986. Hematocrit variation during sleep apnea in elephant seal pups. *American Journal of Physiology* 251: R429–R431.

Castellini, M. A., R. W. Davis, and G. J. Kooyman. 1988. Blood chemistry regulation during repetitive diving in Weddell seals. *Physiological Zoology* 61: 379–386.

Castellini, M. A., G. L. Kooyman, and P. J. Ponganis. 1992. Metabolic rates of freely diving Weddell seals: Correlations with oxygen stores, swim velocity, and diving duration. *Journal of Experimental Biology* 165: 181–194.

Castellini, M. A., B. J. Murphy, M. Fedak, K. Ronald, N. Gofton, and P. W. Hochachka. 1985. Potentially conflicting demands of diving and exercise in seals. *Journal of Applied Physiology* 251: R429–R431.

Guppy, M., R. D. Hill, R. C. Schneider, J. Qvist, G. C. Liggins, W. M. Zapol, and P. W. Hochachka. 1986. Microcomputer-assisted metabolic studies of voluntary diving of Weddell seals. *American Journal of Physiology* 250: R175–R187.

Harris, R. C., J. C. Harman, D. J. Marlin, and D. H. Snow. 1986. Acute changes in the water content and density of blood and plasma in the thoroughbred horse during maximal exercise: Relevance to the calculation of metabolic concentrations in these tissues and muscles. In *Equine Exercise Physiology*, ed. J. R. Gillespie and N. E. Robinson, 464–475. Davis: ICEEP.

Hill, R. D., R. C. Schneider, G. C. Liggins, A. H. Shuette, R. L. Elliot, M. Guppy, P. W. Hochachka, J. Qvist, K. J. Falke, and W. M. Zapol. 1987. Heart rate and body temperature during free diving of Weddell seals. *American Journal of Physiology* 253: R344–R351.

Hubbard, R. C. 1968. Husbandry and laboratory care of pinnipeds. In *The Behavior and Physiology of Pinnipeds*, ed. R. J. Harrison, R. C. Hubbard, R. S. Petersen, C. Rice, and R. J. Schusterman, 299–383. New York: Appleton-Century-Crofts.

Huntley, A. C. 1984. Relationships between metabolism, respiration, heart rate, and arousal states in the northern elephant seal. Ph.D. dissertation, University of California, Santa Cruz.

Kooyman, G. L. 1968. An analysis of some behavioral and physiological characteristics related to diving in the Weddell seal. In *Antarctic Research Series*, vol. 11, *Biology of the Antarctic Seas III*, ed. W. L. Schmidt and G. A. Llano, 227–261. Washington, D.C.: American Geophysical Union.

———. 1981. *Weddell Seal: Consummate Diver*. Cambridge: Cambridge University Press.

———. 1989. *Diverse Divers: Physiology and Behavior*. Berlin: Springer Verlag.

Kooyman, G. L., and W. B. Campbell. 1973. Heart rate in freely diving Weddell seals (*Leptonychotes weddelli*). *Comparative Biochemistry and Physiology* 43: 31–36.

Kooyman, G. L., D. H. Kerem, W. B. Campbell, and J. J. Wright. 1973. Pulmonary gas exchange in freely diving Weddell seals. *Respiration Physiology* 17: 283–290.

Kooyman, G. L., E. A. Wahrenbrock, M. A. Castellini, R. W. Davis, and E. E. Sinnett. 1980. Aerobic and anaerobic metabolism during voluntary diving in Weddell seals: Evidence of preferred pathways from blood chemistry and behavior. *Journal of Comparative Physiology* 138: 335–346.

Le Boeuf, B. J., D. P. Costa, A. C. Huntley, and S. D. Feldkamp. 1988. Continuous, deep diving in female northern elephant seals, *Mirounga angustirostris*. *Canadian Journal of Zoology* 66: 446–458.

Le Boeuf, B. J., D. P. Costa, A. C. Huntley, G. L. Kooyman, and R. W. Davis. 1986. Pattern and depth of dives in northern elephant seals, *Mirounga angustirostris*. *Journal of Zoology, London* 208A: 1–7.

Qvist, J., R. D. Hill, R. C. Schneider, K. J. Falke, G. C. Liggins, M. Guppy, R. L. Elliot, P. W. Hochachka, and W. M. Zapol. 1986. Hemoglobin concentrations and blood gas tensions of free-diving Weddell seals. *Journal of Applied Physiology* 64: 1560–1569.

Ridgway, S. H., R. J. Harrison, and P. L. Joyce. 1975. Sleep and cardiac rhythm in the gray seal. *Science* 187: 553–555.

Strohl, K. P., N. S. Cherniak, and B. Gothe. 1986. Physiologic basis of therapy for sleep apnea. *American Reviews of Respiratory Disease* 134: 791–802.

Zapol, W. M. 1987. Diving adaptations of the Weddell seal. *Scientific American* 256: 100–107.

TWENTY

Expenditure, Investment, and Acquisition of Energy in Southern Elephant Seals

Michael A. Fedak, Tom A. Arnbom, B. J. McConnell, C. Chambers,
Ian L. Boyd, J. Harwood, and T. S. McCann

ABSTRACT. Information on the expenditure and investment of energy in southern elephant seals, *Mirounga leonina*, was collected during breeding and molt over four field seasons at South Georgia. Weight and body composition changes of mothers, pups, and breeding males were monitored during the breeding season. These changes were also measured in adult females, before and after the 70-day period when animals fed at sea between breeding and molt. During this period, information on foraging movements and behavior was gathered using purpose-built satellite-relay data loggers. Body composition changes were measured using isotope dilution techniques.

Breeding energetics information is discussed in relation to the evidence for differential investment in male and female pups. Large females produce larger pups, both at birth and weaning. Male pups are born larger than female pups. However, there is no evidence that mothers invest more energy (either relative or absolute) in male pups after birth once female size and birth weight are taken into account.

Foraging movements and diving behavior are discussed in terms of the oceanography of the foraging area and possible constraints placed by prey consumption on the seals' dive behavior. We suggest that the long distance travel of females to distant feeding locations may be advantageous in providing for the requirements for reliable food sources in a long-lived, uniparous mammal. Dive characteristics changed during the different phases of activity in foraging animals in relation to the average daily velocity of the animal, water depth, and undersea topography.

Southern elephant seals divide their year between land and sea. While on land, they breed and molt, expending energy and material that was stored in their bodies while foraging at sea over the remainder of the year. Some energy is invested in the production of young and some in new skin and hair, while the remainder is metabolized to support the animal during the fasts associated with these activities. Southern elephant seals separate these periods of net energy loss, both geographically and temporally, from their foraging efforts. This presumably allows them some freedom in their choice of the location of their breeding and molting areas and can help to insulate them from local changes in the abundance of prey. The sharp distinction between periods of net energy gain and loss and the temporal separation of breeding and molt make it possible to study each of these phases of the life history separately. Because there is no feeding during either the breeding or molting periods and the animals are on land and accessible during these times, the behavior and energy expenditures of these activities can be studied using relatively straightforward techniques.

The same cannot be said of studies of the periods of energy gain while the animals are at sea. Because the southern elephant seals that breed on South Georgia have a very large number of widely scattered breeding and molting sites from which to choose and because travel on the island (and to other islands) is difficult, recovery of time-depth recorders from animals is uncertain; this makes the use of telemetry advantageous. However, the scale and remoteness of the Southern Ocean make this difficult. The study of the dive behavior of the species therefore has lagged behind that of northern elephant seals (*M. angustirostris*). However, a combined data logger/ Argos transmitter has been developed by the Sea Mammal Research Unit (SMRU) to gather detailed information about their movements and behavior while at sea. These devices are now producing a flood of new information on the southern species.

Here, we bring together information collected as part of a joint program involving the SMRU, the British Antarctic Survey, and the University of Stockholm on (1) breeding energetics, (2) parental investment, and (3) energetics and behavior of molting and foraging in this species. The information was collected over four field seasons from 1986 to 1991 at Husvik, South Georgia. It presents work in the process of analysis and publication and is a preliminary attempt to synthesize the energetics of the life history of this species.

Using a combination of techniques (serial weight changes and isotope dilution measurements of body composition), we provide information on the energy expenditure of female southern elephant seals during breeding and molt and relate this to their size, their energy stores, and the sex and growth of pups. We use this information to consider the evidence for differential parental investment in this species. We briefly consider the reproductive effort of harem males and compare this with that of females. We then present information on foraging behavior of females and their weight gain during the period at sea between breeding and molt using serial weight changes and position and dive depth/velocity information provided by the satellite-relayed data logger.

EXPENDITURE AND INVESTMENT WHILE ASHORE

Overview of Reproductive Season and Annual Cycle

Southern elephant seals have a circumpolar distribution in the Southern Ocean. They breed during October and November on a small number of subantarctic islands and mainland sites in South America (Ling and Bryden 1981; Laws 1984). Approximately half of the world population breeds at South Georgia (54°S, 35°W) (McCann and Rothery 1988). Female southern elephant seals may begin to breed at 3 years of age, but the majority do not come ashore to breed until age 4 or 5 (Laws 1960; McCann 1980, 1981). Pregnant female southern elephant seals begin to arrive on beaches on South Georgia in mid-September, after some males have made an appearance. New females continue to appear until early November. They give birth about one week after arrival and nurse their pups for 18 to 23 days. Mating occurs after an average of about 22 days. Pups are weaned when females leave the beach but then remain ashore for 3 to 6 weeks. After breeding, females spend around 70 days at sea, then come ashore again for approximately one month to molt (Laws 1956). Males spend more time ashore while breeding and molt about one month later than females.

An Important Proviso

Breeding female elephant seals vary greatly in size: the largest females may weigh three times more than the smallest. We have tried to include in the study significant numbers of the smallest and largest animals on the beaches. Very large and small animals are probably somewhat overrepresented in the sample, and, therefore, the distributions about the means for some of the variables of interest are probably not representative of the population norms. Rather, the sample emphasizes the potential range of values the variables can take and the relationships possible over the size range of females in the population.

Growth of Pups

Table 20.1 summarizes information on weight changes for a sample of females and pups taken in 1986 and 1988 at Husvik, South Georgia. Pups are born weighing an average of 43 kg; males average 6 kg heavier than females (McCann, Fedak, and Harwood 1989). Overall, birth weight is positively correlated with the maternal weight, although this correlation is significant in female pups but not in males when the sexes are treated separately. Growth during lactation follows a roughly sigmoid trajectory (fig. 20.1). It may begin during the first day after birth, but in some pups, it may be negligible or even negative for the first 1 to 10 days. A period of constant growth rate follows this postpartum lag. This is in turn followed in

TABLE 20.1 Weight changes of pups and lactating females and duration of lactation (see proviso in text).

	Male pups		Female pups		Pups combined	
	N	Mean ± SD	N	Mean ± SD	N	Mean ± SD
Mother's weight change (kg)	13	130 ± 43	15	142 ± 29	28	137 ± 36
Mother's rate of weight loss (kg/day)	13	8.0 ± 2.3	15	8.2 ± 1.5	28	8.1 ± 1.9
Pup's weight change (kg)	13	79 ± 26	15	81 ± 16	28	80 ± 22
Pup's rate of weight gain (kg/day)	14	3.5 ± 1.3	16	3.4 ± 0.8	30	3.5 ± 1.0
Birth–weaning duration (days)	14	23	16	23	30	23
Pup's weaning weight (kg)	14	124 ± 28	16	117 ± 23	30	121 ± 25

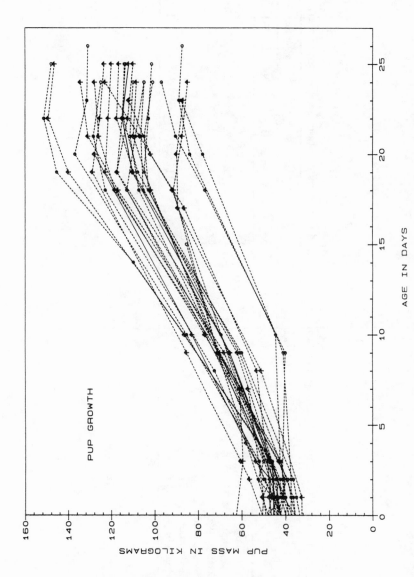

Fig. 20.1. Weight changes in male (arrows) and female (circles) pups. Growth follows a roughly sigmoid trajectory with a variable lag and some indication of a growth slowdown near weaning.

many cases by a decreasing rate just prior to weaning. Size at weaning is not highly correlated with weight at birth, and the size advantage held by males at birth is often lost by the time of weaning. Average weight at weaning of male and female pups is not significantly different if the size of their mothers and the size of pups at birth are taken into account. This results from the fact that larger females produce larger pups and that the average weight of mothers of male pups was significantly larger than that of female pups, primarily because few small females in the sample had male pups (see discussion below).

Weight Loss of Females

The rate at which females lose weight during lactation depends, in part, on their size; large females lose weight more rapidly than small ones (fig. 20.2a), but they also tend to produce larger pups (fig. 20.2b). Overall, there is no clear difference in the growth of male and female pups (table 20.1) or the weight loss of their mothers when one accounts for differences in the size of the mothers and the birth weight of their pups. However, some small females that produced male pups lose more weight than might be expected, and their male pups seem to grow somewhat faster than average for pups of females of this size (fig. 20.2b). However, there are few small females with male pups in the sample (in spite of the effort to sample small females), and this trend would require much larger sample sizes to establish.

Changes in Body Composition

After weaning their pups, some females leave the beach looking extremely thin, while others (often the largest) look quite fat. This suggested to us that in spite of the fact that larger females produce larger pups at both birth and weaning, there might still be a differential investment with respect to the size or age of the mother. There might be large differences in resources available in small and large females, and although larger females tend to produce larger pups, the amount invested relative to the amount available (relative investment) might vary as a function of size. To measure this investment in terms of energy and materials and to compare it to the resources that mothers of different sizes had available, the body compositions of 45 females were measured at the beginning and end of lactation during 1986 and 1988 (Fedak et al. 1989) using isotope dilution techniques (Reilly and Fedak 1990).

Table 20.2 gives the average use of energy, fat, and protein for the nursing females in the sample, subject to the proviso noted above. On average, approximately 40% of the energy, 47% of the fat, and 17% of the protein in the body is utilized in producing a pup. However, these amounts vary

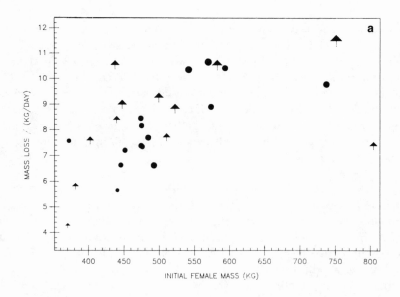

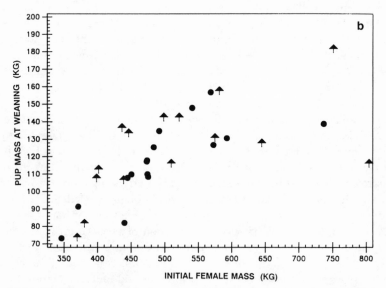

Fig. 20.2. *(A)* The relationship between initial female mass and daily weight loss split by male (arrows) and female (circles) pups. Large females lose weight more rapidly than small females; some small mothers of male pups lose weight rapidly. *(B)* The relationship between initial female mass and male (arrows) and female (circles) pup mass at weaning. Large females produced larger pups. The relation is significant (p < .01) for female pups but not for males when sexes are considered separately. See fig. 20.3 for explanation of symbol sizes.

TABLE 20.2   Utilization of energy, fat, and protein in females nursing male and female pups (see proviso in text).

Component	N	Mean total used ± SD	Mean percentage used ± SD
Energy (MJ)	27	3222 ± 161	40 ± 2.2
Males	12	3198 ± 258	39 ± 3.4
Females	15	3242 ± 212	40 ± 2.1
Fat (kg)	27	72 ± 4	47 ± 3
Males	12	72 ± 6	46 ± 15
Females	15	72 ± 5	47 ± 12
Protein (kg)	27	15 ± 1	17 ± 1
Males	12	14 ± 2	17 ± 1
Females	15	15 ± 1	18 ± 1
Energy density of weight loss (MJ/kg)	27	23.1 ± 0.6	

widely among individual females (fig. 20.3). Some very large females used as little as 30% or less of the total energy or fat available. In contrast, some of the smallest females that produced large pups used almost 60% of the total energy in their bodies and up to almost 70% of the fat. Some of the smallest females used little of their reserves, but these were mothers who produced pups in the lowest quartile of weaning weight.

One way of viewing this distribution of data is to consider females in order of decreasing size. The largest females can produce even the largest pups with small relative investment; indeed, in theory, some have sufficient reserves to produce two. Some midsize females can produce large pups with no more than average relative investment. But small females face a problem: if they produce pups of average size or greater, they will use a very large fraction of their reserves, possibly with deleterious effects on their subsequent reproduction or survival (Huber 1987; Reiter and Le Boeuf 1991). If they produce small pups, these pups may have a reduced probability of survival to adulthood. The best option for small females might be to abort their pup as early in gestation as possible if their own size at parturition will fall below that necessary to produce a viable pup.

Thus, we could view the distribution of investments in figure 20.3 as reflecting a solution to this dilemma. As female size decreases, relative investment increases up to a point; at weights below the median of female size, females produce either normal pups at very high relative investment or very small pups, well below median size. Indeed, the small number of small females with pups of any size in the sample may well suggest that many small females either do not become pregnant or abort before term.

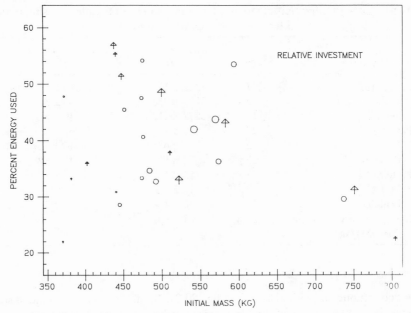

Fig. 20.3. The relationship between initial female mass and the percentage of energy stores used in nursing male (arrows) and female (circles) pups. Symbol size reflects pup size in quartiles: the largest symbol indicates pups in the highest quartile of weaning weight, and so forth. Note the marked variation in relative investment with changes in female size.

Energetics of Reproduction in Males

The information presented below on weight changes of males is an interim and incomplete presentation of our data. We include it here to complement the data on northern elephant seal bulls (Deutsch et al., this volume).

Male elephant seals on South Georgia may become sexually mature at 5 to 6 years of age, although most harem bulls are probably 9 to 12 years old (McCann 1981). As is true among females, a very wide size range of animals is present on the breeding beaches. Bulls occupying positions within harems typically weigh 1,500 to 3,000 kg, and weights up to 3,700 kg have been reported (Ling and Bryden 1981). Bulls peripheral to harems may weigh as little as 1,000 kg. The duration of the bulls' stay in and around harems is also very variable. In our sample, the longest stay was 73 days; the average was 58 days. Males lost weight at 12 to 15 kg per day while ashore, although one male lost weight at a prodigious 20 kg/day. Typical males lose around 700 kg during their breeding fast, and some lose up to 1,000 kg.

When compared to lactating females, the rate of weight loss and total weight loss are much greater and more variable. Females lose around 8 kg/ day for 23 days or about 180 kg. Much of this loss is material, largely fat and protein, transferred to their pup. Male losses are largely metabolic, probably mostly fat consumed to provide energy for breeding activity. While males are 2 to 8 or more times the size of breeding females, their total losses are only about 4 times as great, even though they may spend double the time breeding and fasting.

These rates of weight loss are larger than similar estimates for northern elephant seal bulls (Deutsch et al., this volume). This is not unexpected, since southern elephant seal bulls are somewhat larger. The spatial organization of harems on South Georgia is also quite different from that on most northern elephant seal beaches. Harems on South Georgia tend to be linearly arrayed along the shoreline and are often separated by unoccupied beach or rocky headlands. This distribution may require a swim if males from one harem are to visit another. The number of interactions between animals could therefore be quite different when compared to northern elephant seals among which harems often spread back from the beach and are more crowded. This might well influence the energy used during breeding.

Behavior and Energetics of the Molt in Females

Less is known about the behavior of animals during the annual molt or the energetic and material requirements of this phase of the life cycle of southern elephant seals. However, we have begun work to find out how much energy animals use during the molt and how much energy they gain while at sea between breeding and molt (Boyd, Arnbom, and Fedak, 1993). During the breeding season of 1990–1991 at Husvik, we captured 20 females late in lactation, weighed them, determined body composition isotopically, and marked or radio tagged them. Some were recaptured when they made landfall at the start of the molt and again 2 to 4 weeks into the molt and the procedures repeated. During the molting period, individual animals were located every few days until they left the Stromness Bay area. Reports of this study are now in press (ibid.), and we can now make some statements about the behavior and weight changes during the molt and about the weight gains while the animals are at sea between breeding and molt.

Adult females begin to arrive in significant numbers at molting sites around Husvik in mid-January (Laws 1960). Roughly one-third of the females tagged at breeding returned to the Stromness Bay area to molt. One female molted in another bay 10 km away in an area inaccessible to us, but we could monitor her presence via a VHF radio. At least two of the radio-tagged seals are known to have molted away from South Georgia on islands near the Antarctic Peninsula.

Several phases seem to occur during the molt in terms of both the morphological changes (Worthy et al. 1992) and, from our observations, the behavior of the animals. When radio-tagged animals first come ashore, they may make landfall at one site and then reenter the water and move to other sites, sometimes many kilometers away. They are often quite mobile for several days before settling down in a "wallow" or other place not immediately adjacent to the water's edge. Once in such a spot, they often remain there for 7 to 14 days while their old hair and skin loosens and falls away. Once this stage is complete and new hair begins to be detectable by touch, they often move out of the wallow to areas where their skin remains relatively clean and dry. During this period, hair growth continues and animals lose the velvet feel and appearance they have when leaving the wallows. They often move close to shore during this time and, on warm days, enter the water occasionally. Once at the shore, they may later move along it to more exposed locations on points and rocky headlands. Some animals were found along the shore for many weeks after leaving wallows and after hair seemed completely regrown. These changes in behavior may be related to underlying thermoregulatory or mechanical needs associated with shedding and regrowth of hair and skin.

During this period, animals lost 4 to 5 kg/day, a rate slightly greater than that of northern elephant seals (Worthy et al. 1992), but the animals in that study were somewhat smaller. This loss is roughly half that during lactation. If animals remain ashore for 30 days, the total loss may approach two-thirds to three-fourths of the lactational loss. The chemical composition or the weight lost is slightly different from that lost during lactation, with 39% of the loss being fat and 16% protein (Boyd, Arnbom, and Fedak 1993) compared with 47% fat and 17% protein during lactation. Thus, each year, females invest half as much energy in their skin as they do in their offspring. It seems fair to say that elephant seals look after their own skin.

ACQUISITION OF ENERGY: MOVEMENTS AND BEHAVIOR AT SEA

Little is known about where southern elephant seals go after leaving breeding and molting beaches. But as this and other chapters in this volume demonstrate, we are starting to acquire information on foraging movements and diving behavior to complement the serial data on individuals gathered before and after they go to sea. Gathering such serial data is difficult at South Georgia because animals may breed and molt on widely separated islands and because of the large number of inaccessible breeding and molting sites available to animals that do return over several years.

Animals that bred and returned to the Husvik area to molt tended to arrive for the molt in the same order in which they left the beach after breed-

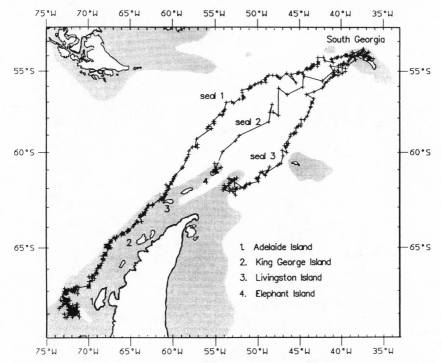

Fig. 20.4. The tracks of three postbreeding females obtained from Argos-compatible transmitter packages. The cross-hatched area is the continental shelf. Locations with a Location Quality Index (LQI) = 0 are shown as a cross and those with LQI > 0 as a circle. All three seals swam southwest from South Georgia to areas of the continental shelf.

ing. Their periods at sea ranged from 66 to 75 (mean = 72) days, during which time they gained 70 to 153 kg (mean = 107 kg, n = 8). Daily weight gains ranged from 1.1 to 2.3 (mean = 1.5) kg while at sea. Foraging success seems to vary significantly from individual to individual, even though the size range of animals recaptured was small (breeding beach departure weights of 340–457 kg). There was no relationship between size on leaving the beach and weight gain while at sea. The sample is, however, very small.

Four postbreeding female southern elephant seals at South Georgia were fitted with Argos compatible transmitter packages (Argos 1989) in November 1990 (McConnell, Chambers, and Fedak 1992). One failed when bitten by a copulating male. The movements of the remaining three animals are shown in figure 20.4. All traveled southwest to sites on the Antarctic continental shelf.

Seal 1 provided detailed information over 70 days. Its track was divided

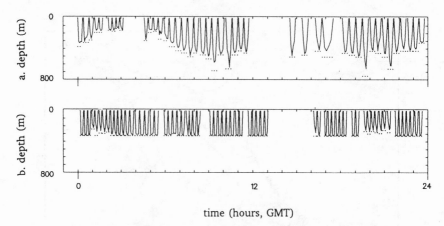

time (hours, GMT)

Fig. 20.5. Typical dive depth profiles over 24 hours from seal 1 while *(A)* in phase 1 and *(B)* in phase 3. Four depth values were transmitted for each dive at intervals of one-fifth of the duration of the dive. The maximum depth of each dive was also transmitted and is shown as a dotted line in the figure. Preceding surface intervals are represented by a horizontal line at zero depth. Breaks in the continuous plot indicate missing data, due primarily to diurnal variation in satellite visibility. Note the variability of dive depth and duration in phase 1 compared with phase 3.

into three phases based on location and movement. During phase 1 (23 days), it swam 1,845 km (average daily velocity [ADV] across the sea surface 0.93 m/sec.) to Livingston Island. In phase 2 (17 days), it hauled out at Livingston Island for 18 hours and then swam a further 805 km (ADV .55 m/sec.) to the southwest, following the continental shelf margin of the Antarctic Peninsula to a location 110 km west of Adelaide Island where water depth was 300 to 400 m. During this phase, it spent several periods of up to 12 hours at the surface. During phase 3 (29 days), it remained within 20 km of this location.

Seal 2 swam 1,420 km in 16 days (ADV 1.02 m/sec.) to Elephant Island where it hauled out. The transmitter failed one day later. It was captured, reweighed, and the transmitter removed at a molting site at South Georgia on February 8, 1991. Seal 3 swam 1,435 km in 16 days (ADV 1.03 m/sec.) to the continental shelf 110 km southeast of Elephant Island. It remained within a 60-km radius of this area for the next 4 days, after which the transmitter failed. It was sighted on January 19, 1991, at a molting site on King George Island.

Figure 20.5 provides a detailed view of dive profiles from seal 1 on two days typical of phases 1 and 3. Such dive-by-dive data were available from 65% of the animal's track: 6-hour dive summaries covered the entire period. We suggest that most of phase 1 was spent in transit to feeding grounds, although the variability of maximum depths and dive durations suggests that there was some opportunistic feeding. Dives made during phase 3, when the animal remained in the small area off Adelaide Island, were shallower and less variable (fig. 20.6a) than during phases 1 and 2. The seal tended to swim directly to the bottom and remain there, sometimes swimming slowly, until it returned to the surface (fig. 20.6d). These dives involved less swimming activity than those in phases 1 and 2 (fig. 20.6e), yet they tended to be shorter (fig. 20.6b) with a smaller proportion of time spent underwater (fig. 20.6c). Between days 63 and 65, the seal moved 8 km north, 40 km west and then returned to its starting location. Over these 3 days, all dive parameters (fig. 20.6a–e) shifted to values similar to those in phase 1.

Seal 2 gained 141 kg over 78 days (equivalent to 1.8 kg/day) between breeding and molt. By analogy with female northern elephant seals whose weight gain is similar (Le Boeuf et al. 1989; Sakamoto et al. 1989), we estimate that South Georgia seals require 9 to 20 kg of prey per day; that is, 630 to 1,400 kg over the approximately 70-day interval between breeding and molt.

Our interpretation of the activity of seal 1 in phase 3 is of targeted benthic or demersal feeding. This is consistent with data from stomach samples taken on land in which cephalopods predominate. These cephalopods include species of demersal squid and benthic octopods (Rodhouse et al. 1992; Murphy 1914; Laws 1960; Clarke and MacLeod 1982). However, if, as we have demonstrated, feeding areas are far from breeding sites, these stomach samples may underrepresent the consumption of fish whose remains are retained in the stomach for shorter periods. Samples from seals hauled out on sites near foraging areas could help to assess this bias.

Dives in phase 3 were shorter by a factor of 1.5 than in phase 1, and thus more time was spent traveling to and from the surface than would be the case for longer dives to the same depth. The proportion of time spent underwater in these dives was 6% (81% vs. 87%) less on average than those in the first two phases. The following argument suggests that the seal was not merely resting during these dives. These animals spend 80 to 90% of their time underwater and might be better referred to as "surfacers" rather than "divers." If these were rest dives, there would seem to be little reason why they should be more frequently interrupted by traveling to and from the surface to breathe given that oxygen stores should last longer at rest.

But if these were feeding dives, why should they be shorter? Extended periods at the surface were rare during phase 3, implying that the processes of digestion and assimilation were combined with diving activity on a steady state basis and not delayed to breaks in diving activity. Assimilation of food is known to increase basal metabolic rate up to 1.7 times in another

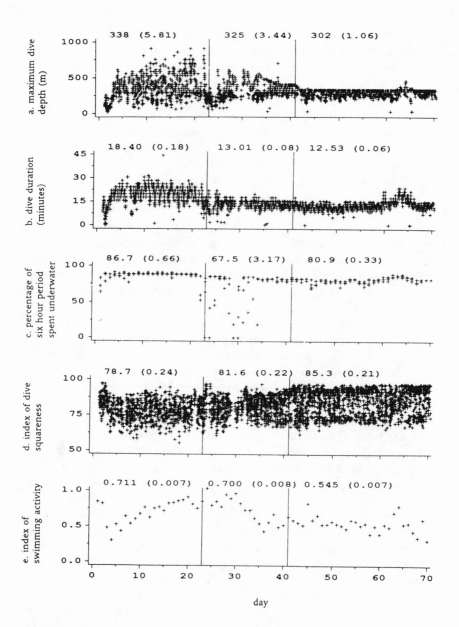

Fig. 20.6. Dive parameters from seal 1 over the 70-day tracking period. Vertical lines demarcate the three phases described in the text. Mean and standard error values, for each phase, are shown for each parameter. The parameters refer either to individual dives, to individual dives averaged over a day, or to 6-hour summary periods. Note that during days 63 to 65 there was a temporary change in all dive parameters.

phocid, the harp seal, *Phoca groenlandica* (Gallivan and Ronald 1981). We therefore suggest that the seal's aerobic dive limit was reduced during this phase, even though swimming activity was also reduced (fig. 20.6e), because the specific dynamic effect (SDE) (Kleiber 1961) of food assimilation increased the rate at which O_2 stores were used.

An additional component of this metabolic increase could also be the energy cost of warming the prey from sea temperature to body temperature, some 37°. The amount of heat this would require depends on the specific heat of the prey. If this is assumed to be equivalent to that of water (i.e., 4.19 J/g (and this assumption maximizes the likely effect since all prey will have specific heats less than this), then warming 20 kg of prey would account for the equivalent of about 10% of the standard metabolic rate of a 500-kg animal. However, this is not to say that warming food would require an increase in metabolic rate. This would depend on whether or not the animal was at or below the lower limit of its operational thermoneutral zone under the conditions prevailing while foraging. If it was not, it could decrease its heat loss via other avenues to the extent necessary to make up for the additional loss to the food. Given that SDE will cause a much larger increase in metabolism than 10%, the requirement for heating food in the stomach may not be important.

⊲————————————————————————————————————

(A) The maximum depth attained in each dive. Depth values are accurate to within 5 m at 100 m, to 50 m at 1,600 m. The maximum depth of approximately 910 m was obtained during phase 1. The variability of maximum depths declined markedly through phases 1 to 3 (SD = 171.2, 105.9, and 45.2). The constant upper limit of depth in phase 3 corresponds to the depth of the seabed as determined from Admiralty Charts.

(B) The duration of each dive. Dive durations were shorter in phase 3 than in phase 1 and less variable (SD = 2.5 and 5.4).

(C) The percentage of time within each 6-hour summary period that the seal spent underwater. This was least in phase 2, due primarily to extended periods on the surface. The time spent underwater in phase 3, when we suggest the seal was feeding, was less than during the transit in phase 1.

(D) Index of dive depth profile squareness of each dive. Four depth readings at intervals of one-fifth of the dive duration were transmitted in addition to the maximum depth attained. The average of these four values are expressed as a percentage of maximum depth. Dives were more flat bottomed in phase 3 than in phase 1.

(E) Daily averages of mid-dive swimming activity for each dive. A velocity turbine provided an index of swimming activity over five equally spaced intervals within each dive. The average of the middle three values is shown here to exclude descent and ascent activity. Mid-dive activity was less in phase 3 than in phase 1.

INTERACTIONS BETWEEN ACQUISITION AND EXPENDITURE: LINKING DATA FROM LAND AND SEA

Implications of Weight Change for Diving

Consideration of the huge weight losses of males (and, indeed, females as well) raises the question of how seals maintain effective control of buoyancy during diving throughout the annual cycle. Taking an extreme case, if a 3,000-kg male loses 1,000 kg while ashore and if most of the weight loss is lipid, then a thin animal would be much less buoyant than the same animal just prior to arrival on the breeding beach. Given that lipid is only .9 times the density of water (note that it is the density of the lipids used rather than blubber density that is important here), if the animal was neutrally buoyant on arrival, it could leave the beach 100 kg negatively buoyant, which amounts to a force equivalent to 3 to 5% of the weight of the animal.

Clearly, this is an extreme example, and the actual effect could be reduced, both because materials other than lipid are lost during the fast and because animals may come ashore positively buoyant and leave negatively buoyant, halving the imbalance. The imbalance could nonetheless amount to many kilograms upward or downward. While the imbalance in buoyancy would be much less in smaller animals, even small imbalances could be significant. Most human divers will be aware of the dramatic effect even small imbalances in buoyancy can have on maintaining their position in the water column and on the time they can remain submerged on a tank of air. This proportion of buoyancy change would amount to 4 kg extra of lead on a 70-kg human diver's weight belt, which would make for very hard work while operating free in midwater.

While an imbalance might have less energetic consequence during ascent and descent (losses in the energy required to swim up being partially repaid on the way down, or vice versa), it would have profound effects on the effort required to remain at a working depth in midwater or to remain on the bottom while buoyancy is positive. How might the animals compensate for the change in buoyancy? Changing the amount of gas taken down from the surface in the respiratory system would have little or no effect at the depths used by these species. While rocks and gravel are often found in the stomachs of elephant seals, it is hard to imagine enough being taken on board to compensate, and these might not be available in midocean. Could the physical properties (density and its relation to pressure and temperature) of fat stores of these seals be such that changes in the amount stored would not adversely affect buoyancy? Do animals change diving behavior, for example, switching from benthic to midwater foraging, as weight is regained to mitigate or take advantage of the effects of buoyancy changes? The questions remain open.

Foraging in Relation to Oceanography

The animals we tracked traveled more than 85 km each day for 16 to 23 days in a directed way to areas of the continental shelf. The use of distant foraging areas (associated with the Antarctic Polar Front, continental shelf margin, or ice edge) has also been inferred from water temperature data for elephant seals breeding on Macquarie Island (54°S, 157°E) (Hindell, Burton, and Slip 1991). Why do they adopt this strategy?

Elephant seal females are long-lived animals that invest large amounts of resources in a single pup each year over many years. They must, therefore, locate food reliably each year for many years in succession. Movement away from South Georgia may be explained by the fact that the local shelf area contains insufficient prey items to sustain the local elephant seal breeding population (McCann 1985). There may be an advantage, therefore, in adopting strategies that minimize the risk of yearly failure at the expense of the energetic costs of long-distance transits. The narrow Antarctic continental shelf, ice edges, and the Antarctic Polar Front are highly productive and attract many top predators (Ainley and De Master 1990) and fisheries (Itchii 1990). This contrasts with the open reaches of the Antarctic Ocean where concentrations of prey are both spatially and temporally variable and may be associated with unpredictable hydrographic conditions (El-Sayed 1988). If an animal has the energy storage capability (as elephant seals, with their prodigious potential for blubber accumulation, do), the benefit of using distant foraging areas where food is reliably associated with readily relocatable oceanographic features, such as the continental shelf and the Antarctic Polar Front, may outweigh the costs of transport to these areas. That is, a long swim on an empty stomach may, in the long term, be more productive than pelagic meandering.

Variability in Size and Success

Yet, in spite of these capabilities, some animals gain more weight than others while at sea. And energy stores and the size at age of both males and females seem likely to be important to reproductive success. What factors of an animal's movements, experience, and foraging techniques are important in determining how successful it is?

Of particular importance is information on all aspects of the variability of foraging parameters: how they vary year to year in the same animal and change with age, sex, and size between animals. How do the patterns observed develop in naive animals when they first go to sea, and how does the variability of such early experience affect later development? No less important are the effects of the choice of breeding and molting sites in relation to near and distant oceanography.

Though the information to answer such questions is not available for southern elephant seals, the techniques to do so are at hand. What is needed is their continued application and particularly their effective combination in land- and sea-based studies. These techniques will make new demands on the resolution, the accuracy, and, crucially, the availability of oceanographic data.

ACKNOWLEDGMENTS

This work was carried out as part of the Sea Mammal Research Unit (Natural Environment Research Council) Open Oceans Programme in conjunction with the British Antarctic Survey and the University of Stockholm. We are particularly grateful to Neil Audley, Colin Hunter, Kevin Nicholas, Tim Barton, Ash Morton, and David Davies-Hughes for technical assistance and essential help in the field. We thank John Croxall for his usual unusually useful comments on the manuscript given good-naturedly with the usual short notice.

REFERENCES

Ainley, D. G., and D. P. DeMaster. 1990. The upper trophic levels in polar marine ecosystems. In *Polar Oceanography, Pt. B: Chemistry, Biology, and Geology*, ed. W. O. Smith, 599–630. San Diego: Academic Press.

Argos. 1989. *Guide to the Argos System*. Toulouse: Argos CLS.

Boyd, I. L., T. Arnbom, and M. A. Fedak. 1993. Water flux, body composition, and metabolic rate during molt in female southern elephant seals (*Mirounga leonina*). *Physiological Zoology* 66: 43–60.

Clarke, M. R., and N. MacLeod. 1982. Cephalopods in the diet of elephant seals at Signy Island, South Orkney Islands. *British Antarctic Survey Bulletin* 57: 27–31.

El-Sayed, S. Z. 1988. Seasonal and inter-annual variabilities in Antarctic phytoplankton with reference to krill distribution. In *Antarctic Ocean and Resources Variability*, ed. D. Sahrhage, 101–119. Berlin, Heidelberg, and New York: Springer Verlag.

Fedak, M. A., I. L. Boyd, T. Arnbom, and T. S. McCann. 1989. The energetics of lactation in southern elephant seals, *Mirounga leonina*, in relation to the mother's size. Abstract from the Eighth Biennial Conference on the Biology of Marine Mammals, Monterey, Calif., December 7–12, 1989.

Gallivan, G. J., and K. Ronald. 1981. Apparent specific dynamic action in the harp seal (*Phoca groenlandica*). *Comparative Biochemistry and Physiology*. 69A: 579–581.

Hindell, M. A., H. R. Burton, and D. J. Slip. 1991. Foraging areas of southern elephant seals, *Mirounga leonina*, as inferred from water temperature data. *Australian Journal of Marine and Freshwater Research* 42: 115–128.

Huber, H. R. 1987. Natality and weaning success in relation to age of first reproduction in northern elephant seals. *Canadian Journal of Zoology* 65: 1311–1316.

Itchii, T. 1990. Distribution of Antarctic krill concentrations exploited by Japanese krill trawlers and minke whales. *Proceedings of the NIPR Symposium on Polar Biology* 3: 36–56.

Kleiber, M. 1961. *The Fire of Life*. New York: John Wiley and Sons.

Laws, R. M. 1956. The elephant seal (*Mirounga leonina* Linn). II. General, social, and reproduction behaviour. *Falkland Islands Dependencies Survey, Scientific Reports* 13: 1–88.

———. 1960. The southern elephant seal (*Mirounga leonina* Linn.) at South Georgia. *Norsk Hvalfangst-Tidende* 49: 466–476, 520–542.

———. 1984. Seals. In *Antarctic Ecology*, vol. 2, ed. R. M. Laws, 621–715. London: Academic Press.

Le Boeuf, B. J., Y. Naito, A. C. Huntley, and T. Asaga. 1989. Prolonged, continuous, deep diving by northern elephant seals. *Canadian Journal of Zoology* 67: 2514–2519.

Ling, J. K., and M. M. Bryden. 1981. Southern elephant seal, *Mirounga leonina* Linnaeus, 1758. In *Handbook of Marine Mammals, 2, Seals*, eds. S. H. Ridgway and R. J. Harrison, 297–327. London: Academic Press.

McCann, T. S. 1980. Population structure and social organisation of southern elephant seals, *Mirounga leonina* (L.). *Biological Journal of the Linnaean Society of London* 14: 133–150.

———. 1981. The social organization and behaviour of the southern elephant seal, *Mirounga leonina* (L.). Ph.D. dissertation, University of London, England.

———. 1985. Size, status, and demography of southern elephant seal (*Mirounga leonina*) populations. In *Sea Mammals in South Latitudes: Proceedings of a Symposium of the 52d ANZAAS Congress in Sydney—May 1982*, ed. J. K. Ling and M. M. Bryden, 1–17. Northfield: South Australian Museum.

McCann, T. S., M. A. Fedak, and J. Harwood. 1989. Parental investment in southern elephant seals, *Mirounga leonina*. *Journal of Behavioural Ecology and Sociobiology* 25: 81–87.

McCann, T. S., and P. Rothery. 1988. Population size and status of the southern elephant seal (*Mirounga leonina*) at South Georgia, 1951–85. *Polar Biology* 8: 305–309.

McConnell, B. J., C. Chambers, and M. A. Fedak. 1992. Foraging ecology of southern elephant seals in relation to the oceanography of the Southern Ocean. *Antarctic Science* 4: 393–398.

Murphy, R. C. 1914. Notes on the sea elephant, *Mirounga leonina* (Linn.). *Bulletin of the American Museum of Natural History* 33: 63–78.

Reilly, J. J., and M. A. Fedak. 1990. Measurement of the body composition of living gray seals by hydrogen isotope dilution. *Journal of Applied Physiology* 69: 885–891.

Reiter, J., and B. J. Le Boeuf. 1991. Life history consequence of variation in age at primiparity in northern elephant seals. *Journal of Behavioral Ecology and Sociobiology* 28: 153–160.

Rodhouse, P., T. R. Arnbom, M. A. Fedak, J. Yeatman, and A. W. A. Murray. 1992. Cephalopod prey of the southern elephant seal, *Mirounga leonina* L. *Canadian Journal of Zoology* 70: 1007–1015.

Sakamoto, W., Y. Naito, A. C. Huntley, and B. J. Le Boeuf. 1989. Daily gross energy requirements of female northern elephant seal, *Mirounga angustirostris*, at sea. *Nippon Suisan Gakkaishi* 55: 2057–2063.

Worthy, G. A. J., P. A. Morris, D. P. Costa, and B. J. Le Boeuf. 1992. Molt energetics of the northern elephant seal. *Journal of Zoology, London* 227: 257–265.

TWENTY-ONE

Hormones and Fuel Regulation in Fasting Elephant Seals

Vicky Lee Kirby and C. Leo Ortiz

ABSTRACT. This chapter summarizes current knowledge about fasting physiology in the northern elephant seal, *Mirounga angustirostris*. Changes in metabolic fuel distribution and plasma hormone levels as well as changes in insulin secretion and peripheral tissue sensitivity to plasma insulin are addressed. Pups at weaning and during an eight-week postweaning fast were hyperglycemic, hyperlipidemic, hypo-insulinemic with impaired glucose tolerance, and relatively insulin insensitive. Fasting northern elephant seal weanlings did not closely regulate their blood glucose.

It is suggested that the suckling elephant seal pup is preadapted to the postweaning fasting period because of the lack of carbohydrate in the milk, its high fat content (85–95% of the calories), and the large increase in body fat (up to 50% of the mass at weaning). All of these contribute to impaired insulin secretion and action in other mammals. Blood glucose could only be maintained by hepatic gluconeogenesis because of the lack of dietary carbohydrate in this species at all stages of its life history. Adaptations to a low carbohydrate, high fat diet are similar to those necessary for adaptation to fasting. Low plasma insulin and relative tissue insensitivity to insulin are normal adaptations to low carbohydrate diets and fasting and would not be clinically abnormal for carnivores.

The northern elephant seal, *M. angustirostris*, provides the physiologist with a model for studying the basic physiological, biochemical, and anatomical mechanisms underlying the ability to undergo natural extended periods of complete food and water abstinence in large nonhibernating mammals. With the exception of nursing pups, individuals of all ages and both sexes fast entirely during the terrestrial phase of their life cycle, notably, the reproductive and molting phases. This rigorous life-style begins early in life when pups of the year are weaned abruptly at one month of age (Le Boeuf, Whiting, and Gantt 1972; Reiter, Stinson, and Le Boeuf 1978). During this period, young animals not only cope with the rigors of zero nutritional and water input but do so while continuing normal neonatal development. This entails substantial intertissue reorganization of protein, minerals, and other cellular components.

Over the past several years, we have investigated aspects of the physiology of spontaneous fasting in these young weanlings with two basic objectives. First, we wanted to understand the physiology of integrated biochemical processes underlying these prolonged fasts. Second, we wanted to determine the control mechanisms that simultaneously integrate catabolic processes involved in meeting energy needs with the anabolic processes required for protein recruitment and synthesis during development.

Our initial studies focused on the major regulatory hormones, insulin and glucagon, in fasting weanlings (Kirby and Ortiz 1989; Kirby 1990). In this chapter, we summarize our current understanding of fasting physiology during the postweaning fast, including a discussion of (1) fuel depots and fuel turnover studies; (2) plasma metabolite and hormone levels during the postweaning fast; and (3) changes in pancreatic and peripheral tissue responsiveness to glucose and insulin tolerance tests.

FUEL DEPOTS: STORAGE AND UTILIZATION

Northern elephant seals undergo dramatic changes in body composition during their first year of life. Unlike terrestrial mammals, the accumulation of adipose tissue occurs early in the neonatal period (Bryden 1968). Throughout the nursing period, pups gain on average 2 kg of adipose and 1 kg of lean tissue daily, and at weaning, the fat mass averages 48% in healthy pups (Ortiz, Costa, and Le Boeuf 1978). The fat mass gained during nursing is important for survival during the postweaning fast, as evidenced by the correlation between duration of the postweaning fast and the relative level of body fat at weaning (Kirby 1992).

In fasting animals, changes in body mass compartments can be used to calculate how much lean and adipose tissue contribute to metabolism. Newly weaned pups lose 1 to 2 kg of tissue/day in the first 2 weeks of fasting as compared to .65 kg/day during the rest of the 8- to 10-week fast (Kretzmann 1990; Rea 1990). This progressive sparing of tissue reserves is accomplished by a reduction of resting metabolic rate during the postweaning fast (Rea 1990). Although fasting seals catabolize both lean and adipose tissue at an equivalent rate, these tissues do not have equal energy content. Hydrated proteinaceous tissue has a lower energy content than an equivalent weight of adipose tissue. Thus, the size of the fat depot is important because energy mobilized from adipose tissue has to supply more than 85% of the total energy needs of the pup.

Normally, fuel stores important in carbohydrate metabolism are muscle and hepatic glycogen, amino acid stores in lean tissue, and triglycerides in adipose tissue. Although glycogen levels have not been measured specifically in elephant seal tissues, it has been shown in other species of pinnipeds that glycogen is not an important energy store. Therefore, glucose must be made from noncarbohydrate precursors, such as the glycerol moiety derived by triacylglycerol oxidation and the glucogenic amino acids derived from lean tissue. However, the relative contributions of lean and fat tissue to glucose formation cannot be determined from just monitoring changes in fuel depot size.

FUEL TURNOVER STUDIES

The relative contributions of lean and fat tissue to total energy needs and to glucose formation have been examined by isotope-labeled fuel metabolite studies in fasting weanlings. Fatty acid oxidation studies confirmed that lipid is the main energy source and suggested that sufficient glycerol was liberated by lipolysis to meet all glucose precursor needs (Castellini, Costa, and Huntley 1987). In fact, the direct contribution of glucose to the total metabolic rate was shown to be less than 1% in seals fasting longer than one month (Keith and Ortiz 1989). Although glucose turnover rates were within mammalian norms, most of the glucose carbon appeared to be recycled, possibly by futile cycling, and was not oxidized. This has also been observed in harbor seals (Davis 1983) and grey seals (Nordoy and Blix 1991).

Similar results from urea, albumin, and leucine turnover studies in fasting pups confirmed that protein oxidation contributed to less than 3% of total energy needs (Ortiz 1990; Pernia 1984; Pernia, Hill, and Ortiz 1980). Since it appears that protein and glucose oxidation, together, provide less than 10% of total energy needs, the physiological role of active turnover of these substrates remains unclear. It is possible that recycling protein and glucose carbon may serve as an important carbon shuttle mechanism, for example, glucose synthesis or synthesis of nonessential amino acids or other cellular components.

PLASMA METABOLITE LEVELS

Plasma levels of metabolites from carbohydrate, protein, and lipid metabolism are common parameters in characterizing in vivo fuel homeostasis. These metabolites usually include glucose, blood urea nitrogen (BUN), creatinine, nonesterified fatty acids (NEFA), and β-hydroxybutryate (BOHB).

Early studies reported that elephant seal pups may be hyperglycemic

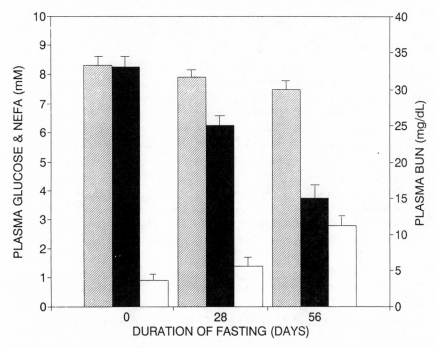

Fig. 21.1. Plasma levels (mean ± SD) of glucose (cross-hatched bars), blood urea nitrogen (solid bars), and nonesterified fatty acids (open bars) in eight elephant seal pups before fasting (at weaning), after 28 days of fasting, and after 56 days of fasting.

because plasma glucose levels were rarely seen below 6 mM, which is substantially higher than predicted by body size (Umminger 1975; Costa and Ortiz 1982). Plasma levels of glucose as high as 200 mg/dL have been reported for other species of pinnipeds (Englehardt and Ferguson 1980; Hochachka et al. 1979; Worthy and Lavigne 1982). Subsequent studies (Kirby 1990) confirmed that plasma glucose levels remain elevated above 6.5 mM throughout the postweaning fast, although there was a small but steady decrease in plasma glucose from 8 mM at the start of the fast to 7 mM by the end of the fast (fig. 21.1).

Plasma BUN levels are an indirect measure of protein oxidation and demonstrate predictable changes associated with fasting. They were highest at weaning (33 mg/dL) and decreased significantly over the fasting period to 15 mg/dL, as shown in figure 21.1 (Adams 1991; Kirby 1992; Costa and Ortiz 1982). Another index of lean tissue catabolism is creatinine (C), which is released into plasma when muscle tissue is catabolized. The decrease in the BUN:C ratio from 37 to 16 by the end of the fast was consistent with protein sparing adaptations.

Conversely, plasma levels of NEFA progressively increased from
< 1.0 mM at the beginning of the fast (at weaning) to ≥ 2.6 mM after
8 weeks of fasting (fig. 21.1; Kirby 1992). The ketone body BOHB, a by-
product of fat oxidation, also increased slowly during the fast (Castellini and
Costa 1990). After 8 weeks of fasting, there was a sharp decrease in BOHB,
and the seals departed within 10 days. This suggests that the seals termi-
nated their land-based fast because adipose reserves had become too
low to support ketogenesis and still provide enough insulation to aid
thermoregulation while feeding at sea.

All of these changes in plasma fuel levels were consistent with the general animal fasting model (Felig et al. 1969; Cahill 1970) in which there is a
metabolic shift from protein oxidation to lipid oxidation early in the fasting
period. Similar adaptations have been observed in other animals in which a
spontaneous fast is part of their life cycle, for example, the king penguin,
Aptenodytes patagonica (Cherel, Stahl, and Le Maho 1987).

PLASMA HORMONE LEVELS

The shift to a lipid-based metabolism occurs in conjunction with specific
changes in insulin and glucagon levels (Cahill et al. 1966; Felig et al. 1979).
A decrease in insulin secretion is vital for fuel homeostasis during fasting
because insulin normally inhibits lipolysis. In fasting elephant seals, mean
plasma insulin levels decreased throughout the fast from 11 ± 4 μU/ml at
weaning, to 9 ± 3 μU/ml after 4 weeks of fasting and 8 ± 2 μU/ml after
8 weeks. Low plasma insulin levels (5–15 μU/ml) were also observed in
lactating and molting adult females, molting yearlings, and weanlings
captured just after their return from feeding at sea (Kirby 1990; Kirby and
Ortiz 1990). Similar low levels have also been reported for feeding harbor
seals (Robin et al. 1981) and Weddell seals (Hochachka et al. 1979).

Glucagon levels were highly variable, especially in nursing pups. Different glucagon antibodies in the glucagon radioimmunoassay suggested the
possibility that nonpancreatic glucagon cross-reacted with the antibody for
pancreatic glucagon, thus overestimating glucagon. Extrapancreatic alpha
cells have been observed in the gut of the South African fur seal, *Arctocephalus pusillus* (Van Aswegan and Viljoen 1987). The lowest total glucagon
levels measured contributed to a molar I:G ratio ≤ 1.0 in all animals
studied. In other mammals, an I:G ratio of less than 1 is correlated with
high rates of hepatic gluconeogenesis (Unger, Eisentraut, and Madison
1963; Unger 1985).

In general, changes in plasma levels of insulin and glucagon as well as in
their ratios were consistent with fasting adaptations observed in animals as
diverse as humans and penguins (Marliss et al. 1970; Cahill and Aoki 1977;
Cherel, Leloup, and Le Maho 1988). Decreased levels of insulin and a low

I:G ratio allowed for the increased mobilization of lipid from fat stores as protein and glucose were conserved. It was not expected that the prefasting insulin levels in newly weaned seals would be so low. However, potential explanations are that (1) the high fat milk diet (Le Boeuf and Ortiz 1977; Riedman and Ortiz 1979; Kretzmann 1990) suppressed insulin secretion; (2) the high body fat levels impaired insulin secretion (Kirby 1990); (3) 1-month-old seals were still too young developmentally to regulate glucose normally (Tieran 1970); or (4) insulin secretion was low because of its minor role in glucose homeostasis in this species (Kirby 1992).

GLUCOSE TOLERANCE TESTS

Glucose tolerance tests (GTT) can provide an index of pancreatic islet sensitivity to changes in plasma glucose. If insulin does not regulate blood glucose levels in these animals, one would expect an impaired response in which animals do not closely regulate their blood glucose levels. To test this, glucose was administered intravenously (IV) as a 50% solution over a 2- to 4-minute period in a standard dose of 25 g or 0.5 g glucose/kg body weight, and the disappearance rate (K) of glucose was measured over time. Feeding mammals normally have K values of 2.3 or greater in contrast to fasting animals, which have K values of 1.0 or less. The glucose tolerance profile for five seals at weaning and after 8 weeks of fasting is shown in figure 21.2. K values were ≤1.0 for all seals. This profile is similar to that seen in obese and diabetic humans (Salans, Knittle, and Hirsch 1983). Normal (nonobese) mammals would restore euglycemia within 60 to 90 minutes of the glucose injection due to an acute insulin response.

Impaired glucose clearance (K ≤ 1.0) and the total lack of an insulin response to a glucose injection were virtually identical for pups at weaning and after fasting 8 to 11 weeks. These data indicated that these seals were adapted for fasting, that is, they switched to fat oxidation metabolism, *prior* to weaning.

In other mammalian neonates, the transition from the glucose-based metabolism of the fetus to the high fat milk diet of the neonate is accompanied by an increase in gluconeogenesis and fatty acid oxidation and a decrease in hepatic lipogenesis in the perinatal period (Kalhan 1992). As carbohydrate is introduced into the diet, the neonate decreases its reliance on de novo synthesis of glucose and increases insulin secretion. Elephant seal milk does not contain carbohydrate; in fact, the fat content is at its highest just prior to weaning. Thus, the suckling elephant seal neonate cannot develop the ability to secrete or utilize insulin for carbohydrate uptake into tissues.

Impaired insulin secretion and tissue insensitivity to insulin are correlated with high body fat levels (and high fat diets) in other mammals. We

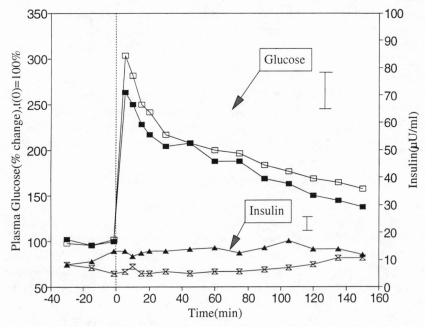

Fig. 21.2. Glucose tolerance test for five elephant seal pups just prior to weaning (closed symbols) and after 8 weeks of fasting (open symbols). The mean glucose response is represented by the square symbols (solid line). The mean insulin response is represented by the triangle and hourglass symbols. At time zero, a bolus injection of 25 g glucose in a 50% saline solution was injected intravenously. For purposes of clarity, the largest standard deviations for glucose and insulin are shown as vertical bars.

investigated the effect of body fat on the insulin response to a glucose challenge by comparing glucose clearance rates (K) in pups with known body weights, body composition, and age (Kirby and Ortiz 1989). It proved to be very difficult to separate changes in body fat levels from changes in age. Virtually all seals larger than 100 kg body weight at weaning had a fat mass of 48 to 50%. Seals smaller than 90 kg at weaning had lower body fat levels (<40%). However, all seals, regardless of weaning mass, age, and duration of fasting, had impaired glucose clearance values less than 1.0 with little or no insulin response to the glucose injection. Glucose clearance values estimated from figures in harbor seal and Weddell seal studies were also less than 1.0 (Hochachka et al. 1979; Robin et al. 1981).

INSULIN TOLERANCE TESTS

As a mammal adapts to fasting, there should be a decrease in tissue sensitivity to insulin concomitant with decreased insulin secretion (Cahill et al.

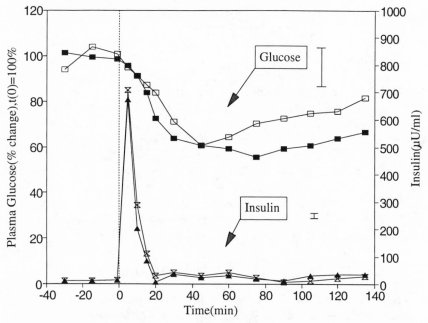

Fig. 21.3. Insulin tolerance test for five elephant seal pups just prior to weaning (closed symbols) and after 8 weeks of fasting (open symbols). The mean glucose response is represented by the square symbols (solid line). The mean insulin response is represented by the triangle and hourglass symbols. At time zero, a bolus injection of 0.1 U insulin/kg body weight was injected intravenously. For purposes of clarity, the largest standard deviations for glucose and insulin are shown as vertical bars. Note change of scale from fig. 21.2.

1966). Changes in peripheral tissue sensitivity to insulin were assessed by intravenous injections of 0.05 to 0.1 U insulin per kg body weight. In response to insulin, plasma glucose levels decreased in all seals studied. Although the exogenous insulin was cleared from the blood within 20 to 30 minutes postinjection, blood glucose levels did not reach a nadir until 40 to 90 minutes postinjection.

A typical insulin tolerance test (ITT) profile is shown in figure 21.3 for five seals prior to and after 8 weeks of fasting. Blood glucose was not restored within 135 minutes postinjection even though some seals tolerated blood glucose levels as low as 3.5 to 4 mM for almost an hour without any behavioral changes. There was no difference in glucose recovery rates for nursing and fasting pups. This pattern of impaired response has been observed in fasting and feeding obese humans as well as in diabetics (Drenick et al. 1972). In contrast, most mammals restore euglycemia within two hours of the insulin injection. The maximal hypoglycemia in response to insulin is reached within 30 minutes of the injection.

Elephant seal pups tolerated a reduction of plasma glucose of 50% of basal level glucose. This suggests that plasma glucose levels were not closely regulated by the standard insulin-glucagon push-pull model. It is possible that glucose control was not as important due to adipose tissue insensitivity to insulin. We measured changes in the by-products of lipolysis, plasma NEFA and BOHB, in response to an insulin injection in four seals after an 8-week fast. Although NEFA and BOHB both decreased in response to insulin, plasma NEFA levels began to recover immediately and returned to preinjection levels 20 minutes after the exogenous insulin was cleared from the plasma. Ketone levels did not start to recover until 60 minutes after the injection and, like glucose, did not return to preinjection levels within the 150-minute experimental period. Thus, insulin had an effect on lipolysis, although in this experiment we could not determine whether fatty acid levels decreased due to insulin suppression of adipocyte lipolysis or due to facilitated muscle cell uptake of fatty acids.

SUMMARY AND CONCLUSIONS

Overall, fasting elephant seal pups conformed to the general mammalian model of fuel homeostasis for fasting animals in which protein conservation is paralleled by increased mobilization and utilization of lipid. It is surprising that these same adaptations are also seen in suckling pups just prior to weaning. However, when we examine the nutritional life history of elephant seals, we find that fat is the major energy source throughout development whether individuals are consuming high fat milk, high fat fish, or body fat stores. Since elephant seals consume only a minimal amount of dietary carbohydrate, all glucose must be made from precursors derived from the diet or from tissue stores. To date, changes in fuel distribution, fuel turnover rates, and plasma metabolites confirm that lipolysis is the major source of energy—and possibly glucose—in northern elephant seals.

The insulin and glucose studies show that nursing elephant seals are preadapted for fasting. The lack of pancreatic islet response and the impaired glucose clearance ($K \leq 1$; fig. 21.2), the tissue insensitivity to insulin (fig. 21.3), and low plasma insulin and low ratio of I:G all contributed to maximizing the release of lipid from adipose tissue and were not a function of age, body fat, or nutritional status in seals. This emphasis on lipid mobilization concurrent with minimal oxidation of glucose and protein suggests that glucose may be synthesized from glycerol while the fatty acids are oxidized to meet the total energy needs of feeding and fasting seals. The important difference between the feeding and fasting state would be the partitioning of amino acids into protein synthesis or into gluconeogenesis.

Despite the fact that insulin is considered essential to protein synthesis in muscle tissue, suckling pups can gain a kilogram of lean tissue daily, even with plasma insulin levels of less than 12 μU/ml. We suggest that this same anabolic condition exists for young fasting seals who are actively recycling protein and glucose as they reorganize their tissues. The observation that active synthesis and reorganization of tissue protein occurs throughout the postweaning period of fasting and development is supported by the previously mentioned protein turnover studies. Although we do not understand how these seals balance anabolic and catabolic processes during the postweaning fast, the hormonal data suggest that we reevaluate the importance of insulin and glucagon in fuel regulation in these animals in particular and in other carnivores in general.

If insulin is not important in fuel homeostasis in this species, then other factors may play a more important role in glucose homeostasis. The influence of diving hypoxia on various tissues as well as the brain may change the influence that insulin has on muscle. It has been observed that hypoxia enhances muscle tissue sensitivity to insulin, reducing the amount needed to stimulate tissue growth (King et al. 1987). Diving may also enhance plasma fuel availability because exercise induces a release in epinephrine, which then increases glucose and free fatty acid release (Hamburg, Hendler, and Sherwin 1979). Increases in epinephrine and glucocorticoid plasma levels can also inhibit glucose-stimulated insulin release (Ploug, Galbo, and Richter 1984; Porte, Smith, and Ensinck 1979).

Alternative speculations suggest that insulin may not be necessary for protein synthesis in muscle tissue because seals may have high plasma levels of growth hormone or growth factors. The cellular requirements for energy are also regulated in part by thyroid hormone and glucocorticoids. These hormones are important for phocid molting (Ashwell-Erickson et al. 1986) and could be important to overall fuel homeostasis in seals. However anabolic processes are regulated, the glucose needs of the brain and other glucose obligate tissues must first be met so that seals can survive as diving carnivores who routinely experience apnea, tissue hypoxia, extreme exercise (as in deep diving), and periodic fasting. While fuel homeostasis in fasting seals is consistent with the general mammalian model of fasting, adipose tissue may be far more important to the elephant seal for glucose homeostasis than it is in other mammals.

REFERENCES

Adams, S. H. 1991. Changes in protein metabolism and water conservation in northern elephant seals during the postweaning fast. M.Sc. thesis, University of California, Santa Cruz.

Ashwell-Erickson, S., F. Fay, R. Elsner, and D. Wartzok. 1986. Metabolic correlates of molting and regeneration of pelage in Alaskan harbor and spotted seals. *Canadian Journal of Zoology* 64: 1086–1092.

Bailey, B. A., R. G. H. Downer, and D. M. Lavigne. 1980. Neonatal changes in tissue levels of carbohydrate and lipid in the harp seal. *Comparative Biochemistry and Physiology* 67B: 179–182.

Bryden, M. M. 1968. Growth of the southern elephant seal, *Mirounga leonina* (Linn.). *Growth* 33: 69–82.

Cahill, G. F., Jr. 1970. Starvation in man. *New England Journal of Medicine* 282: 668–675.

Cahill, G. F., and T. T. Aoki. 1977. The role of glucagon in amino acid homeostasis. In *Glucagon: Its Role in Physiology and Clinical Medicine*, ed. P. Foa, J. Bajaj, and N. Foa, 487–494. New York: Springer Verlag.

Cahill, G. F., M. G. Herrera, A. P. Morgan, J. S. Soeldner, J. Teinke, P. L. Levy, G. A. Reichard, and D. M. Kipnis. 1966. Hormone-fuel interrelationships during fasting. *Journal of Clinical Investigation* 45: 1751–1769.

Castellini, M. A., and D. P. Costa. 1990. Relationships between plasma ketones and fasting duration in neonatal elephant seals. *American Journal of Physiology* 259: R1086–R1089.

Castellini, M. A., D. P. Costa, and A. C. Huntley. 1987. Fatty acid metabolism in fasting elephant seal pups. *Journal of Comparative Physiology* 157B: 445–449.

Cherel, Y., J. Leloup, and Y. Le Maho. 1988. Fasting in king penguins. II. Hormones and metabolite changes during molting. *American Journal of Physiology* 254: R178–R184.

Cherel, Y., J. Stahl, and Y. Le Maho. 1987. Ecology and physiology of fasting in king penguin chicks. *Auk* 104: 254–262.

Costa, D. P., and C. L. Ortiz. 1982. Blood chemistry homeostasis during prolonged fasting in the northern elephant seal. *American Journal of Physiology* 242: R591–R595.

Davis, R. W. 1983. Lactate and glucose metabolism in the resting and diving harbor seal. *Journal of Comparative Physiology* 153: 275–288.

Davis, R. W., M. A. Castellini, T. H. Williams, and G. L. Kooyman. 1991. Fuel homeostasis in the harbor seal during submerged swimming. *Journal of Comparative Physiology* 160B: 627–635.

Drenick, E. J., L. C. Alvarez, G. C. Tamasi, and A. S. Brickman. 1972. Resistance to symptomatic insulin reactions after fasting. *Journal of Clinical Investigation* 51: 2757–2762.

Englehardt, F. R., and J. M Ferguson. 1980. Adaptive hormonal changes in harp seals and grey seals during the post-natal period. *General Comparative Endocrinology* 40: 434–445.

Felig, P., O. E. Owen, J. Wahren, and G. F. Cahill, Jr. 1969. Amino acid metabolism during prolonged starvation. *Journal of Clinical Investigations* 48: 584–594.

Felig, P., R. S. Sherwin et al. 1979. Hormonal interactions in the regulation of blood glucose. *Recent Progress in Hormone Research* 35: 501–511.

Hamburg, S., R. Hendler, and R. S. Sherwin. 1979. Epinephrine: Exquisite sensitivity to its diabetogenic effects in normal man. *Clinical Research* 27: 252A.

Hochachka, P. W., B. Murphy, G. C. Liggins et al. 1979. Unusual maternal-fetal blood glucose changes in the Weddell seal. *Nature* 277: 388–389.

Kalhan, S. C. 1992. Metabolism of glucose and methods of investigation in fetus and newborn. In *Fetal and Neonatal Physiology*, vol. 1, eds. R. A. Polin and W. W. Fox, 357–372. Philadelphia: W. B. Saunders.

Keith, E. O., and C. L. Ortiz. 1989. Glucose kinetics in neonatal elephant seals during postweaning aphagia. *Marine Mammal Science* 5(2): 99–115.

Kerem, D., D. Hammond, and R. Elsner. 1973. Tissue glycogen levels in the Weddell seal: A possible adaptation to asphyxial hypoxia. *Comparative Biochemistry and Physiology* 45A: 731–736.

King, D. S., G. P. Dalsky, M. A. Staten, W. E. Clutter, D. R. Canhouten, and J. O. Holloszy. 1987. Insulin action and secretion in endurance-trained and untrained humans. *Journal of Applied Physiology* 63: 2247–2252.

Kirby, V. L. 1990. Marine Mammal Endocrinology. In *The CRC Handbook of Marine Mammal Medicine*, ed. L. Dieruf, 303–352. Boca Raton: CRC Press.

———. 1992. The regulation of fuel homeostasis in young northern elephant seals. Ph.D. dissertation, University of California, Santa Cruz.

Kirby, V. L., and C. L. Ortiz. 1989. Body fat and glucose insulin response in fasting and feeding northern elephant seals during the first year of life. In *Proceedings of Eighth Biennial Conference on the Biology of Marine Mammals*, December 7–11, Pacific Grove, Calif.

———. 1990. Glucose regulation in fasting and feeding northern elephant seals less than one year of age. *Physiologist* 33: A–108.

Kretzmann, M. B. 1990. Milk intake, metabolic rate and mass change in northern elephant seal pups: Evidence for sexual equality. M.Sc. thesis, University of California, Santa Cruz.

Le Boeuf, B. J., and C. L. Ortiz. 1977. Composition of elephant seal milk. *Journal of Mammalogy* 58: 683–685.

Le Boeuf, B. J., R. J. Whiting, and R. F. Gantt. 1972. Perinatal behavior of northern elephant seal females and their young. *Behaviour* 43: 121–156.

Marliss, E. B., T. T. Aoki, R. H. Unger, J. S. Soeldner, and G. F. Cahill, Jr. 1970. Glucagon levels and metabolic effects in fasting man. *Journal of Clinical Investigations* 49: 2256–2270.

Nordoy, E. S., and A. S. Blix. 1991. Glucose and ketone body turnover in fasting grey seal pups. *Acta Physiologica Scandinavica* 41: 565–571.

Ortiz, C. L. 1990. Protein metabolism during the prolonged natural postweaning fast in rapidly developing elephant seal pups. *Physiologist* 33: A–56.

Ortiz, C. L., D. P. Costa, and B. J. Le Boeuf. 1978. Water and energy flux in elephant seal pups fasting under natural conditions. *Physiological Zoology* 51: 166–178.

Pernia, S. D. 1984. Protein turnover and nitrogen metabolism during long-term fasting in northern elephant seal pups. Ph.D. dissertation, University of California, Santa Cruz.

Pernia, S. D., A. Hill, and C. L. Ortiz. 1980. Urea turnover during prolonged fasting in the northern elephant seal. *Comparative Biochemistry and Physiology* 65B: 731–734.

Ploug, T., H. Galbo, and E. A. Richter. 1984. Increased muscle glucose uptake during contraction: No need for insulin. *American Journal of Physiology* 247: E726–E731.

Porte, D., Jr., P. H. Smith, and J. W. Ensinck. 1979. Neurohormonal regulation of the pancreatic islet. *Metabolism* 25 (Suppl. 1): 1453–1460.

Rea, L. 1990. Changes in resting metabolic rate during long-term fasting in northern elephant seal pups (*Mirounga angustriostris*). M.Sc. thesis, University of California, Santa Cruz.

Reiter, J., N. L. Stinson, and B. J. Le Boeuf. 1978. Northern elephant seal development: The transition from weaning to nutritional independence. *Behavioral Ecology and Sociobiology* 3: 337–367.

Riedman, M., and C. L. Ortiz. 1979. Changes in milk composition during lactation in the northern elephant seal. *Physiological Zoology* 52: 240–249.

Robin, E. D., J. Ensinck, A. J. Hance et al. 1981. Glucoregulation and simulated diving in the harbor seal. *American Journal of Physiology* 241: R293–R300.

Salans, L. B., J. R. Knittle, and J. H. Hirsch. 1983. Obesity, glucose tolerance, and diabetes mellitus. In *Diabetes Mellitus: Theory and Practice*, 3 ed., ed. M. Ellenberg and H. Rifkin, 469–479. New York: Medical Examination Publishing Co.

Tieran, J. R. 1970. Insulin in fetal and neonatal metabolism. In *Physiology of the Perinatal Period*, vol. 2, ed. U. Stave. New York: Plenum Medical.

Umminger, B. L. 1975. Body size and whole blood sugar concentrations in mammals. *Comparative Biochemistry and Physiology* 52A: 455–458.

Unger, R. H. 1985. Glucagon physiology and pathophysiology in the light of new advances. *Diabetologia* 28: 574–579.

Unger, R. H., A. M. Eisentraut, and L. L. Madison. 1963. The effects of total starvation upon the levels of circulating glucagon and insulin in man. *Journal of Clinical Investigations* 42: 1031.

Van Aswegen, G., and A. T. Viljoen. 1987. Endocrine cells in the gut of the South African fur seal, *Arctocephalus pusillus*. *Acta Anatomica* 130: 93–95.

Worthy, G. A., and D. M. Lavigne. 1982. Changes in blood properties of fasting and feeding harp seals after weaning. *Canadian Journal of Zoology* 60: 586–592.

TWENTY-TWO

Endocrine Changes in Newborn Southern Elephant Seals

Michael M. Bryden

ABSTRACT. The structure and function of some endocrine organs (hypophysis, thyroid, and pineal gland) of newborn southern elephant seals and their role in maintaining homeothermy are reviewed briefly. The adrenal gland is probably important, particularly in the immediate perinatal period, but it has not been examined yet.

The predominant cell in the hypophysis of the newborn seal is the somatotroph, suggesting that secretion of growth hormone is a major function of the gland during suckling, when pups grow rapidly. The degree of control of other endocrine organs by the hypophysis in newborn seals is unknown.

The thyroid gland is large, with histological evidence of activity at and soon after birth. The thyroid hormones T3 and T4, present in circulating blood at birth, are elevated within the subsequent two hours and remain high for approximately the first week postpartum. As circulating thyroid hormones decline, metabolic rate probably declines also, but direct measurements have not been made.

The pineal gland is very large and active in newborn southern elephant seals and remains so until 7 to 10 days postpartum. Circumstantial evidence suggests it is involved in the control of thermogenesis in early postnatal life.

Newborn phocid seals have a characteristically short lactation period, when the young undergo dynamic change (Bryden 1969; Oftedal, Boness, and Tedman 1987). Growth is rapid; most species at least treble their birth weight in just a few days or weeks (Laws 1959; Bowen, Oftedal, and Boness 1985). Most of the increase in weight is due to deposition of fat, although growth of the musculature and many other organs does occur. There is little increase in bones during this phase of growth (Bryden 1969). Dynamic shifts in the relative size of different tissues and organs occur during this stage; for example, the muscles that are used in terrestrial locomotion grow relatively more quickly than those that are not. This situation changes when the seals begin swimming, when those muscles responsible for aquatic propulsion grow relatively more quickly (ibid.).

Elephant seals are subjected to severe environmental conditions at and soon after birth. Although they experience a less extreme temperature change in passing from the uterus to the external environment than do polar seals such as Weddell and ringed seals, it is conceivable that they experience greater cold stress at birth. The coat of polar seals dries very quickly after birth, as the placental fluids on the coat quickly freeze and fall off or can be shaken off. In addition, the natal fur of these species is thicker than that of the newborn elephant seal and is a more effective insulator.

Southern elephant seals are born on subantarctic islands, where the environmental temperature is usually in the range of −5°C to +5°C. Rain, sleet, or snow is common, and the coat of the newborn seal may remain wet for several hours. This combined with the relatively ineffective insulation of the coat means that the newborn elephant seal can be subjected to extreme cold stress at birth, exacerbated by average wind speed on many subantarctic beaches of 40 km/hour.

As a rule, newborn pups do not adopt behavioral means of reducing heat loss, such as curling up, seeking shelter, or seeking to huddle close to neighbors. Neither does the mother protect the newborn young from heat loss: she does not attempt to shelter it or draw it close to her body.

These observations suggest that physiological mechanisms are predominant in maintaining body temperature in early postnatal life and play an important role in other body mechanisms during this dynamic part of the seal's postnatal life. Shivering is observed rarely (Laws 1953; Little 1989) and does not appear to be an important means of thermogenesis except in the most extreme circumstances.

My group has been studying aspects of the endocrine changes in early postnatal life for several years. Major aims of the work have been to examine the endocrine control of early postnatal development and gain some insight into possible means of thermogenesis in the first hours and days after birth.

THE ENDOCRINE GLANDS IN NEWBORN SEALS

It was first observed about forty years ago that in newborn seals certain organs, notably, the gonads, are very large and appear to be subjected to endocrine stimulus at about the time of birth (Harrison, Matthews, and Roberts 1952; Bonner 1955). E. C. Amoroso, G. H. Bourne, R. J. Harrison, L. H. Matthews, I. W. Rowlands, and J. C. Sloper (1965) described the histological appearance of several endocrine organs in fetal, newborn, and adult gray and common seals and noted particularly that the hypophysis (pituitary gland) was well developed at birth. However, they concluded

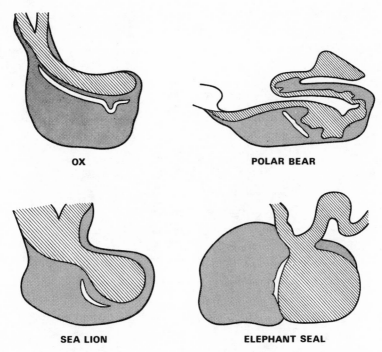

OX   **POLAR BEAR**

SEA LION   **ELEPHANT SEAL**

Fig. 22.1. Comparative median sections of the hypophyses of some mammals. The shaded structure is the adenohypophysis; the cross-hatched structures are the neurohypophysis and adnexa.

that the rapid decrease of endocrine influence on the gonads, which involute rapidly, suggested that the placenta rather than the fetal hypophysis is implicated. The large size of the gonads is due mainly to proliferation of interstitial cells and does not reflect functional status of the organs.

The Hypophysis

The hypophysis of the southern elephant seal conforms to the general mammalian pattern. Its form is compared with that of some other mammalian species in figure 22.1, which shows variations in shape but not of general arrangement. The significance of the intraglandular cleft, expansive in the newborn hypophysis but greatly reduced in the adult, is unknown.

Cell types in the pars distalis were differentiated by staining with a Periodic Acid Schiff–Alcian blue—orange G method modified from M. F. El Etreby, K.-D. Richter, and P. Günzel (1973). Four clearly distinct chromophilic cells were visible following staining, namely, gonadotrophs, thyrotrophs, lactotrophs, and somatotrophs. The relative frequencies of the different cell types in adult cows and young pups are shown in table 22.1.

TABLE 22.1 Frequencies (%) of different cell types in the adenohypophysis (*pars distalis*) of 3 adult cows and 5 neonatal southern elephant seals at Macquarie Island.

	Somatotroph	Thyrotroph	Gonadotroph	Lactotroph	Chromophobe
Adult cows	15.3	22.0	6.3	47.2[a]	9.3
	20.3	14.6	10.4	1.3	53.5
	23.4	28.0	12.3	3.2	33.0
Pups	30.2	8.9	3.1	2.1	55.8
	37.0	7.9	0.9	0.3	53.9
	33.6	11.5	2.7	2.5	49.8
	19.4	9.8	6.0	0.0	64.8
	38.0	20.1	5.1	0.7	36.1

NOTE: From Griffiths 1980.
[a] Lactating.

A preponderance of chromophobic cells in the hypophysis at birth contrasts with the situation in the adult, in which chromophilic cells suggest activity in controlling various physiological events and reproductive events in particular (Griffiths and Bryden 1986). A similar, relatively immature appearance of the adenohypophysis of the newborn animal, with the presence of many chromophobic cells, was observed in the harp seal by J. F. Leatherland and K. Ronald (1979). This contrasts with the finding in the common seal that the maturity of the adenohypophysis at birth is striking (Amoroso et al. 1965).

By far the most abundant chromophilic cell in the hypophysis of the elephant seal pup is the somatotroph, suggesting that the adenohypophysis may be actively secreting growth hormone during the rapid early postnatal growth of the pup (see table 22.1). Attempts to assay growth hormone in the plasma of elephant seal pups were unsuccessful, but the morphological evidence suggests that plasma concentrations of growth hormone may be quite high.

Immunocytochemical methods would improve the sensitivity of identifying the different cell types and would enable one to identify corticotrophic cells. They would also differentiate between luteotrophin-secreting and FSH-secreting cells, not possible with the histochemical staining regime used here.

The Thyroid

The hormones secreted by the thyroid gland, thyroxine (T4) and triiodothyronine (T3), are responsible for normal growth and maintenance of homeothermy through their role in controlling body metabolism. We have examined both histological and endocrinological indicators of thyroid function. A detailed description of the thyroid gland and its function in southern elephant seals has been provided by G. J. Little (1991).

Histology. Squamous thyroid epithelial cells are indicative of low or reduced thyroid activity, whereas cuboidal or columnar epithelium signifies increased activity. In southern elephant seal pups, thyroid epithelial cell height is significantly greater in the first two days of life than subsequently (Little 1991).

Ultrastructurally, dramatic morphological changes occur in the thyroid epithelial cells in the first 24 to 48 hours after birth and have been illustrated by Little (1991). Pseudopodia protrude into the thyroid follicular lumen from the surface of the epithelial cells. The pseudopodia engulf colloid within the follicular lumen and take it up into the epithelial cells. For the first few hours of postnatal life, colloid is confined to the apical portion of the epithelial cell, but after 6 hours, colloid droplets become distributed throughout the cytoplasm. This indicates that thyroglobulin, absorbed from the lumen following engulfment by the pseudopodia, is being hydrolyzed and that T4 and T3 are being released.

Endocrinology.   The plasma concentrations of T4 and T3 increased in the first 24 hours or so of postnatal life (Little 1991). Plasma T4 concentration showed about a twofold increase in the first 2 hours after birth, which is similar to the situation in many other mammals. Plasma T3 concentration, however, increased about eightfold at 24 hours after birth, the most dramatic increase of any newborn mammal. The concentration remained elevated until 7 days of age. Little (1991) concluded from these observations that the hypophysis-thyroid axis is functional at birth in the southern elephant seal and that the thyroid gland is the source of circulating T4 and T3 in the pup.

T3 is physiologically much more active than T4, and in newborn elephant seals, it seems to play an important role in thermogenesis, at least in the first week of postnatal life. It probably does so by increasing the metabolic rate. The growth of fat increases markedly after the first week (Bryden 1969), so the physical insulation it provides probably reduces heat loss, leading to lowered metabolic rate. We have not confirmed this in elephant seals, but it has been shown to be the case in common seals and harp seals (Davydov and Makarova 1964).

In summary, there is morphological and physiological evidence that the thyroid gland is very active within a few hours of birth and is responsible for increased levels of circulating T4 and T3 for the first days of life. It was not possible to confirm whether thyroid function was initiated and maintained by the hypophysis, but the presence of mature thyrotrophs in reasonable numbers in the hypophysis of the newborn seal (table 22.1) suggests this is probably the case.

The Pineal

It has been known for almost twenty years that polar animals tend to have a very large pineal gland. Southern elephant seals are no exception. A remarkable feature of this species, however, like the Weddell seal and other polar seal species, is that the pineal gland is particularly large at birth (Bryden et al. 1986; see fig. 22.2).

Histologically, the pineal contains many pinealocytes at birth, indicating that the gland is potentially functional (ibid.). This contrasts with the gonads, whose enlargement at birth results from the great proliferation of interstitial tissue.

Ultrastructurally, the pinealocytes appear relatively immature (Little and Bryden 1990). This suggests that the pineal, although enlarged, may not be very active at birth. However, the density of pinealocytes is similar

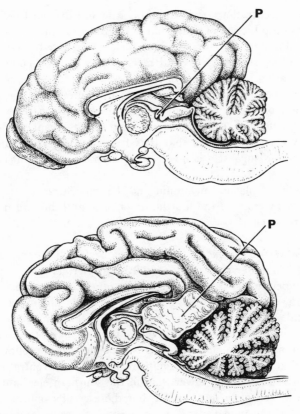

Fig. 22.2. Median section of the brain of a dog (top) and a newborn elephant seal (bottom). The pineal of each is marked (P).

in pups and adults, and the pup pineal contains 50 to 100 times more pinealocytes than the adult (Bryden et al. 1986).

We tested for pineal activity by assay of plasma for the hormone secreted by the pineal gland, melatonin. Details of the assay method are given in D. J. Kennaway, T. A. Gilmore, and R. F. Seamark (1982) and C. R. Earl, M. J. D'Occhio, D. J. Kennaway, and R. F. Seamark (1985).

The assays revealed extremely high concentrations of circulating melatonin in the first days of postnatal life. Concentrations in excess of 60,000 picomoles per liter (pM/l) were measured. (This contrasts with levels in most adult mammals of 100–300 pM/l.) It is possible that at least some of the melatonin may originate from organs other than the pineal, for example, the retina. However, the fact that the pineals of newborn pups also contained high concentrations of melatonin (Little and Bryden 1990) argues for a high level of production of melatonin in the pineal itself.

The pineal and its secretory product, melatonin, have been implicated in the control of several physiological mechanisms, in particular, reproduction, by mediating the influence of the photoperiod on the neuroendocrine-reproductive axis. As currently perceived, the function of the pineal is a rather general one, serving as an intermediary between the external environment (particularly the photoperiod) and the organism as a whole (Reiter 1981). We have observed that it is large and active in the newborn elephant seal and suggest it is involved in thermoregulation at this stage of the seal's life.

The profiles of plasma melatonin and T3 concentration are somewhat similar. They increase in the first hours of postnatal life and decline after about the first week.

Comparison of Southern and Northern Elephant Seals. If our conclusion that the pineal gland in newborn pups is involved in thermoregulation is correct, we should expect to see a lesser role for the pineal in early postnatal life in the northern elephant seal as compared to the subantarctic southern elephant seal. Plasma has been assayed for melatonin concentration in northern elephant seals (fig. 22.3). We see that although the concentration of melatonin in northern elephant seals is very high, and it follows a similar pattern to that in the southern elephant seal, the concentrations for the most part are substantially lower than in the southern elephant seal. This needs to be looked at in more detail, but these results tentatively support the notion that the pineal gland is involved in thermogenesis in elephant seals in early postnatal life.

Adrenal

An important omission so far is the adrenal gland, which has not been examined in elephant seals. F. R. Engelhardt and J. M. Ferguson (1980) reported high concentrations of cortisol and aldosterone immediately after birth in gray seal and harp seal pups. They discussed the possible role of these hormones acting synergistically to promote gluconeogenesis and the role of cortisol in increasing cold tolerance in the newborn. The extreme cold to which newborn southern elephant seals are subjected induces stress almost certainly, and the role of the adrenal needs to be addressed.

CONCLUSIONS

Endocrine glands of newborn southern elephant seals are active and almost certainly vital to survival in the first hours and days of postnatal life. Morphological evidence suggests the hypophysis secretes growth hormone, in-

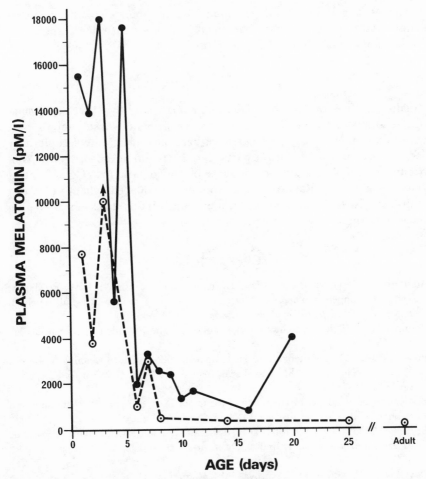

Fig. 22.3. Plasma melatonin concentration (pM/1) in neonatal southern (closed circles, solid line) and northern (open circles, broken line) elephant seals.

volved in the very rapid growth during the lactation period. It has not been possible to quantify the degree of hypophyseal control of the thyroid gland in early postnatal life by the methods used so far.

The thyroid gland is large and active within about 2 hours of birth in the production of thyroid hormones. Increased amounts of thyroid hormones are released into the bloodstream during the first day, and high circulating levels are maintained for the first week of life. The rate of metabolism probably declines in the second week as a consequence of reduction in the levels of thyroid hormones, when the rate of blubber deposition increases and physical insulation is enhanced.

The pineal gland is very large and active in newborn elephant seals. Circumstantial evidence suggests that it plays an important role in thermoregulation in the first hours and days of postnatal life.

ACKNOWLEDGMENTS

I thank the Director, Antarctic Division, Hobart, and the Director, Tasmanian National Parks and Wildlife Service, without whose support over many years this work would not have been possible. Many people have assisted in many ways, but in particular, I wish to acknowledge the contributions and stimulating discussions of David Griffiths, Gerald Little, David Kennaway, Raymond Tedman, and Jean Ledingham. I extend thanks to Ms. B. Jantulik for illustrations and Mrs. L. Hicks for preparing the manuscript.

REFERENCES

Amoroso, E. C., G. H. Bourne, R. J. Harrison, L. H. Matthews, I. W. Rowlands, and J. C. Sloper. 1965. Reproductive and endocrine organs of foetal, newborn and adult males. *Journal of Zoology, London* 147: 430–486.

Bonner, W. N. 1955. Reproductive organs of foetal and juvenile elephant seals. *Nature* 176: 982–983.

Bowen, W. D., O. T. Oftedal, and D. J. Boness. 1985. Birth to weaning in 4 days: Remarkable growth in the hooded seal, *Cystophora cristata*. *Canadian Journal of Zoology* 63: 2841–2846.

Bryden, M. M. 1969. Relative growth of the major body components of the southern elephant seal, *Mirounga leonina* (L.). *Australian Journal of Zoology* 17: 153–177.

Bryden, M. M., D. J. Griffiths, D. J. Kennaway, and J. Ledingham. 1986. The pineal gland is very large and active in newborn antarctic seals. *Experientia* 42: 564–566.

Davydov, A. F., and A. R. Makarova. 1964. Changes in heat regulation and circulation in newborn seals on transition to aquatic form of life. *Fiziol. Zhur. SSSR* 50: 894–897.

Earl, C. R., M. J. D'Occhio, D. J. Kennaway, and R. F. Seamark. 1985. Serum melatonin profiles and endocrine responses of ewes exposed to a pulse of light late in the dark phase. *Endocrinology* 117: 226–230.

El Etreby, M. F., K.-D. Richter, and P. Günzel. 1973. Histological and histochemical differentiation of the glandular cells of the anterior pituitary in various experimental animals. *Excerpta medica* (Amsterdam), *International Congress Series* 288: 270–281.

Engelhardt, F. R., and J. M. Ferguson. 1980. Adaptive hormone changes in harp seals, *Phoca groenlandica*, and gray seals, *Halichoerus grypus*, during the postnatal period. *General and Comparative Endocrinology* 40: 434–445.

Griffiths, D. J. 1980. The control of the annual reproductive cycle of male elephant seals (*Mirounga leonina*) at Macquarie Island. Ph.D. dissertation, University of Queensland, Australia.

Griffiths, D. J., and M. M. Bryden. 1986. Adenohypophysis of the elephant seal (*Mirounga leonina*): Morphology and seasonal histological changes. *American Journal of Anatomy* 176: 483–495.

Harrison, R. J., L. H. Matthews, and J. M. Roberts. 1952. Reproduction in some pinnipedia. *Transactions of the Zoological Society of London* 27: 437–540.

Kennaway, D. J., T. A. Gilmore, and R. F. Seamark. 1982. The effect of melatonin feeding on serum prolactin and gonadotropin levels at the onset of estrous cyclicity in sheep. *Endocrinology* 110: 1766–1772.

Laws, R. M. 1953. The elephant seal (*Mirounga leonina* Linn.). I. Growth and age. *Falkland Islands Dependencies Survey, Scientific Reports* 8: 1–62.

———. 1959. Accelerated growth in seals, with specific reference to Phocidae. *Norsk Hvalfangst-Tidende* 48: 425–452.

Leatherland, J. F., and K. Ronald. 1979. Thyroid activity in adult and neonate harp seals, *Pagophilus groenlandicus*. *Journal of Zoology, London* 189: 399–405.

Little, G. J. 1989. Thermoregulation in the newborn southern elephant seal, *Mirounga leonina* (L.). Ph.D. dissertation, University of Queensland, Australia.

———. 1991. Thyroid morphology and function and its role in thermoregulation in the newborn southern elephant seal (*Mirounga leonina*) at Macquarie Island. *Journal of Anatomy* 176: 55–69.

Little, G. J., and M. M. Bryden. 1990. The pineal gland in newborn southern elephant seals, *Mirounga leonina*. *Journal of Pineal Research* 9: 139–148.

Oftedal, O. T., D. J. Boness, and R. A. Tedman. 1987. The behavior, physiology, and anatomy of lactation in the pinnipedia. In *Current Mammalogy*, vol. 1, ed. H. H. Genoways, 175–245. New York: Plenum.

Reiter, R. J. 1981. The mammalian pineal gland: Structure and function. *American Journal of Anatomy* 162: 287–313.

INDEX

Designer: U.C. Press Staff
Compositor: Asco Trade Typesetting Ltd.
Text: Baskerville 10/12
Display: Baskerville
Printer: BookCrafters, Inc.
Binder: BookCrafters, Inc.